Key Challenges in Geography

EUROGEO Book Series

This book series addresses relevant topics in the wide field of geography, which connects the physical, human and technological sciences to enhance teaching, research, and decision making. Geography provides answers to how aspects of these sciences are interconnected and are forming spatial patterns and processes that have impact on global, regional and local issues and thus affect present and future generations. Moreover, by dealing with places, people and cultures, Geography explores international issues ranging from physical, urban and rural environments and their evolution, to climate, pollution, development and political economy.

Key Challenges in Geography is an initiative of the European Association of Geographers (EUROGEO), an organization dealing with examining geographical issues from a European perspective, representing European Geographers working in different professional activities and at all levels of education. EUROGEO's goal and the core part of its statutory activities is to make European Geography a worldwide reference and standard. The book series serves as a platform for members of EUROGEO as well as affiliated National Geographical Associations in Europe, but is equally open to contributions from non-members.

The book series addresses topics of contemporary relevance in the wide field of geography. It has a global scope and includes contributions from a wide range of theoretical and applied geographical disciplines.

Key Challenges in Geography aims to:

- present collections of chapters on topics that reflect the significance of Geography as a discipline;
- provide disciplinary and interdisciplinary titles related to geographical, environmental, cultural, economic, political, urban and technological research with a European dimension, but not exclusive;
- deliver thought-provoking contributions related to cross-disciplinary approaches and interconnected works that explore the complex interactions among geography, technology, politics, environment and human conditions;
- publish volumes tackling urgent topics to geographers and policy makers alike;
- publish comprehensive monographs, edited volumes and textbooks refereed by European and worldwide experts specialized in the subjects and themes of the books;
- provide a forum for geographers worldwide to communicate on all aspects of research and applications of geography, with a European dimension, but not exclusive.

All books/chapters will undergo a blind review process with a minimum of two reviewers.

An author/editor questionnaire, instructions for authors and a book proposal form can be obtained by contacting the Publisher.

More information about this series at http://www.springer.com/series/15694

Kostis C. Koutsopoulos · Jan H. Stel
Editors

Ocean Literacy: Understanding the Ocean

euro
geo

Springer

Editors
Kostis C. Koutsopoulos
European Association of Geographers
National Technical University of Athens
Pikermi, Greece

Jan H. Stel
Maastricht Sustainability Institute
Maastricht University
Maastricht, The Netherlands
Puurs-Sint-Amands, Belgium

ISSN 2522-8420 ISSN 2522-8439 (electronic)
Key Challenges in Geography
ISBN 978-3-030-70154-3 ISBN 978-3-030-70155-0 (eBook)
https://doi.org/10.1007/978-3-030-70155-0

This Springer imprint is published by the registered company Springer Nature Switzerland AG
The registered company address is: Gewerbestrasse 11, 6330 Cham, Switzerland

We would like to dedicate this book to: the youth of today, the educators and leaders of tomorrow

Drawing Credit: Arthur Verlinden

Foreword

Peace is central to the mission of UNESCO, which was born in 1945 after two devastating world wars. The Preamble to the Constitution of UNESCO states that "since wars begin in the minds of men, it is in the minds of men that the defences of peace must be constructed". Yet peace is much more than the end of an armed conflict. Defences of peace are education, science, culture, communication and information. Science illuminates the path to addressing the global challenges of humanity such as climate change or a pandemic. It is the main factor in changing human behaviour. This is reflected in the vision of UNESCO's climate strategy, "changing minds not the climate". Central to the climate discussion but going far beyond the climate nexus is the ocean, the earth's life support system.

We have turned the ocean into a massive garbage bin, because either we had thought it was too big to suffer, or simply because we did not think enough! In the divided world, suffering from poverty, hunger and anger, there is no place for care and deep thought. However, we need to change the world. We need to save the ocean, and the ocean will help us to save us. Knowledge about the role of the ocean for our lives and about our society's impact on the ocean is therefore the key to our survival as a species. As was eloquently expressed in the already famous phrase by Sylvia Earle, "far and away the biggest threat to the ocean is ignorance". To know, understand the ocean and make sure that people are emotionally connected to the ocean is the main purpose of ocean literacy—as a concept and as a movement.

The future ocean will be managed based on science. The tools will include marine protected areas, coastal zone management, marine spatial planning, adaptation and mitigation of climate change, fisheries and aquaculture management, assessment of risks, deployment of warning systems and facilitating a new, sustainable ocean economy. Scientific understanding will make this possible. That is why in 2017, the Intergovernmental Oceanographic Commission (IOC) of UNESCO proposed to the United Nations to dedicate the period from 2021 to 2030 to the UN Decade of Ocean Science for Sustainable Development. A proposal that was excepted by the UN, and thus by the world.

Please expect during these 10 years a breakthrough in ocean observations, data management, ecosystem understanding, mapping of the ocean bottom, creating an ocean genetic image, production of more food and renewable energy for people.

One of the Challenges of the Decade is not to leave anyone behind, so that we not only move forward the cutting edge of science but also create conditions for *all* to contribute and to benefit. The Decade plan speaks about "an inspiring and engaging ocean". To incite a behaviour change, there needs to be a step change in society's relationship with the ocean. This can be only achieved through the elevated "ocean literacy" of our society and starting to develop it early for people, by giving the ocean more weight in the school curricula.

This book is about "Sea Change", a very significant and pioneering European project in the area of ocean literacy. I hope that the summary of knowledge acquired and the methodologies developed in the course of the project will be illuminating the path to success of ocean literacy in the future, and that many new projects in all parts of the world will move forward as contributions to the Decade. If we succeed, we will open the door for systematic and welcomed use of ocean science for saving our ocean, for a sustainable future. I would like to invite the readers to join IOC on the road towards "the ocean we need for the future we want".

Dr. Vladimir Ryabinin
Executive Secretary of UNESCO/IOC;
Assistant Director General of UNESCO

Preface

This book is a result of the Sea Change project (March 2015–February 2018), the first of four projects funded by the European Commission under Horizon 2020. It is part of a stimulation by the European Commission, to introduce ocean literacy (OL) in the culturally diverse, educational systems of Europe. The four projects started in 2015. They were supported by the EU Maritime Affairs Directorate, are in alignment with the statement on TransAtlantic Cooperation on Ocean Research, signed in Galway in May 2013 and are part of the EU Blue Growth programs. The three other projects were: Columbus, ResponSEAble and Atlantic Research. Ocean literacy will, play an important role in the new Horizon Europe (2020–2027) program and most likely in the European Green Deal. Additionally, ocean literacy has been strongly supported by UNESCO's Intergovernmental Oceanographic Commission, which leads the upcoming UN Decade of Ocean Science for Sustainable Development (2021–2030). At the European level, ocean literacy now is part of the activities of the European Marine Board and is advocated by most of its members.

The main aim of the Sea Change project was to encourage citizens to take direct and sustainable action to protect healthy and biodiverse seas, by increasing their ocean literacy. Which is defined as an understanding of the ocean's influence on a person and that person's impact on the ocean. Sea Change initiated a number of citizen-focused education initiatives with influential networks like the European Network of Science Centers and Museums, the European Association of Geographers and the European Marine Science Educators Association. The project builds on cooperation with Canada and the USA, which have substantial OL experience. The ocean is a crucial part of the earth system, and makes it a habitable place to live. The marine environment is a source of vital human health benefits, that is why the Sea Changes slogan states: Our Ocean, Our Health.

What Is In This Book?

The book is organized into three sections: (i) General, (ii) Education and (iii) Application. The first section provides the reader with a brief overview of the development and global spread of the OL concept. In the second section, a number of interesting and highly successful educational activities are addressed, and an OL assessment structure is described. The last section focuses on projects that deliver a variety of educational innovations.

General on Ocean Literacy

The first chapter "Ocean Literacy: Background, Future Drivers, and Opportunities" is the leading chapter and provides an overview of current trends in ocean sciences, relevant for the development of ocean literacy. The history of oceanography is briefly discussed, a number of characteristics of the ocean are given, and the notion of ocean space in relation to a 4D interpretation of the ocean, is introduced in the section *The ocean as the Planet's life support system.* Next the needs to understand, observe, protect and involve citizens and pupils from primary and secondary schools are discussed.

The second chapter "Ocean Literacy: From a Ripple to a Wave". Payne provides a brief history of ocean literacy, and connects it with insights in ocean science and geography. Ocean literacy was developed in the USA to incorporate ocean science into science education. Later it developed into a global movement to include ocean sciences in many culturally diverse educational systems. She also relates ocean literacy to the UN Sustainable Development Goal 14, life below water, and UNESCO's Intergovernmental Oceanographic Commissions, holistic OL activities. Payne addresses a number of critical challenges to the long-term success of ocean literacy, like the need for centralized communication, inclusion of cultural relevance, a solid foundation of evaluation and education research, nuances of formal and non-formal education and adequate and consistent sources of funding.

The third chapter "A Framework for the Assessment of the Effectiveness of Ocean Literacy Initiatives" discusses how to evaluate the effectiveness of new tools and initiatives in ocean literacy. Molloy, Ashley and McCrossan give a wonderful description of the approaches developed within the ResponSEAble project, one of the four projects funded by Horizon 2020 to focus on increasing ocean literacy in Europe. The project focused on key ocean health issues through key stories on eutrophication and agriculture, ballast water and invasive alien species, sustainable fisheries and aquaculture, microplastics and cosmetics, coastal tourism and marine renewable energy. The authors advocate a system approach by using well-known models like DPSIR and DAPSIWR. The letters in these acronyms mean: Drivers (D), activity (A), Pressures (P), State (S), Impacts (I), Welfare (W) and Responses (R). These models are linked with the environmental literacy ladder and incorporated in an assessment of OL tools within a pilot study.

The fourth chapter "Exploring and Exploiting Deep Ocean Space" provides a detailed and extremely interesting historical review of technological developments and explorations of every corner of the Earth's surface since the British industrial revolution. Stel stresses that the links between pollution, technology and an unprecedented increase of human activities since World War II conflict with natures carrying capacity. CO_2 pollution unwillingly but since the 1960s willingly led to the present anthropogenic climate change. Under a business-as-usual scenario it becomes evident that to solve our urgent demand for resources, there is a need to exploit the unknown treasure troves of deep ocean space. The protection and conservation of ocean space requires an adaptation of the prevailing international legislation. In addition, in a low-carbon society, citizens should be aware of and involved in ocean-related issues. For this ocean literacy is a *sine-qua-non.*

The fifth chapter "Ocean Literacy—In the Context of Naming of Seas: Case Study: The Sea Between Korea and Japan" addresses the problem that countries bordering a sea often named this sea differently. Sometimes this leads to tension between riparian countries, especially as the ownership of marine resources is involved. In this chapter, Dormels discusses the naming conflict of the "Sea of Japan", which has different names in Korea and Japan. In an interesting way he sketches how history through the perception of western cartographers, Japanese imperialism and the admission of the Republic of Korea (South Korea) and the Democratic People's Republic of Korea (North Korea) to the UN in 1991, shaped the discussion. Moreover, the effects of the proclamation of Exclusive Economic Zones sharpens the conflict. At the end of the paper, a number of suggestions are made on how this could offer opportunities for ocean literacy,

Education

The sixth chapter "Design-Based Implementation Research for Exploring the Ocean: A Geographical Perspective" De la Vega discusses five of the seven Ocean Literacy Principles from a geographical perspective. OLP 2, the ocean and life in the ocean shape the features of Earth and OLP 5, the ocean supports a great diversity of life and ecosystems, are not discussed. A global analysis is given of the OL contents and learning standards, and the concept of ocean citizenship is introduced. The methodological framework applied in this chapter is based on Design-based Implementation Research. This approach is based on pedagogical methods like Problem-based Learning and Project-based Learning, learning strategies such as fieldwork and a range of digital and analogical resources. On this basis, the author develops three exciting and doable OL proposals. Two have a focus on classroom activities, being: Meddies in the Mediterranean Sea and North Atlantic and floating lava islands. A third activity concerns a transdisciplinary oceanographical campaign on ancient Mediterranean civilizations.

The seventh chapter "Innovative Education Strategies to Advance Ocean Literacy" stresses the urgent need to raise levels of ocean literacy through education, to make it possible for citizens to mandate politicians and policy makers to manage the ocean sustainably. But ocean literacy is more than just knowledge about the ocean. It should lead to a deeper understanding and change in our behaviour to take care of the health of the ocean. McCauley, Davison, McHugh, Domegan and Grehan describe two innovative ocean education strategies within the Sea Change project, that stimulate the spread of ocean literacy: gaming pedagogy and e-book design. Inclusive participation with a stakeholder group is central to these developments. In developing gaming pedagogy, the authors successfully used CoderDojo, a global movement of free, volunteer-led programming clubs of 7–17-year-old children. In the design of a *Harmful Algal Blooms* e-book, it turns out that tablets are a promising instrument. The authors conclude that the application of new media and methodologies provide and attract a modern education platform, to increase ocean literacy.

The eighth chapter "Sail Training Has Set Sail on a Course Towards Ocean Literacy" introduces the concept of "sail training" as a non-formal, adventurous and experiential, alternative education approach mostly on board large, ocean-going sailing vessels. Lyth notes that today sail training is taking place in some 40 counties and involves more than 200 operators with about 570 ships. This offers an interesting OL opportunity for young people between 12 and 25 years of age. In the chapter these opportunities are related to the seven Ocean Literacy Principles, of which principles 1, 3 and 6 can easily be achieved by most sail training operators. During sail training pupils and students get to know and are immersed in natural elements such as ocean currents and winds. Additionally, they will see the beauty of sea life, and the consequences of human activities on land and sea. Lyth also relates sail training to the UN Sustainable Development Goals quality education and life below water.

Applications

The nineth chapter "The Importance of Ocean Literacy in the Mediterranean Region—Steps Towards Blue Sustainability" examines the complex Mediterranean Sea region in terms of the main Anthropogenic pressures, being: coastal urbanization, tourism, overfishing and marine aquaculture and pollution. These pressures are then related to OL activities. Human-induced climate change highly affects the region through climate warming and sea-level rise. Mokos, Cheimonopoulou, Koulouri, Previati, Realdon, Santoro, Mogias, Boubonari, Satta and Loakeimidis extensively describe the role of the UN Environment's Regional Seas Programme and other regional bodies in the protection of the marine environment, as well as the application of Integrated Coastal Zone Management and the establishment of Marine Protected Areas in the Mediterranean Sea. Again, these activities are related to ocean literacy, and to achieve UN Sustainable Development Goal 14, *life below water*.

The tenth chapter "Fostering Ocean-Literate Generations: The Portuguese Blue School". Costa, Mata, Silva, Conceição and Guimarães is present the Portuguese

Blue School program. Portugal as a pre-eminent maritime nation and having an extensive Exclusive Economic Zone was the first country to translate and adapt the Ocean Literacy Principles in Europe. In this chapter, the development of both the Blue School and the Portuguese Blue School concept is discussed. The authors provide a fascinating and clear picture of the way how the Blue School concept is embedded in the Portuguese national policy, what a Blue School is, which criteria apply, how the application process functions and the sort of partnerships playing a role at the local, national, European and international level. Moreover, the Blue School concept relates to the IOC-UNESCO ocean literacy perspectives, the Ocean Literacy Principles and the UN Sustainable Development Goals 4, 14 and 17. Currently upgrading of the concept is taking place in cooperation with other countries in Africa, Europe and South America, and within the European Commission.

The eleventh chpater "Two Ocean Aquarium Academy: An Introduction to Ocean Literacy Programmes and a Marine Sciences Curriculum" describes the activities of the privately run Two Oceans Aquarium in Cape Town, South Africa. The aquarium is educating the general public about human impacts on the ocean, and is located in the famous and very popular Waterfront area. Russel notes that the Western term ocean literacy has a negative connotation in South Africa. He sketches the development of the aquarium and focuses both on its role in creating public awareness, and developing focused educational efforts for young people. For this, a separate Environmental Education Centre with discovery classrooms was constructed. In these discovery rooms Gr 3 to Gr 12 pupils and students are offered a wide range of lessons related to animals in the Aquarium. Additionally, an Aquarium Marine Sciences Academy Course and a Young Biologist Course are developed successfully. These activities are supported by the government as they perfectly fit in the South African Blue Economy or Operation Phakisa, which means "hurry up" in the Sesotho language, initiative.

Target Audience

The book *Ocean Literacy: Understanding the Ocean*, is giving a wealth of information on ocean literacy as it is today, and as it can be in the near future. Ocean literacy is an emerging field of a broad focused awareness-raising activity through the educational systems. An interesting challenge will be to fine-tune OL activities at the culturally diverse national and regional levels, as is demonstrated so well in the last chapter of this book.

It is our strong belief that this book will be a valuable tool for educational leaders, teachers, lifelong learners, ICT developers, etc., as well as for scientists, policy makers and politicians. In addition, we think that the book will be highly relevant to schools, universities and national and international organizations, with respect to sustainability.

Finally, this book is also important for the general public, as it provides insights into how we can ensure that citizens are better informed to decide how we can address the global challenge to improve the health of the ocean. And that is crucial for our survival as a species.

July 2020 The Editors

Acknowledgments

The editors would like to acknowledge a group of people, a professional organization and one person without whose efforts and support the book you have in your hands and your browser could not have reached you.

First and foremost, the editors would like to thank the authors for not only providing their expertise, but mainly for doing so under difficult circumstances. Of course, the same holds true for the Reviewers for their efforts and time within a very tight timeline.

Second, we would like to thank EUROGEO (the European Association of Geographers) for helping this endeavor becoming a reality. EUROGEO by dealing with places, people and cultures and exploring international issues ranging from physical, urban and rural environments and their evolution, to climate, pollution, development and political-economy, is greatly interested in scientific issues like ocean literacy. EUROGEO believes that a book on ocean literacy is a fundamental contribution to scientific knowledge. Therefore, EUROGEO has been committed to bring to a successful end the publication of this ocean literacy book by Springer.

Third, we need to thank and express our gratitude to Karl Donert who participated in the project this book is based on, proposed and wrote the proposal for the publication of the book by Springer and supported us through all the difficulties and tribulations.

Prof. Kostis C. Koutsopoulos
National Technical University
Athens and EUROGEO
Pikermi, Greece

Prof. em. Jan H. Stel
Maastricht University
Maastricht, The Netherlands
Puurs-Sint-Amands, Belgium

Contents

Applications

About the Editors

Professor Kostis C. Koutsopoulos has taught at the University of Colorado, the University of Iowa and the National Technical University of Athens where he served as director of the Geography and Spatial Analysis Lab., as Chairman of the Geography and regional planning Department and as Dean of the Rural and Surveying Engineering School. He is a member of several scientific and academic associations as well as journal boards. In addition, he is VP of EUROGEO and Chief Editor of the European Journal of Geography and the book series titled "Key challenges in geography" published by Springer. He has organized numerous congresses, meetings and seminars and he has participated in many more. He has presented more than 170 papers in various meetings, has published more than 80 papers in refereed journals and has edited or written 70 books and chapters. Finally he would like to dedicate his work in this book to the newest member of his family, his granddaughter ELLI.

Prof. em. Jan H. Stel studied geology and paleontology. His thesis concerned a paleo-biological study of Silurian favositid corals of the Swedish island Gotland. As an ocean science manager, he organized ocean-going expeditions, developed a European consortium of small European countries to participate in the Ocean Drilling Program, initiated capacity building programs for IOC UNESCO, developed the Dutch Antarctic research program, was initiator of "Das schwimmende Klassenzimmer" outreach project with RV *Polarstern* in 1998, participated in the development of an air link between Cape Town in South Africa and Dronning Maud Land through the DROMLAN-project, Antarctica, was a facilitator in UNESCO's Training-through-Research (TTR) program, and organized projects at the interface with the ocean industry. From 2000, he was a professor in "Ocean Space and Human Activities" at the University of Maastricht, The Netherlands. Jan wrote some 350 (popular) science papers and blogs and made a number of TV documentaries, to inform the public at large why ocean space is important for us.

Abbreviations

AAAS	American Association for the Advancement of Science (USA)
ACCOBAMS	Agreement on the Conservation of Cetaceans in the Black Sea, the Mediterranean Sea and the contiguous Atlantic Area
AMEA	Asia Marine Educators Association
ASTO	Association of Sail Training Organisations
AUV	Autonomous Underwater Vehicle
BASF	German Chemical Company
BBC	British Broadcasting Corporation
BBNJ	Biodiversity Beyond National Jurisdiction
CaNoe	Canadian Ocean Literacy Network
CBD	Convention on Biological Diversity (UN)
CFP	Common Fisheries Policy
CHM	Common Heritage of Mankind (UNCLOS)
CIA	Central Intelligence Agency (USA)
Ciimar	Interdisciplinary Centre of Marine and Environmental Research
CLAMER	Climate Change and European Marine Ecosystem Research
CoE	College of Exploration
COSEE	Centers for Science Education Excellence
COVID-19	Corona Virus Disease 2019
DAPSIWR	Drivers-Activities-Pressures-State-Impact-Welfare-Responses
DBIR	Design-Based Implementation Research
DG MARE	European Commission's Directorate-General for Maritime Affairs and Fisheries
DGPM	Directorate General for Maritime Policy (Portugal)
DPRK	Democratic People's Republic of Korea
DPSIR	Drivers-Pressures-State-Impact-Responses
DTU	Technical University of Denmark
ECSA	European Citizen Science Association
EEZ	Exclusive Economic Zone (UNCLOS)
EL	Environmental Literacy
ELL	Environmental Literacy Ladder
EMB	European Marine Board

EMSEA	European Marine Science Educators Association
EMSEA-Med	European Marine Science Educators Association—Mediterranean
ENSO	El Niño–Southern Oscillation
ESA	European Space Agency
EU	European Union
EUROGEO	European Association of Geographers
EuroGOOS	European Global Ocean Observing System
FAO	Food and Agriculture Organization (UN)
FEE	Foundation for Environmental Education
FIFA	Fédération Internationale de Football Association
GCOS	Global Climate Observing System
GEBCO	General Bathymetric Chart of the Oceans
GEO	Group on Earth Observations
GEOMAR	Helmholtz Centre for Ocean Research Kiel (Germany)
GEOSS	Global Earth Observation Systems of System
GES	Good Environmental Status
GFCM	General Fisheries Commission for the Mediterranean (FAO)
GIS	Geographical Information Systems
GMES	Global Monitoring for Environment and Security (ESA)
GOC	Governmental Oceanographic Commission
GOOS	Global Ocean Observing System
GOS	Global Observing System
GTOS	Global Terrestrial Observing System
GVA	Gross Value Added
HAB	Harmful Algal Bloom
HIPPO	Habitat destruction, Invasive Alien, Population, Pollution, Overexploitation
HMS	His or Her Majesty's Ship
IAS	Invasive Aquatic Species
ICCAT	International Commission for the Conservation of Atlantic Tunas
ICT	Information and Communications Technologies
ICZM	Integrated Coastal Zone Management
IEA	International Association for the Evaluation of Educational Achievement
IGC	Intergovernmental Conference of States
IGU	International Geographical Union
IHE	Institute of Higher Education
IHO	International Hydrographic Organization (UN)
IMBBC	Institute of Marine Biology, Biotechnology and Aquaculture (Greece)
IMO	International Maritime Organization (UN)
IOC	Intergovernmental Oceanographic Commission (UNESCO)
IOLR	Israel Oceanographic and Limnological Research
IOLS	International Ocean Literacy Survey

IPBES	Intergovernmental Science-Policy Platform on Biodiversity and Ecosystem Services
IPCC	Intergovernmental Panel on Climate Change (UN)
IPMEN	International Pacific Marine Educators Network
ISA	International Seabed Authority (UNCLOS)
IUCN	International Union for Conservation of Nature
JSON	JavaScript Object Notation
K-2	From kindergarten to second grade in the USA education system
LED	Light-Emitting Diode
LHS	The Lawrence Hall of Science
LTK	Local and Traditional Knowledge
MARPOL	International Convention for the Prevention of Pollution from Ships (IMO)
MBA	Marine Biological Association
MH	Malaysia Airlines
MIO-ECSDE	Mediterranean Information Office for Environment, Culture and Sustainable development
Mofat	Ministry of Foreign Affairs and Trade (Republic of Korea)
MPA	Marine Protected Area
MSFD	Marine Strategy Framework Directive (EU)
MSL	Mediterranean Sea Literacy
MSP	Maritime Spatial Planning
NAHF	The Northeast Asian History Foundation
NASA	National Aeronautics and Space Administration (USA)
NEARGOOS	North East Asia GOOS
NGO	Non-Governmental Organization
NGS	National Geographic Society
NGSS	Next Generation Science Standards
NMEA	National Marine Educators Association
NOAA	National Oceanic and Atmospheric Administration (USA)
NOS	National Ocean Strategy
NSF	National Science Foundation (USA)
OECD	Organisation for Economic Co-operation and Development
OHI	Ocean Health Index
OL	Ocean Literacy
OLF	Ocean Literacy Framework
OLP	Ocean Literacy Principles
OOI	US Ocean Observatories Initiative
PAHs	Polycyclic Aromatic Hydrocarbons
PBL	Problem-Based Learning
POPs	Persistent Organic Pollutants
PRB	Population Reference Bureau (Republic of South Africa)
PRECEDE	Predisposing, Reinforcing, and Enabling Constructs in Educational Diagnosis and Evaluation

PROCEED	Policy, Regulatory, and Organizational Constructs in Educational and Environmental Development
PT	Portugal
RMS	Royal Mail Ship
ROK	Republic of Korea
ROV	Remotely Operated Vehicle
SDG	Sustainable Development Goal
STEAM	Science, Technology, Engineering, Arts and Mathematics
STEM	Science, Technology, Engineering and Mathematics
STI	Sail Training International
STS	Sail Training Ship
TEK	Traditional Ecological Knowledge
THC	ThermoHaline Circulation
TIMSS	Trends in International Mathematics and Science Study
TSA	Tall Ships America
TV	Television
UGOT	University of Gothenburg (Sweden)
UK	United Kingdom
UN	United Nations
UNCLOS	UN Convention on the Law of the Sea
UNCSGN	United Nations Conference on the Standardization of Geographical Names
UNESCO	United Nations Educational, Scientific and Cultural Organization
UNEP	United Nations Environmental Programme
UNEP/MAP	UNEP Mediterranean Action Plan
UNGEGN	United Nations Group of Experts on Geographical Names
US / USA	United States of America
VLIZ	Vlaams Instituut voor de Zee/Flanders Marine Institute
WAZA	World Association of Zoos and Aquariums
WFD	Water Framework Directive (EU)
WHOI	Woods Hole Oceanographic Institution (USA)
WMO	World Meteorological Organization (UN)
WWF	World Wildlife Fund
YB	Young Biologist

General on Ocean Literacy

Ocean Literacy: Background, Future Drivers, and Opportunities

Jan H. Stel

Abstract In this introductory chapter a brief overall introduction is given in which current trends in ocean sciences, relevant for the development of ocean literacy, are sketched. In *The ocean as the planet's life support system*, the history of oceanography is briefly discussed, and a number of characteristics of the ocean are mentioned. In four *The need to…* sections trends in ocean sciences understanding, observing, protecting, and involving citizens and pupils from primary and secondary schools, are discussed.

Keywords Perceptions · Ocean characteristics · Water cycle · Monitoring · Census of Marine Life · UN Decade of Ocean Science for Sustainable Development · UN Sustainable Development Goals

The Ocean as the Planet's Life Support System

We live in a human-made world on a planet we call ours, and with an ocean that we often also stubbornly call ours. That is our first fallacy. Today, March 10, 2021 at 10.00 hours, the World Population Clock indicates that we humans number 7,851,162,073. The clock is ticking fast. Since the 1950s, human activities have increased manyfold, as has the associated pollution. We have, over time, created a wasteful and unsustainable society in which we abuse nature. And, we assume that this can go on forever. That is our second fallacy.

Like all life around us, we are just part of a billion years evolutionary process, forming the present web of life. Although we previously shared the planet with other human species, we are now the only surviving one. Some say we exterminated the other human species, as we probably also did with the large Pleistocene mammals, like mammoths. We are the only human species who thousands of years ago, domesticated plants, wild mammals, and birds for food. Additionally, we started to build

J. H. Stel (✉)
Maastricht Sustainability Institute, Maastricht University, Maastricht, The Netherlands
e-mail: janstel@skynet.be

Puurs-Sint-Amands, Belgium

K. C. Koutsopoulos and J. H. Stel (eds), *Ocean Literacy: Understanding the Ocean*, Key Challenges in Geography, https://doi.org/10.1007/978-3-030-70155-0_1

societies dating back thousands of years, at the various continents and at various times at those continents.

For tens of thousands of years we have seen the ocean just as a surface, which we could cross to explore the other side. The voyage (Fig. 1) of the British corvette HMS *Challenger* marks the beginning of ocean sciences. The vessel was a sixty-meter sailing ship with an engine of 1200 horsepower, one of the new achievements of the booming British Industrial Revolution. With 220 crew and seven scientists on board, she left port in December 1872 and returned on May 24, 1876, after traveling 68,890 nm and obtaining a wealth of new data. Unique to this expedition was a standardized series of measurements, which were taken at regular distances. The voyage of the *Challenger* led to a new and overwhelming view of the seas and ocean: a place full of life, with a complex physical structure, and unknown resources, such as manganese nodules, on the deep-sea floor.

Oceanographic research quickly changed from this two-dimensional perspective into a 3D one. The scientific quest for a better understanding of the ocean began to blossom after the Second World War. New technology including robots and artificial intelligence, modeling, and synoptic observation by dedicated satellites played a main role in this development. This again changed our perspective, as we learned about deep ocean currents, interactions between the ocean and the atmosphere such as the El Niño phenomena, as well as interactions with the ocean floor, seen at hydrothermal vents. Today, it is well known that the ocean has a fourth dimension: time (EMB 2019). As a consequence, the notion of ocean space was introduced by myself (Stel 2002, 2003, 2013). Moreover, operational oceanography matured through, among others, the international Global Ocean Observing System (GOOS) and the European Global Monitoring for Environment and Security, GMES (see websites).

The ocean covers almost 71%, being 361 million km^2, of the earth's surface. A fact we realize when we see pictures taken from space. Some people think that naming the planet Earth was a mistake. It should have been 'Planet Ocean' or 'Blue Planet' (Earl 2009). Yet, in the early Middle Ages, when the notion Earth was framed, this fit in very well with the land-based peasant society. But from a modern geological perspective the ocean is just a thin veneer on the earth's surface (Fig. 2).

Fig. 1 HMS *Challenger* sailing in the Southern Ocean (left) and its officers and scientific crew (right). © Wikipedia

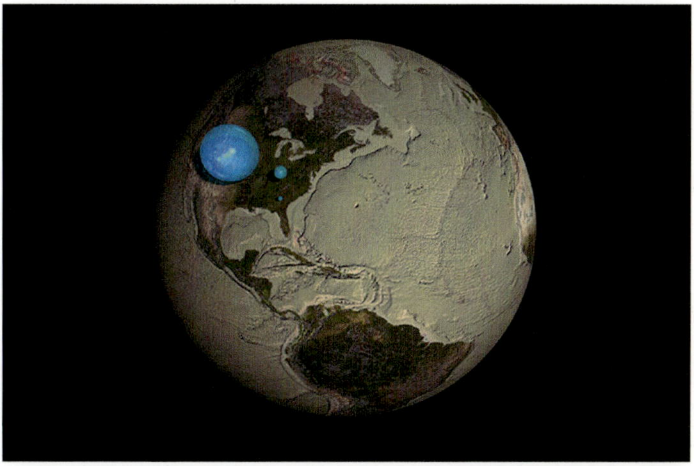

Fig. 2 Bubble sizes represent all water on the planet (largest; diameter 1,384 km), all fresh liquid water in lakes, rivers, swamps, and the ground (mid-sized; diameter 272.8 km), and fresh water in lakes and rivers (smallest; diameter 56.2 km). Credit: Howard Perlman, USGS; globe illustration by Jack Cook, Woods Hole Oceanographic Institution and Adam Nieman

The ocean is a crucial part of the earth system, and interacts with the other subsystems: air, land, and life, at various timescales. Through earth systems thinking we now know that human activities also affect that system, mainly by pollution, by the extraction of raw material like wood, oil and gas, and fish, and by building infrastructures, such as dams. We are changing the world we live in, and we are now even threatening our own lives.

Water is essential to life on earth. It connects the major parts of the earth's climate system. The present ocean contains 1.39 billion km^3 of salty water, circling the present five continents. In earth's geological history, the shape of the ocean has continuously changed, as has its size. Sometimes the ocean covered up to 80% of the earth surface. For billions of years, water has been ceaselessly circulating and recycling.

The water cycle (Fig. 3) is driven by the sun. Water is evaporated from the ocean, which moves through the atmosphere and precipitates as rain and snow, being temporarily on loan from ocean space. Fresh water flows across the land and is stored in glaciers and ice sheets, lakes, rivers, and the ground. Just a few tenths of one percent is directly available for drinking water. That resource we have to share with all other life. Moreover, it has been unevenly distributed across the land surfaces.

Due to the present anthropogenic climate change and population growth, fresh water is rapidly becoming a scarce resource. According to the 2018 edition of the United Nations (UN) World Water Development Report, the global demand for water is increasing at a rate of about 1% per year. This is caused by population growth, economic development, and changing consumption.

The vast majority of the growing demand for water occurs in countries with developing or emerging economies. Yet, in developed countries like Belgium,

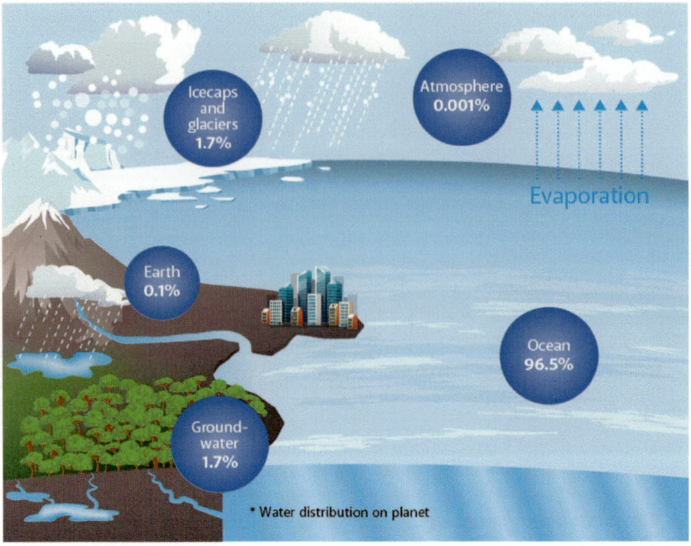

Fig. 3 Water cycle. Every living organism is connected with ocean space through the water cycle. © Flanders Marine Institute, Ostend, Belgium

droughts are increasingly causing problems in agriculture, while adequate preservation measures are not taken. Moreover, water wars might again be lurking in the near future. Vulnerable hotspots are rivers like the Nile, Ganges–Brahmaputra, Indus, and Tigris–Euphrates.

At the same time, the global water cycle is intensifying due to climate change, with wetter regions generally becoming wetter and drier regions becoming even drier. The consequences of climate change are well explained in the three landmark special reports of the IPCC, being *Global Warming of 1.5°C,* above preindustrial levels, *Climate Change and Land,* and *The Ocean and Cryosphere in a Changing Climate*, published in 2018 and 2019. Just as the 2019 Global Assessment Report on Biodiversity and Ecosystem Services, they jointly paint a grim picture, if we do not change our Western lifestyle of extreme consumerism, and reharmonize with nature.

The ocean is everywhere in our lives. The ocean touches you with every breath you take, with every drop of water you drink, with the food you consume. It is our life support system, and it begins to falter due to human activities on land. We need to forget our deep-rooted perception of the ocean as too vast and too resilient to be affected by human activities. It also is crucial to realize that the ocean provides a vast number of free ecosystem services like regulating climate and weather, providing food and leisure, absorbing both 93% of the excess heat released over the past 50 years by human activities, and 30% of the CO_2 we produce. But this heat and CO_2 is not absorbed uniformly. It varies in both space and time, with the largest changes in polar regions. The ocean acts as a lifeline for sustaining our economies, with more than 90% of all transport occurring across its surface (UNCTAD 2019).

A Need to Understand

From space the ocean is an eye-catching feature of the planet. On July 19, 2013, NASA's Cassini spacecraft captured the Earth from a distance of 1.45 billion kilometers beneath Saturn's imposing rings, as a faint blue speck in the vast blackness of outer space (Fig. 4). Planet earth is a unique feature within the solar system.

Yet, planets like Mars are better mapped than the earth's ocean floor, of which only a bare 15% has been mapped in detail, meaning with a resolution high enough to identify features the size of an autobus. It is no wonder that we can't find the lost Malaysian Airlines flight MH 370. If we look at the ocean waters, just 0.0001% of the deep-sea, the waters below 200 m, is explored. Even today, the ocean is relatively unknown. It is the last frontier of the earth to be explored in the twenty-first century.

The first international ocean project of the twenty-first century was the Census of Marine Life, which started in 2000 and ended ten years later. Finally, more than 2,700 scientists from 80 countries recorded, in one of the largest scientific collaborations so far, the diversity, distribution, and abundance of life in the ocean. The Census addressed three basic questions: What has lived in the ocean, what does live in the ocean, and what will live in the future ocean? During a decade, scientists explored and sampled the ocean from the polar regions to the tropics, and from deep-sea hydrothermal vents to coastal systems. Within the Census, scientists studied species as large as a blue whale and as small as a microbe. More than 6,000 potentially new ocean species were discovered.

The easily accessible nearshore zone of the ocean, with a depth of up to 20 m, forms an interface with the land. It is the most-studied region of the ocean. Yet, even here we still don't know how many species live at our doorstep. Estimates range from 178,000 to more than ten million species. As there are many ecosystems in the coastal zone, the Census focused on the rocky bottoms, dominated by kelp forests

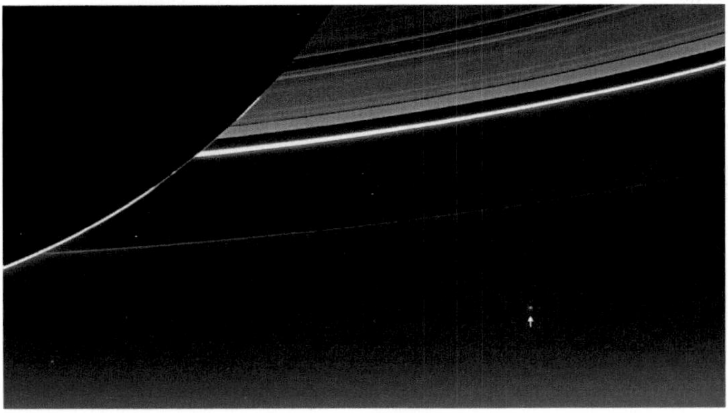

Fig. 4 On July 19, 2013, Cassini spacecraft captured Saturn's rings and the Earth which is 1.45 billion kilometers away, as a blue dot. © NASA Jet Propulsion Laboratory Caltech/Space Science Institute

and other seaweeds, and soft-bottom areas covered by seagrasses. Both show a global distribution, a high biodiversity, and are among the most productive ecosystems on earth. They also provide a range of ecosystem goods and services on which human populations depend.

The ocean may seem featureless to us, but not to the organisms that dwell in it. The Census found that many predatory animals, like sharks, sea turtles, and whales, congregate off the coast of California in the California Current. Here, cold, nutrient-rich water moves from deep ocean space toward the surface. It brings blooms of phytoplankton and food like squid, sardines, and krill.

Another remarkable outcome was the discovery of a remote mid-Pacific Ocean area, with a 250 km radius, halfway between Mexico and Hawaii, acting as a winter and spring habitat of coastal great white sharks. Scientist dubbed it the "White Shark Café." Satellite tracking data showed that white sharks in a Serengeti-like fashion travel from diverse rookeries along the North American coast to the "Café." Here they loiter for several months.

On average these sharks travel around a hundred days, while diving up to almost thousand meters for food. At the "Café" they dive, once every ten minutes, just up to 460 m when feeding (Schmidt Ocean Institute 2018). Based upon satellite information it was hitherto assumed that this area was an ocean desert. Yet, research in 2018 showed it teeming with life in the deeper twilight zone layers of the ocean. These zones are undetectable by satellites, but can be explored by remotely operated vehicles and robotic underwater drones, such as gliders.

The Census also studied microbes, life in the ocean we can't see with the naked eye, including microscopic viruses, bacteria, phytoplankton, and zooplankton that drift with the ocean currents. An average liter of ocean water holds around 38,000 microbial bacteria most of which are supporting ocean life. Moreover, there are approximately ten million viruses in every drop of surface seawater. Marine microbes are the dominant life forms in ocean space. They comprise more than 90% of the living biomass in the ocean. Microbial communities thrive in places where one wouldn't expect life to survive, such as at deep-sea hydrothermal vents or deep in the ocean's crust.

Some hundred scientists analyzed historical population data of marine species to learn about how the number of animals changed over time, and how their traits, such as the size of caught fish, have changed. This research was done by archeological digs and studying waste pits, by reading historical documents and old menus and by reviewing trophy-fishing pictures. They are all vital indicators to understand how human activities have affected marine populations over the last 500–2000 years. As such, they provide a baseline, which, among others, can be used for future conservation efforts.

We are just beginning to understand the ocean's complex system, while at the same time we are heedlessly and irreparably damaging its ecosystems by mining, oil and gas exploitation, and fisheries. We now know that there are limits to the waste, like heat, CO_2, and plastics, which we can dump into the ocean, without changing the stability of the system. Actually we are already changing the ocean on an unprecedented scale, due to our polluting activities in a consumer-focused and throw-away society, which we are apparently not willing to change. The present corona crisis has shown us our vulnerability, and gives us a glimpse of how our life in a less polluting world would look like. No more traffic jams, beautiful blue skies without the polluting white trails of planes, less nitrogen pollution, etc. There is an urgent need to transform our world into a sustainable one.

In 2015 the world agreed to the UN 2030 Agenda for Sustainable Development, with its 17 Sustainable Development Goals (SDGs; Fig. 5; see websites) and 169 targets. As such, the Agenda is a blueprint for a sustainable world. The interconnected goals and targets address the present global challenges, such as those related to poverty, inequality, climate change, environmental degradation, and peace and justice. All goals have to be achieved in 2030, just over a decade away. Yet, human activities increasingly affect the health of ocean space. CO_2 pollution and most of the marine pollution by chemicals and plastics are reaching alarming levels. They are mostly derived from land-based sources. SDG 14 is related to the conservation and sustainable use of the ocean, seas, and marine resources. There is an urgent need for a healthy ocean.

The ocean, the largest biome on earth, is under threat from both the ever-increasing human population and the related rapidly diversifying human activities, and a Western lifestyle. Together, they put an enormous pressure on marine ecosystems goods and services, which are fundamental to our well-being. In 2016 the OECD calculated that the ocean economy generated US$1.5 trillion in 2010, and could grow to US$3 trillion in 2030 (OECD 2016). In 2015 the First World Ocean Assessment or First Global Integrated Marine Assessment was published by the UN (UN 2015). This report notes that the oceans' carrying capacity is near or at its limit.

In 2017 the IOC-UNESCO assessed the status and trends in the ocean science capacity around the world for the first time. This Global Ocean Science Report

Fig. 5 Sustainable Developments Goals for 2030. © UN

demonstrates that national investment in ocean science is low, on average just 1% of national science budgets. As a result, ocean observation networks are less sustainable than the meteorological ones (IOC-UNESCO 2017). In 2019 the Intergovernmental Science-Policy Platform on Biodiversity and Ecosystem Services (IPBES) published its Global Assessment Report on Biodiversity and Ecosystem Services. The report shows overwhelming evidence that nature is declining globally at rates unprecedented in human history. It sketches a grim future (IPBES 2019).

The upcoming UN Decade of Ocean Science for Sustainable Development (2021–2030) aims to ensure that ocean science will support, assist, and guide the SDGs with a focus on SDG 14, life below water. The Decade comes at a time in which society strongly depends on the health of the ocean for its future. For that future, we need to make decisions based on solid science in a time with insufficient scientific knowledge. Peter Thomson, the United Nations Secretary-General's Special Envoy for the Ocean, is rather outspoken in this matter (Voices of the Ocean Decade 2020). "*I believe that what's happening in the ocean will determine the survival of our species. The IPCC special report on global warming tells us that once we go over the dreaded line of 2 °C above pre-industrial levels, we lose what's left of the planet's living coral reefs, which are home to around 30% of the ocean's biodiversity. If you take 30% of the ocean's biodiversity away, do you have a healthy ocean? Surely not. Can you have a healthy planetary ecosystem without a healthy ocean ecosystem? No, you can't. The ocean is the most important element of this blue planet, and that's why I say that our fate may be closely linked with that of coral.*"

The Decade offers a possibility to build scientific capacity and the potential to develop sustainable ways to use ocean resources in line with SDG 14. Additionally, there are numerous interactions between this SDG and the achievement of many other SDGs. For example, under optimistic projections the ocean has the potential to supply up to six times more food than it does today (SDG 2, zero hunger). New technologies in renewable energy or carbon storage could increase the capacity of the ocean to mitigate the worst effects of climate change (SDG 7, affordable and clean energy; SDG 13, climate action). New knowledge and tools for coastal nature-based solutions could increase the adaptive capacity of hundreds of millions of the most vulnerable people (SDG 3, good health and well-being; SDG 10, reduced inequalities), (see websites; IOC 2020b).

The Decade is much broader than a traditional multidisciplinary oceanographic research project. It builds on IOCs holistic interdisciplinary approach to ocean sciences, in which the human element is an important aspect too. In the coming Decade, ocean science is defined broadly and includes: social sciences and human dimensions; the infrastructure that supports ocean science (observations, data systems, etc.); the application of those sciences for societal benefit, including knowledge transfer and applications in regions that are lacking science capacity; ocean literacy; and the science–policy/user interface. Ocean science also integrates local and indigenous knowledge. The leading principle in the Decade is to move from the "ocean we have" to the "ocean we want," that being a healthy ocean (IOC 2020a, b).

A Need to Observe

Most of the major discoveries in ocean sciences have occurred within the last 50 years. We have opened the well-preserved archives in the ocean floor by the International Ocean Discovery Program and its various predecessors since 1968 (see Chapter 4). Through this we have learned that the ocean is shaped by tectonic processes at a hundreds of million years time scale, and that the size of the ocean varies during the geological history of the planet. We also learned that climate change is a feature of the earth, as is life. Exploration and technological innovations, such as the developments of satellites and modeling, go hand in hand. The latter is continuously improving in concert with rapidly increasing computer power. This academic research transitioned into becoming part of operational oceanography in the 1990s (Stel et al. 1997).

The satellite revolution began with the launch of the Sputnik on October 4, 1957 by the former Soviet Union, soon followed by the Explorer I on January 2, 1958 by the US. Advances in computers and space technology at the end of the 1950s and the beginning of the 1960s, led to this revolution. This forever changed the way people are observing the planet. In the mid-1960s, this led to the development of the WMO World Weather Watch for the lower atmosphere, and its Global Observing System (GOS) as well as the Global Data Processing System and the Global Telecommunication System. GOS is an extremely complex undertaking, and one of the most successful international partnerships of the last sixty years (see websites). Yet, there is a looming observational gap in middle and higher atmosphere forming geospace (Mlynczak et al. 2021). Today, a fleet of satellites provides data to different user communities, in the field of meteorology, oceanography, and climate.

Space-based observations contributed strongly to a dawning understanding that human activities are without doubt taking their toll on the environment. The 1972 United Nations Conference on the Human Environment in Stockholm, Sweden, was the first international conference to address these environmental problems directly. In 1992 it was followed by the United Nations Conference on Environment and Development, or the Earth Summit, in Rio de Janeiro, Brazil. This world summit produced the Rio Declaration on Environment and Development, the Statement of Forest Principles, and adopted Agenda 21, an unprecedented global action plan for sustainable development. It also led to the establishment of the Convention on Biological Diversity, and the United Nations Framework Convention on Climate Change. For the ocean, the acceptance of an IOC proposal for the development of a Global Ocean Observing System was crucial. It stimulated the development of that observing system, as well as the transition toward operational oceanography.

In 2005 at the second World Summit on Sustainable Development in Johannesburg, South Africa, the Group on Earth Observations (GEO) was accepted and established. The main goal of GEO is to create the Global Earth Observation Systems of System, GEOSS, to bring together various specialized earth observing systems. In Europe, this resulted in the establishment of the Global Monitoring for Environment and Security (GMES) earth observation program. This program was launched by the European Space Agency (ESA), and funded by the EU (Fig. 6). During its

Fig. 6 ESA's Sentinel 3 satellite is observing land and ocean within the European Copernicus program. © ESA

development there were strong sentiments between ESA and EuroGOOS, which was taking care of the development of the crucial in situ observation infrastructure, which ESA could not provide. GMES became a success story and is now known under the name Copernicus, the second flagship program of the EU (see websites). The Global Ocean Observing System is briefly discussed in Chapter 4.

Observations from space do not have the capability to look into the ocean. As a consequence, novel in situ observing systems have to be developed. Today, a wide variety of underwater robots and underwater drones are available to monitor ocean space. The Argo-network, discussed in Chapter 4, is an excellent example of this. Autonomous underwater drones have become a common feature in ocean research and monitoring. Rutgers University Center for Ocean Observing Leadership (RU COOL) gives a glimpse into the future with their daring gliders programs (see websites). They, among others, organized the first ever crossing of the Atlantic Ocean, and repeated the famous *Challenger* expedition with a number of gliders (Fig. 7). Another glimpse into the future are the development of wave-powered surface robots that can transport and release gliders, and the development of autonomous networks on the ocean floor to explore ocean space. These developments should be an attractive element of future ocean literacy activities.

Fig. 7 A Challenger glider dives off the Brazilian coast, into ocean space. Credit the Rutgers University Center for Ocean Observing Leadership

A Need to Protect

The ocean is the largest ecosystem on Earth and it is the planet's life support system. Yet, the first global holistic assessment of ocean space ever was published only six years ago (UN 2015). A large part of the ocean still is unexplored. Therefore, we should be careful in using it. In 2015 the world also agreed to SDG 14 and by this, committing itself to conserve and sustainably use ocean resources. This SDG is raising awareness about the crucial role the ocean is playing in protecting us from, for example, climate change. On the other hand, the ocean is providing us with a series of valuable ecosystem services from oxygen and food to climate regulation. Estimates of the economic value of the carbon storage by the High Seas range from US$74 billion to US$222 billion per year. Yet, the ocean remains chronically undervalued, poorly managed, and is inadequately protected and governed. This is particularly true for the High Seas, which cover more than 60% of the ocean surface and more than 70% of its space (Rogers et al. 2014).

The key international regime governing the ocean is the UN Convention on the Law of the Sea (UNCLOS), which was adopted in 1982 (see websites). Then the key driver was access to, and use of living resources for marine fisheries and the rapidly developing offshore oil and gas sector. This led to the proclamation of the Exclusive Economic Zone, EEZ. During the 1970s it was also thought that the technological feasibility of deep-sea mining was imminent. There was a genuine concern about a fair distribution of the potential profits from the mining industry, and it was considered that these should not exclusively go to the technologically advanced countries. As a solution, the innovative and sustainable notion of the "Common Heritage of Mankind" (CHM) was strongly advocated but not accepted by the Maltese ambassador Arvid Pardo (1914–1999). Today, UNCLOS provides rules regarding the freedom of navigation, the extent of territorial seas, and deep-sea mining. On the other hand, it gives minimal guidance on environmental conservation, while the pressure of deep-water fisheries, shipping, deep-sea mining, uses of new marine resources (bioprospecting) in the near future are rapidly increasing. Moreover, the impacts of climate change and ocean acidification on marine biodiversity are not addressed.

After over a decade of discussions and negotiations at the UN, the General Assembly decided on December 24, 2017 to start a negotiation process for a new treaty and a legally binding framework on the conservation and sustainable use of marine biological diversity within areas beyond national jurisdiction, being in the High Seas. This will be the Biodiversity Beyond National Jurisdiction, or BBNJ Treaty. The new treaty will come under the umbrella of UNCLOS. These negotiations are an important opportunity to fill a number of gaps in the international legal framework governing marine biodiversity. It will address emerging threats to, and use of, biodiversity, for example, and the potential threat of commercialization of marine genetic resources (see Chapter 4).

An intergovernmental conference of states (the IGC) has met three times since September 2018. A difference of opinion is once again noted between countries like the US, Japan, South Korea, and some European countries with advanced marine capabilities, and the emerging economies and developing countries with less advanced or without adequate marine capabilities. Again discussions on sharing the benefits from genetic resources and deep-sea mining and the CHM show the need for capacity building, a fair sharing of the profits, and conservation and protection of key marine areas. The fourth and final session of the IGC was scheduled to take place at the UN Headquarters in New York from March 23 to April3, 2020, but was postponed due to COVID-19.

One of the most effective tools to ensure the sustainability of the ocean is establishing a Marine Protected Area (MPA) or a Marine Protected Volume (EMB 2019), if one looks at it from an ocean space perspective. MPAs are places where nature comes first, where conservation of marine resources have been secured in the long-term perspective, and typically in a marine park or reserve. They range from small sites, often established by local communities, to vast tracts of ocean space, like the Ross Sea MPA established in 2016. It covers an area of 1.570.000 km^2 in the Southern Ocean and offers protection for 35 years. MPAs vary depending on the types of activities which are permitted within its boundaries, from multiple use to no access at all. MPAs can also vary in terms of how long the area will be protected, from permanent to seasonal and rotating.

When the BBNJ treaty will be approved and comes into force, it will allow the designation of more MPAs in the High Seas, as well as developing an MPA-network. A current concern in the negotiations is a lobby to exclude existing agreements on commercial fisheries and other human activities from the treaty. An analysis published in Nature Ecology & Evolution in August 2019 insists that a failure to ensure the coverage of the full scope of fish biodiversity could result in thousands of species continuing to slip through the cracks of a fragmented global ocean governance framework (Crespo et al. 2019). Visalli et al. outline high-priority regions for MPAs and present a top ten of MPA-hotspots within the High Seas. Additionally, the 30% ocean protection target of the International Union for Conservation of Nature was supported, and leads to the protection of 27,3% of the High Seas, i.e., an area of

52,545,634 km^2, which is three times the surface area of Russia and more than five times that of the US (Visalli et al. 2020).

A Need to Explain and Involve

One of the questions frequently posed during the 1998 International Year of the Ocean was: How to protect the ocean if one does not know it? Ocean awareness and outreach were the main tools to advocate its central theme "Our Common Heritage," which was a tribute to Arvid Pardo and Elisabeth Mann Borgese (1918–2002). Jointly with my German colleagues from the Alfred Wegener Institute for Polar and Marine Research in Bremerhaven, I then organized an international competition for schools in Europe. Students had to write a research proposal for evaluation. The winners were offered a visit to the 1998 Lisbon World Exposition (May–September 1998) with a focus on the ocean, and the participation in "Das Schwimmende Klassenzimmer. Eine *Polastern*-Expredition für die Schule" ("*Polarstern* Cruise for Schools") from Lisbon to Bremerhaven, in June 1998 (Fig. 8).

On the fringe of the thirtieth *Pacem in Maribus* conference in Kiev, Ukraine, in the summer of 2003, an exhibition was organized for students from a drawing academy. The conference itself was organized by the International Ocean Institute, the brainchild of Elisabeth Mann Borgese who was widely known as "The mother of the Ocean." When I met her, she was a brave, intelligent, endearing old lady who

Fig. 8 Participant "*Polarstern* cruise for Schools" in June 1998 during the International Year of the Ocean. © Author

Fig. 9 Children are attracted by the sea. Why does the current education system not respond to this? © Arthur Verlinden

also participated in the preparatory meeting of the International Year of the Ocean. She invited me to the conference on "A Year after Johannesburg. Ocean Governance and Sustainable Development: Ocean and Coasts - a Glimpse into the Future." I was pleasantly touched by the beautiful drawings of enthusiastic teenagers, giving explanations during a reception. Today, I would call this a successful ocean literacy activity.

Young children like the sea and the beach, where they go for pleasure and bathing, for collecting seashells, for building sandcastles, etc. As soon as the weather is good enough, hundreds of thousands of families pay a visit to our, often very crowded, Belgian North Sea coast. All over the world, the coast is the number one tourist attraction. During my six-monthly check-up at the dentist, I see (Fig. 9/Frontpage) a beautiful drawing in shades of blue of the sea. A yellow starfish with glasses, a light blue smiling fish, and an exotic red crab make you smile, and makes it clear that you are in ocean space. The newspaper boat in the middle refers to one of the oldest uses of the sea. A way to travel to the other side or an adventurous journey to an unknown destination. Arthur's masterpiece illustrates a seascape on a sunny, cloudy day. Why this seven-year-old boy colored the sun green, is a mystery to me. Perhaps he just run out of yellow paint. His and countless other children's drawings show a warm interest in the sea. Why does the current education system not respond to this clear interest? Why don't they use the vast existing literature and handbooks (Santoro et al. 2017)? Why don't they immerse our kids and students in the many secrets of ocean space to empower them as future citizens and leaders in the more sustainable world of tomorrow?

Browsing through the beautifully illustrated coffee table books *The Blue* (Taylor et al. 1999) or *Smithsonian Ocean: Our Water Our World* (Cramer 2008) or watching and reading BBC's *Blue Planet II* (Honeyborne and Brownlow 2017) brings the spectator and reader into a world of awe and wonder. According to some newspapers, BBC's *Blue Planet II* series has been cited as the greatest nature series which the BBC has ever produced, and was viewed by millions of people around the world. To create this masterpiece, the BBC spends US $10 million on a seven-year shoot. Action, wonder, and wave after wave of exquisite photography are the characteristics of this fascinating journey through ocean space. It has also affected the behavior of millions of viewers who watched the episode about plastic pollution. Almost nine out of ten people (88%) have reportedly changed their habits since then, and 60% choose to use a refillable water bottle and coffee cup instead of the disposable plastic bottles and cups (Waitrose and Partners 2018). What stays with the viewer after watching *Blue Planet II* is that one realizes that ocean space is a world which is completely alien to us, as we are a terrestrial species. In the ocean there is no horizon for orientation. There, above is below and below is above. It is a magic world.

This unexplored world is attractive for young children, schools, students, and citizens (Fauville et al. 2019). It is no wonder then that ocean-related citizen or community science projects are rapidly increasing in number and size. Citizen science is the involvement of the public in scientific research. It includes a broad range of activities, from mapping natural phenomena to counting seashells at a beach. Through citizen science, people share and contribute to data monitoring and collection programs. Citizen science is blooming, as it basically brings a wide variety of benefits to researchers, citizens, policy makers, and society. It can, for instance, accelerate and realize the production of new scientific knowledge. It can assist policy makers to monitor the implementation and compliance with regulations. And, last but not least, it is also an instrument to increase public awareness about science as well as a feeling of ownership of policies. Finally, this approach offers opportunities to raise awareness on ocean issues, to empower citizens, and to increase ocean literacy. All of this is leading to a more environmentally and ocean-friendly behavior (EMB 2017).

The Citizen Science Global Partnership is a network-of-networks that promotes and advances citizen science for a sustainable world. It was launched in December 2017 at the UN Science-Policy-Business Forum on the Environment. The partnership brings together existing regional, national, and local networks of citizen science researchers and practitioners. The European Citizen Science Association (ECSA) was launched in 2013 and is a nonprofit association. Its aim is to encourage the growth of the citizen science movement in Europe. It draws on more than 200 individual and organizational members from over 28 countries across the European Union and beyond. Many citizen science projects have, however, a national or local focus. It is interesting that even in the Antarctic Peninsula, one of the fastest warming regions on earth, the tourism industry has developed a citizen science program. In the FjordPhyto project (see websites), tourists collect environmental data and biological samples that are sent to scientists for research (Cusick et al. 2020).

In 2008, the UN General Assembly announced that 8 June would be the "World Oceans Day" from 2009 onward. This initiative became a successful and interesting ocean awareness instrument, together with the booming of (marine) citizens science projects, and the rapid emergence of ocean literacy. Additionally, at a European level, the Commission has ambitious plans related to the European Green Deal for increasing ocean and water literacy, citizen science, and engagement activities in the EU (European Commission 2020) in the Member States. The aim is to create and upskill a new blue workforce through a European Ocean and Water education, training, and culture program. By 2025 this should include "*the creation of a pan-European ocean literacy network, the provisions of ocean literacy activities and curricula for schools as well as a coherent Europe-wide curriculum for marine science.*"

Improving ocean literacy will lead to a key societal change toward a sustainable ocean society. It is this new generation of ocean-informed citizens and leaders that will secure a healthy ocean for the generations to come.

References

Cramer D (2008) Smithsonian Ocean: Our Water, Our World. HarperCollins 196p

Crespo GO, Dunn DC, Gianni M, Gjerde K, Wright G, Halpin PN (2019) High-seas fish biodiversity is slipping through the governance net. Nature Ecology & Evolution August 2019, pp 1273–1276. https://doi.org/10.1038/s41559-019-0981-4

Cusick AM, Gilmore R, Bombosch A, Mascioni M, Almandoz GO, Vernet M (2020) Polar tourism as an effective research tool: Citizen science in the Western Antarctic Peninsula. Oceanography 33(1):50–61. https://doi.org/10.5670/oceanog.2020.101

Earl SA (2009) The world is blue how our fate and the Ocean's are one. National Geographic Society, Washington, USA, 320p

European Commission (2020) Regenerating our Ocean and Waters by 2030 Interim report of the Mission Board Healthy Oceans, Seas, Coastal and Inland Waters. European Commission, Directorate-General for Research and Innovation, Directorate C—Healthy Planet, Unit C.4—Healthy Oceans & Seas, 56p

European Marine Board (2017) Advancing Citizen Science for Coastal and Ocean Research. Position Paper 23 of the European Marine Board, Ostend, Belgium, 109p. ISBN: 978-94-92043-30-6

European Marine Board (2019) Navigating the Future V: Marine Science for a Sustainable Future. Position Paper 24 of the European Marine board, Ostend, Belgium, 89p. ISBN: 9789492043757. ISSN: 0167-9309. https://doi.org/10.5281/zenedo.2809392

Fauville G, Payne DL, Marrero ME, Lantz-Andersson A, Crouch F (Eds.) (2019) Exemplary Practices in Marine Science Education. A Resource for Practitioners and Researchers. Springer International Publisher AG, Cham, Switzerland, 452p. ISBN 978-3-319-90777-2. ISBN 978-3-319-90778-9. https://doi.org/10.1007/978-3-319-90778-9

Honeyborne J, Brownlow M (2017) Blue Planet II. BBC Books, London, UK, 312p

IOC (2020a) Advancing Science for Sustainable Ocean Business. IOC 02/20, 24p. Downloaded on June 20 from: https://ungc-communications-assets.s3.amazonaws.com/docs/publications/Advancing-Science-for-Sustainable-Ocean-Business.pdf

IOC (2020b) United Nations Decade of Ocean Science for Sustainable Development. Zero draft, 18 March 2020, 48p. Downloaded on June 20 from: file:///C:/Users/Jan/Downloads/Implementation_Plan_Zero_Draft_March_2020.pdf

IOC-UNESCO (2017) Global Ocean Science Report—The current status of ocean science around the world. L. Valdés et al (eds) Paris, UNESCO Publishing, 277p. ISBN 978-92-3-100226

IPBES (2019) Summary for policymakers of the global assessment report on biodiversity and ecosystem services of the Intergovernmental Science-Policy Platform on Biodiversity and Ecosystem Services. Díaz S, Settele J, Brondízio ES, Ngo HT, Guèze M, Agard J, Arneth A, Balvanera P, Brauman KA, Butchart SHM, Chan KMA, Garibaldi LA, Ichii K, Liu J, Subramanian SM, G. Midgley GF, Miloslavich P, Molnár Z, Obura D, Pfaff A, Polasky S, Purvis A,

Razzaque J, Reyers B, Roy Chowdhury R, Shin YJ, Visseren-Hamakers Ij, Willis, KJ, Zayas CN (eds) IPBES secretariat, Bonn, Germany. 56p. https://doi.org/10.5281/zenodo.3553579
Mlynczak MG, Yue J, McCormack J, Liebermann RS, Livesey, NJ (2021) An observational gap at the edge of space. Eos. 102. https://doi.org/10.1029/2021EO155494
OECD (2016) The Ocean Economy in 2030. OECD Publishing, Paris, 256p
Rogers AD, Sumaila UR, Hussain SS, Baulcomb C (2014) The high seas and us: understanding the value of high seas ecosystems: Global Ocean commission. Oxford, UK, 22p
Santoro F, Santin S, Fauville G, Scowcroft G, Tuddenham PD (2017) Ocean literacy for all: a toolkit. IOC Manuals and guides; Vol. 80, 136p., UNESCO Office Venice and Regional Bureau for Science and Culture in Europe, Printed in Venice, Italy
Schmidt Ocean Institute (2018) Scientists Voyage To The White Shark Café, June 4, 2018. Press release, obtained on June 20, from: https://schmidtocean.org/scientists-voyage-to-the-white-shark-cafe/
Stel JH (2002) Mare Nostrum—Mare Liberum—Mare sit Aeternum, duurzaam gebruik van de oceanische ruimte. Maastricht University (Maastricht), 47p
Stel JH (2003) Society and sustainable use of the Exclusive Economic Zones. In: Dahlin H, Flemming NC, Nittis K, Petersson SE (eds) Building the European capacity in operational oceanography. Proceedings of the Third International Conference on EuroGOOS. Elsevier Oceanography Series 69, pp 592–597
Stel JH (2013) Ocean Space and the Anthropocene, new notions in geosciences?—An essay. Netherlands Journal of Geosciences—Geologie en Mijnbouw 92(2/3):193–211
Stel JH, Behrens HWA, Borst JC, Droppert LJ, van der Meulen JP (1997) Operational oceanography the challenge for European co-operation. Proceedings of the First International Conference on EuroGOOS, 7–11 October 1996, The Hague, The Netherlands. Elsevier Oceanography Series, Vol. 62, 757p
Taylor V, Earle SA, Bellamy D, Halsted B, Cousteau J-M, Holmes M, Severin T, Aldrin B, Partridge J (1999) The Blue. EM International Ltd, London, UK, 220p
UN (2015) Summary of the first global integrated marine assessment UN General Assembly, 22 July 2015, A/70/112, Seventieth session, Item 80 (a) of the provisional agenda, Oceans and the law of the sea, 60p. Downloaded on June 20 from: https://issuu.com/uniccanberra/docs/n1518709
UNCTAD (2019) Review of Maritime Transport. Untied Nations, New York, USA, 109p
Visalli ME, Best BD, Cabral RB, Cheung WWL, Clark NA, Cristina Garilao C, Kristin Kaschner K, Kesner-Reyes K, Lam VWY, Maxwell SM, Mayorga J, Moeller HV, Morgan L, Crespo GO, Pinsky ML, White TD, McCauley DJ (2020) Data-driven approach for highlighting priority areas for protection in marine areas beyond national jurisdiction. Marine Policy, 14p. Available online 28 March 2020, 103927
Voices of the Ocean Decade (2020) Q&A with Peter Thomson: Our future health depends on the health of the ocean. 30/04/2020, 2p. Downloaded on June 20 from: https://en.unesco.org/news/voices-ocean-decade-qa-peter-thomson-our-future-health-depends-health-ocean
Waitrose & Partners (2018) Food & Drink Report 2018–19. 12p. Downloaded on June 22 from: https://www.waitrose.com/content/dam/waitrose/Inspiration/Waitrose%20&%20Partners%20Food%20and%20Drink%20Report%202018.pdf

Websites

Copernicus https://www.copernicus.eu/en
ESA https://www.esa.int/
EuroGOOS http://eurogoos.eu/
FjordPhyto https://fjordphyto.ucsd.edu/
GEO https://www.earthobservations.org/index2.php

GOOS http://www.goosocean.org/
IOC UNESCO https://ioc.unesco.org/
RU COOL https://rucool.marine.rutgers.edu/
SDGs https://sdgs.un.org/goals
UNCLOS http://www.un.org/depts/los/
WMO https://public.wmo.int/en

Jan H. Stel studied geology and paleontology. His thesis concerned a paleo-biological study of Silurian favositid corals of the Swedish island Gotland. As an ocean science manager, he organized ocean-going expeditions like the Dutch-Indonesian Snellius II Program in the mid-1980s, developed a European consortium of small European countries to participate in the Ocean Drilling Program, initiated capacity building programs for the IOC-UNESCO, developed the Dutch Antarctic Research Program, organized and executed a visit of the present Dutch king and queen to Antarctica in 2007, and organized projects at the interface with the European ocean industry. From 2000, he was a professor in 'Ocean Space and Human Activity' at the Maastricht University in the Netherlands. Here he coined the notion of 'ocean space' and 'ocean states'. Jan has written some 350 scientific and popular science papers and blogs, to inform the public at large why ocean space is important for us.

Ocean Literacy: From a Ripple to a Wave

Diana L. Payne and Meghan E. Marrero

Abstract This chapter provides a brief history of ocean literacy and the Ocean Literacy Framework, beginning in the United States and expanding around the world. With connections to the "big ideas" in science, geography, and United Nations Sustainable Development Goal 14, the authors illustrate how the one world ocean transcends both disciplinary and geographical boundaries. Simultaneously, a global campaign reaching beyond such limits has many inherent challenges including communication, cultural relevance, evaluation, education, and funding. The authors acknowledge these limitations but also suggest opportunities and recommendations for a more unified transdisciplinary approach culminating in improved communication and ocean-literate citizens across the globe.

Keywords Geography · Global ocean literacy network · Ocean literacy · Traditional ecological knowledge (TEK) · Transdisciplinary · UN Sustainable Development Goal (SDG) 14

Introduction

Released in 2005, the original impetus for the creation of the Ocean Literacy Essential Principles and Fundamental Concepts Guide was to incorporate ocean science content into United States (US) curriculum and instruction, in particular the science education standards (Schoedinger et al. 2010). To assist educators in implementing the content outlined in the Guide, the Scope and Sequence (2009) and alignment to the US Next Generation Science Standards (NGSS) (2015) were developed. These

D. L. Payne
University of Connecticut, Groton, CT, USA

M. E. Marrero (✉)
Mercy College, Dobbs Ferry, NY, USA
e-mail: mmarrero3@mercy.edu

K. C. Koutsopoulos and J. H. Stel (eds), *Ocean Literacy: Understanding the Ocean*, Key Challenges in Geography, https://doi.org/10.1007/978-3-030-70155-0_2

three documents—the Ocean Literacy Guide, Scope and Sequence, and NGSS alignment—comprise the Ocean Literacy Framework. Although originally intended for a US ocean science and education audience, the ocean literacy campaign has grown significantly to influence global initiatives in ocean and coastal education, science, and policy.

Due in part to its original intent, empirical evidence supporting the impact of the ocean literacy campaign is limited. Therefore, this chapter sets the historical context for the chapters that follow, which is critical to evaluate the present situation and future developments. This chapter briefly outlines the history of the ocean literacy campaign, including key milestones in the expansion of ocean literacy to different regions worldwide, and provides an overview of the connections between ocean literacy and related disciplines and frameworks. The authors also identify critical challenges, limitations, and opportunities for preparing ocean-literate global citizens.

A Brief History of Ocean Literacy

Ocean Literacy Definition and the Essential Principles

Prior to 2002, there was no cohesive structure or consensus regarding what people should know about the ocean. In 2002, a group of experts in the United States met through virtual meetings and developed *Oceans for Life,* a publication linking ocean topics with standards for Geography Education. Soon afterward multiple entities, including the Centers for Ocean Science Education Excellence (COSEE), College of Exploration (CoE), National Geographic Society (NGS), National Marine Educators Association (NMEA), and US National Oceanic and Atmospheric Administration (NOAA), worked together through online and face-to-face discussions to ignite the ocean literacy campaign. For the purpose of this chapter, the focus will be on the definition of ocean literacy, the Essential Principles, and resulting significant events.

Box 1 Definition of ocean literacy

Ocean literacy is an understanding of the ocean's influence on you and your influence on the ocean. An ocean-literate person:

- understands the Essential Principles and Fundamental Concepts about the ocean;
- can communicate about the ocean in a meaningful way; and
- is able to make informed and responsible decisions regarding the ocean and its resources.

Box 2 The Essential Principles of ocean literacy

Principle 1: Earth has one big ocean with many features.
Principle 2: The ocean and life in the ocean shape the features of Earth.
Principle 3: The ocean is a major influence on weather and climate.
Principle 4: The ocean makes Earth habitable.
Principle 5: The ocean supports a great diversity of life and ecosystems.
Principle 6: The ocean and humans are inextricably interconnected.
Principle 7: The ocean is largely unexplored.

Significant Developments in Ocean Literacy

The ocean literacy campaign evolved to make notable impacts in education, science, and policy around the world as summarized in Box 3.

Box 3 Brief ocean literacy timeline

2002: *Oceans for Life* virtual conference identifies ocean-related concepts for geography literacy. Hosted by the CoE and NGS.
2004: Ocean literacy virtual conference to develop initial draft of the definition and the Essential Principles and Fundamental Concepts. Hosted by the COSEE, CoE, NGS, NMEA, and NOAA.
2005: *Ocean Literacy Essential Principles and Fundamental Concepts* Guide published.
2006: Work begins on the Scope and Sequence.
2007: International Pacific Marine Educators Network (IPMEN) forms; *Marine education white paper* published in Taiwan.
2008: *Ocean Literacy* Guide published in Japanese.
2009: *Scope and Sequence* published; US-based *Great Lakes Literacy* Guide published.
2010: *NMEA special report #3: The ocean literacy campaign* published; COSEE China forms; *Ocean Literacy* Guide published in Chinese.
2011: *Ocean Literacy* Guide published in Portuguese.
2012: European Marine Science Educators Association (EMSEA) forms; Conference on ocean literacy in Bruges, Belgium; ocean literacy campaign begins in Bangladesh.
2013: *Ocean Literacy* Guide version 2 published; Canada, EU, and the United States sign the *Galway Statement on Atlantic Ocean cooperation*; Canadian Network for Ocean Education (CaNOE) forms.
2014: *The vision statement for transatlantic ocean literacy* is published; First International Marine Science Communication conference.

2015: *Ocean Literacy/Next Generation Science Standards alignment* published; EU funds Sea Change, ResponSEAble, and Atlantic Ocean Research Alliance projects; Asia Marine Educators Association (AMEA) forms; EMSEA-Med forms.
2016: Korean Marine Educators Association forms; CommOCEAN 2016.
2017: Ocean Literacy for All conference; *Mediterranean Sea Literacy* published; *Korean Ocean Literacy* Guide published; UNESCO's *Ocean literacy for all: A toolkit* published.
2018: African Ocean Literacy Network forms; Ocean Literacy Italia forms; CommOCEAN 2018.
2019: *Exemplary practices in marine science education: A resource for practitioners and researchers* published; Relato océano forms.
2020: *Ocean Literacy* Guide version 3 published; UNESCO's Intergovernmental Oceanographic Commission hosts Virtual Ocean Literacy Summit
2021–2030: United Nations Decade of Ocean Science for Sustainable Development.

Recent projects have demonstrated promise for the future of international collaboration to advance ocean literacy. For instance, the EU-funded *Sea Change* project united 17 partners from nine countries. *Sea Change* sought to shift the ways the citizens of Europe see their relationship with the ocean, encourage specific actions to promote a healthy ocean, and empower an ocean-literate citizenship. Another EU-funded project, *ResponSEAble*, produced diverse tools to educate different audiences. From children to fishers to the public, the project stressed the importance of the ocean in many aspects of human life including the global economy, recreation, and health. UNESCO published *Ocean Literacy for All: A Toolkit* (2017) highlighting ocean literacy activities and featuring vignettes of global stakeholders. The toolkit envisions people taking more personal responsibility for the ocean via ocean literacy experiences, networks, and partnerships. Another international collaboration, the book *Exemplary Practices in Marine Science Education* (2019), showcases 24 chapters highlighting education projects around the world. Ranging from research studies to citizen science initiatives, the chapters in this volume feature the different approaches marine educators take in a variety of contexts toward greater goals of improved ocean literacy and a healthy ocean.

While these advances have been impressive and influential, issues and challenges remain. Because global communities share one ocean, how can successful practices be disseminated appropriately given the subtleties within culture and language? How will the traditional knowledge of indigenous people around the world be honored and respected? These and other concerns speak to the importance of broader awareness and education for a more ocean-literate global citizenry.

Ocean Literacy and Related Disciplines

Science

Most components of the Ocean Literacy Framework are based on the fundamental concepts of science in part because of the original intent to bring ocean science content into curriculum and instruction. Additionally, many of the individuals and organizations involved in drafting the documents are trained in science and science education. Therefore, making connections to science, including alignment to the NGSS, was more easily completed than connections to other disciplines. Table 1 is set in the "big ideas" in science as delineated in the NGSS Disciplinary Core Ideas and Crosscutting Concepts.

Geography

The planet we inhabit transcends boundaries. History, policy, culture, science, and other topics typically studied as stand-alone disciplines do in fact interact over space and time, affecting one another and the entire planet. In other words, our influence

Table 1 Ocean literacy and science

Ocean literacy essential principle	Big ideas in science[a]
1. Earth has one big ocean with many features	Scale, proportion, and quantity; Systems; Matter and its interactions; Forces and interactions; Ecosystems
2. The ocean and life in the ocean shape the features of Earth	Energy and matter; Forces and interactions; Earth's place in the universe; Systems; Ecosystems; Matter and its interactions
3. The ocean is a major influence on weather and climate	Patterns; Systems; Energy; Earth and human activity; Structures and processes; Ecosystems; Matter and its interactions; Waves
4. The ocean makes Earth habitable	Energy and matter; Stability and change
5. The ocean supports a great diversity of life and ecosystems	Systems; Structure and function; Earth and human activity; Structures and processes; Ecosystems; Evolution; Heredity; Forces and interactions; Energy and matter
6. The ocean and humans are inextricably interconnected	Systems; Stability and change; Earth and human activity; Ecosystems; Structures and processes; Earth's place in the universe; Evolution; Energy and matter; Engineering Design; Matter and its interactions
7. The ocean is largely unexplored	Engineering design; Earth and human activity; Waves; Earth's place in the universe; Ecosystems

[a]From NGSS crosscutting concepts and the alignment of OL and NGSS

on the ocean and the ocean's influence on us can be viewed through the lens of geography. The National Geographic Society defines geography as:

> …the study of places and the relationships between people and their environments. Geographers explore both the physical properties of Earth's surface and the human societies spread across it. They also examine how human culture interacts with the natural environment and the way that locations and places can have an impact on people. Geography seeks to understand where things are found, why they are there, and how they develop and change over time.
>
> https://www.nationalgeographic.org/education/what-is-geography/

As in other disciplines, standards depicting what an informed person should know have been developed for geography by the National Council for Geographic Education (Gallagher and Downs 2012) as depicted in Box 4.

Box 4 National Geography Standards

The geographically informed person knows and understands…

Essential Element I. THE WORLD IN SPATIAL TERMS

Standard 1. How to use maps and other geographic representations, geospatial technologies, and spatial thinking to understand and communicate information.
Standard 2. How to use mental maps to organize information about people, places, and environments in a spatial context.
Standard 3. How to analyze the spatial organization of people, places, and environments on Earth's surface.

Essential Element II. PLACES AND REGIONS

Standard 4. The physical and human characteristics of places.
Standard 5. That people create regions to interpret Earth's complexity.
Standard 6. How culture and experience influence people's perceptions of places and regions.

Essential Element III. PHYSICAL SYSTEMS

Standard 7. The physical processes that shape the patterns of Earth's surface.
Standard 8. The characteristics and spatial distribution of ecosystems and biomes on Earth's surface.

Essential Element IV. HUMAN SYSTEMS

Standard 9. The characteristics, distribution, and migration of human populations on Earth's surface.
Standard 10. The characteristics, distribution, and complexity of Earth's cultural mosaics.
Standard 11. The patterns and networks of economic interdependence on Earth's surface.

Standard 12. The processes, patterns, and functions of human settlement.
Standard 13. How the forces of cooperation and conflict among people influence the division and control of Earth's surface

Essential Element V. ENVIRONMENT AND SOCIETY

Standard 14. How human actions modify the physical environment.
Standard 15. How physical systems affect human systems.
Standard 16. The changes that occur in the meaning, use, distribution, and importance of resources.

Essential Element VI. THE USES OF GEOGRAPHY

Standard 17. How to apply geography to interpret the past.
Standard 18. How to apply geography to interpret the present and plan for the future.

Geography education and general geographical awareness can help advance understanding of the need for a global ocean-literate citizenry. Table 2 provides an alignment of the Ocean Literacy Essential Principles and National Geographic Society's Geography Standards.

Table 2 National geography standards essential elements and ocean literacy essential principles

Ocean literacy essential principle	Geography standards essential elements
1. Earth has one big ocean with many features	I. The world in spatial terms II. Places and regions III. Physical systems V. Environment and society
2. The ocean and life in the ocean shape the features of Earth	I. The world in spatial terms II. Places and regions III. Physical systems V. Environment and society
3. The ocean is a major influence on weather and climate	III. Physical systems V. Environment and society
4. The ocean makes Earth habitable	III. Physical systems V. Environment and society
5. The ocean supports a great diversity of life and ecosystems	II. Places and regions III. Physical systems
6. The ocean and humans are inextricably interconnected	II. Places and regions IV. Human systems V. Environment and society VI. The uses of geography
7. The ocean is largely unexplored	I. The world in spatial terms V. Environment and society

UNESCO's Education for Sustainable Development

In contrast to the discipline-specific strategies outlined above, the Intergovernmental Oceanographic Commission (IOC) of the United Nations Educational, Scientific and Cultural Organization (UNESCO) outlines a multidisciplinary approach to ocean literacy. Based on the UN Sustainable Development Goal (SDG) 14: "Conserve and sustainably use the ocean, seas, and marine resources for sustainable development," the publication *Ocean Literacy for All: A Toolkit* (2017) frames the discussion in terms of UNESCO's Education for Sustainable Development, which:

> …aims to improve access to quality education on sustainable development at all levels and in all social contexts, to transform society by reorienting education and help people develop knowledge, skills, values and behaviors needed for sustainable development. (Executive Summary)

While the focus of the UNESCO publication is holistic, there are numerous connections with the Ocean Literacy Essential Principles and Fundamental Concepts as shown in Table 3 and Box 5. The Ocean Literacy Framework does not include the actionable component of the UNESCO SDG 14 means of implementation of the global indicator framework, and is thus aligned by content only.

Box 5 Ocean literacy fundamental concepts implicitly relevant to UNESCO SDG 14

Principle 5: The ocean supports a great diversity of life and ecosystems.

Concept F: Ocean ecosystems are defined by environmental factors and the community of organisms living there. Ocean life is not evenly distributed through time or space due to differences in abiotic factors such as oxygen, salinity, temperature, pH, light, nutrients, pressure, substrate, and circulation. A few regions of the ocean support the most abundant life on Earth, while most of the ocean does not support much life.

Concept G: There are deep ocean ecosystems that are independent of energy from sunlight and photosynthetic organisms. Hydrothermal vents, submarine hot springs, and methane cold seeps rely only on chemical energy and chemosynthetic organisms to support life.

Concept I: Estuaries provide important and productive nursery areas for many marine and aquatic species.

Principle 6: The ocean and humans are inextricably interconnected.

Concept D: Humans affect the ocean in a variety of ways. Laws, regulations, and resource management affect what is taken out and put into the ocean. Human development and activity leads to pollution (point source, nonpoint source, and noise pollution), changes to ocean chemistry (ocean acidification), and physical modifications (changes to beaches, shores, and rivers). In addition, humans have removed most of the large vertebrates from the ocean.

Table 3 UNESCO Sustainable Development Goal (SDG) 14 and ocean literacy Essential Principles and Fundamental Concepts

SDG 14 targets	OL principles and concepts
14.1. By 2025, prevent and significantly reduce marine pollution of all kinds, in particular from land-based activities, including marine debris and nutrient pollution	Principle 6, Concepts D, G
14.2. By 2020, sustainably manage and protect marine and coastal ecosystems to avoid significant adverse impacts, including by strengthening their resilience, and take action for their restoration in order to achieve healthy and productive oceans	Principle 5, Concepts F, G, I Principle 6, Concept G
14.3. Minimize and address the impacts of ocean acidification, including through enhanced scientific cooperation at all levels	Principle 6, Concept D
14.4. By 2020, effectively regulate harvesting and end overfishing, illegal, unreported, and unregulated fishing and destructive fishing practices, and implement science-based management plans, in order to restore fish stocks in the shortest time feasible, at least to levels that can produce maximum sustainable yield as determined by their biological characteristics	Principle 6, Concepts D, G
14.5. By 2020, conserve at least 10 percent of coastal and marine areas, consistent with national and international law and based on the best available scientific information	Principle 6, Concept G
14.6. By 2020, prohibit certain forms of fisheries subsidies which contribute to overcapacity and overfishing, eliminate subsidies that contribute to illegal, unreported, and unregulated fishing and refrain from introducing new such subsidies, recognizing that appropriate and effective special and differential treatment for developing and least developed countries should be an integral part of the World Trade Organization fisheries subsidies negotiation	Principle 6, Concepts D, G
14.7. By 2030, increase the economic benefits to Small Island developing States and least developed countries from the sustainable use of marine resources, including through sustainable management of fisheries, aquaculture, and tourism	Principle 6, Concepts D, G

Concept G: Everyone is responsible for caring for the ocean. The ocean sustains life on Earth and humans must live in ways that sustain the ocean. Individual and collective actions are needed to effectively manage ocean resources for all.

Challenges

Numerous challenges face the global ocean literacy community. It is difficult to ensure all voices are heard and included in mapping the way forward in the field and that ocean literacy initiatives are culturally relevant and locally applicable. The body of research on ocean literacy is small, in part because the definition of ocean literacy makes it difficult to measure. Key challenges to the long-term success of the ocean literacy campaign include the need for centralized communication, inclusion of cultural relevance, a solid foundation of evaluation and education research, recognition of the nuances of formal and nonformal education, and adequate and consistent sources of funding.

Challenges of Communication

Although the ocean literacy community is relatively small, it lacks centralized communication. Advances and expansions in ocean literacy are not necessarily well-communicated among different groups (e.g., scientists and educators), and different locations (e.g., Asia, Africa, North America). Some groups have worked particularly hard to connect with others. For example, at the first formal meeting of the Asia Marine Educators Association (AMEA) in 2016, leaders from NMEA and EMSEA were invited to present and participate.

In late 2017 UNESCO, with support from the Swedish government, held an "Ocean literacy for all" conference in Venice, Italy, inviting ocean literacy leaders from 30 countries around the world representing all continents. Representatives from various sectors, including education, art, and science, were invited to the meeting to discuss collaborations and a more inclusive global framework for ocean literacy moving forward.

Such meetings are key examples of international communication and collaboration. In-person meetings offer extended time to discuss ideas, forge partnerships, and enhance programs. However, travel among global communities is time-consuming and expensive, particularly so for those from developing nations and indigenous groups, and impossible in the midst of a global pandemic. Although advances in technology provided a greater platform for virtual meetings, it remains difficult to ensure that collaborative initiatives include *all* individuals and groups working toward improving ocean literacy.

Cultural Relevance and Responsiveness Around a Global Ocean

When the original Essential Principles and Fundamental Concepts of Ocean Literacy were written, the authors took a western science perspective as the goal at the time was to ensure that ocean content was included in US science standards. As the ocean literacy campaign has expanded globally and is used as a framework for other purposes, the role of cultural perspectives on the ocean must be considered. Culture includes ways of knowing, language, historical perspectives, beliefs, arts, and shared values of a particular community. Ensuring that ocean literacy efforts are culturally relevant and responsive is an ongoing challenge.

The US-centric Ocean Literacy Framework does not encompass world cultures, and the ways in which the Essential Principles and Fundamental Concepts are stated and described may not fit well with all cultures. For instance, while many in the United States refer to the global ocean and ocean basins such as the Atlantic and Pacific, Europeans tend to focus on regional seas, such as the North Sea, Mediterranean, and Baltic.

Indigenous people around the world have sustainably used ocean resources for thousands of years. Indigenous people have observed and studied their environments, harvested and cultivated resources such as seaweed, fish, and shellfish, and passed down the traditional ecological knowledge (TEK) of their elders through stories, songs, dance, art, and other culturally important means. TEK is "a cumulative body of knowledge, practice, and belief, evolving by adaptive processes and handed down through generations by cultural transmission, about the relationship of living beings (including humans) with one another and with their environment" (Berkes et al. 2000, 1252).

Only recently has western science cultivated relationships with traditional knowledge practitioners. Some western ocean scientists are conducting studies to document, from a western science perspective, examples of TEK including animal migration patterns, the effectiveness of ocean and atmospheric processes such as tides and winds on species, and changes observed in coastlines (Thornton and Maciejewski Scheer 2012). Other scientists are taking a more collaborative approach toward ecosystem management and ensuring sustainability for future generations by working with TEK practitioners studying environmental changes and resource management. Some collaborative projects manage to narrow the gap using approaches such as reciprocal training, shared monitoring, workshops, and symposia.

In a review of scientific collaborations with local and traditional knowledge (LTK) in marine environments, Thornton and Maciejewski Scheer (2012) note, "there is considerable room for constructive engagement of LTK as part of marine research, monitoring, spatial planning, and conservation" (p. 11). The same can be said for ocean literacy and education. As ocean literacy continues to drive policy, education standards, and practices, along with Sustainable Development Goals, TEK and LTK practitioners should be included in decision making. TEK and LTK as relating to the ocean must also be included in students' education. In the Melanesian nation of Vanuatu, educators are piloting a project to redesign the curriculum to teach TEK and western science together. A description of the initiative explains:

> By presenting indigenous and scientific knowledge systems side-by-side, it is hoped that the youth of Vanuatu will feel pride in their indigenous cultural heritage, creating the space for them to grasp the complexity and sophistication of their own traditional understanding of the environment, as well as empowering them to make their own choices for a sustainable future using both local and scientific knowledge as they see fit. (UNESCO 2017)

The same is true for other local cultural perspectives around the world. The Ocean Literacy Framework should be flexible enough to allow for local adaptation as appropriate. For example, the Essential Principles and Fundamental Concepts have been translated into Chinese, Italian, Japanese, Polish, Portuguese, and Spanish https://www.marine-ed.org/ocean-literacy/translations. The fundamental concepts in particular can be modified to include locally relevant information.

Challenges in Evaluation and Education Research

When the definition of ocean literacy was crafted, the goal was not to measure the ocean literacy of individuals, communities, or global citizens. The three-pronged definition—knowledge, communication, and behavior—is extremely difficult to measure. How does one appropriately measure, for instance, whether one can "communicate about the ocean in a meaningful way?"

While education research studies on marine education have been published since the environmental movement of the 1960s, most have focused on small, local projects and are frequently Master's theses or doctoral dissertations, not peer-reviewed research articles (Marrero 2009; Fortner 1991). Studies in marine education frequently focus on the efficacy of a particular program, not the development of ocean literacy in individuals or groups, because the programs themselves, not the research, are typically funded. Some research has been published regarding ocean science content knowledge, including the International Ocean Literacy Survey (IOLS) (Fauville et al. 2018b) but much less on the more difficult-to-measure aspect of behavior change.

Similarly, evaluators examine whether the programs are meeting the stated goals and objectives of the project and are not focused on studying ocean literacy for the sake of studying it. The most significant evaluations to date have been of the original COSEE Centers, funded primarily by the US National Science Foundation (NSF), and the EU-funded *Sea Change* and *ResponSEAble* projects. Components of evaluation reports are available online. The COSEE network did strive for more cohesive evaluation across all Centers, although each Center was funded individually. In 2012, a report was published regarding the influence of participation in COSEE programs on practicing scientists (Anderson et al. 2012). Most findings reveal changes in individual scientists' college-level teaching and increased receptiveness to education and outreach, although one-third of the scientists surveyed also indicated a change in the way they think about research questions—specifically a shift in focus toward more societally relevant questions (p. 4).

The ocean literacy community recognized long ago that evidence-based, empirical education research and evaluation would be critical to the progress and acceptance of ocean literacy (Payne and Zimmerman 2010). This book and other recent publications (e.g., Fauville et al. 2019) are steps in the right direction, but the campaign still requires more peer-reviewed, empirical publications and direct input from and partnerships with education researchers and evaluators to become truly relevant.

Inherent Differences in Formal and Informal/Nonformal Education

The inclusion of ocean and coastal topics in curriculum and instruction in formal and nonformal settings continues to be an issue, although progress has been made.

Science learning in traditional school curriculum has sought to improve students' scientific literacy. While the definition of scientific literacy has been debated and has its nuances (DeBoer 2000), many science educators agree that educating for scientific literacy can prepare students with an understanding of science and how it is practiced so they can make informed decisions about science-based issues. In the publication *Science for All Americans*, The American Association for the Advancement of Science (AAAS) (1989) explained that:

> The science-literate person is one who is aware that science, mathematics, and technology are interdependent human enterprises with strengths and limitations; understands key concepts and principles of science; is familiar with the natural world and recognizes both its diversity and unity; and uses scientific knowledge and scientific ways of thinking for individual and social purposes. (Introduction section, para. 2)

The definition of ocean literacy is grounded in this view of scientific literacy, as ocean-literate citizens are those who can "make informed and responsible decisions regarding the ocean and its resources." Because the Ocean Literacy Framework was originally designed to inform formal education curriculum, this perspective is appropriate. Even so, the ocean is not typically included in schooling in many regions (Fauville et al. 2018; Marrero 2009; Hoffman and Barstow 2007), although there are certainly examples of successful programs that have been shown to improve students' ocean literacy (Marrero and Mensah 2011; Plankis and Marrero 2010; Lambert 2006). A review by Donert et al. (2015) highlighted many examples of promising practices and programs in different regions, and made recommendations for promoting ocean literacy in informal settings, such as using hands-on and problem-based approaches to promote ocean literacy and using an inclusive approach that encourages learners to take action.

There are also examples of successful partnerships between institutions of higher education (IHE) and/or nonformal education environments and primary and secondary schools to support marine education. For example, Niedoszytko et al. (2019) found that long-term engagement between school pupils and an aquarium in Poland resulted in higher interest levels about the ocean and its resources than students who did not participate, although 40% of students reported they did not learn about the subject in school (p. 129). Teachers participating in the study overwhelmingly felt that this partnership provided a useful learning experience for students and could lead to better environmental awareness. Similarly, Frederick et al. (2019) led a collaboration between universities, secondary students, and teachers in several countries. Students examined authentic biofouling data and made connections to water quality, thus improving their use of scientific practices and connections to the ocean.

Learning in nonformal or free-choice environments, such as aquariums, zoos, and museums, is also critical. Such institutions work with people of all ages focusing on educating through engagement. Visitors can develop knowledge of the natural world as well as about socioscientific issues, and this learning can lead to improved decision making (Ballantyne and Packer 2005). However, learning outcomes in these settings are not as defined, i.e., they typically include much broader learning than a traditional classroom lesson or unit, and efforts to measure learning, e.g., through tests on science concepts, have been controversial (Robinson and Murray 2019; National Research Council 2009).

In terms of developing ocean literacy, both nonformal/free choice and formal learning settings are important and show promise. For instance, Greely (2008) examined an ocean camp for girls and found that the participants had increased environmental awareness and understanding of marine science content. The ¡Youth & the Ocean! (¡YO!) Program provided secondary students with an opportunity to work with graduate students and employ authentic scientific practices, leading to a stronger appreciation for the ocean and conservation of marine resources (Weiss and Chi 2019). Because of the inherent differences between informal and formal educational settings, the ocean literacy community must consider the role of the Essential Principles and Fundamental Concepts, with their focus on specific ocean science content, in developing ocean literacy in other learning contexts.

Lack of Consistent and Sufficient Funding

Many programs and projects are woefully underfunded or rely on grants to sustain programming. Few entities providing grants consider multidisciplinary projects or the continuation of successful projects.

A lack of sustained funding is nearly the only constant for many ocean literacy-related programs and projects. Funding often depends on governmental policies, initiatives, and of course, the economy. While US-based (NSF-COSEE; NOAA Environmental Literacy; NOAA Bay Watershed Education and Training) and EU-based (Horizon 2020) funding has been available, the disciplinary mission or focus of most agencies makes the funding of multidisciplinary projects even more difficult. More importantly, there is limited funding dedicated to international collaboration and what exists is often reserved for new collaborations. Most government-funded grants for ocean literacy initiatives are programmatic in nature, supporting a specific project aimed at a specific audience. The finite length of funding, usually in periods of three to five years, makes sustaining programs difficult. Even the most successful projects struggle to continue without fiscal support.

Given the small education research base in the ocean literacy field, it is challenging to secure large education research grants to study ocean literacy. At the same time, education research is sorely needed to advance the field (see "Challenges in Evaluation and Education Research").

Limitations

While the ocean literacy campaign can claim multiple high-level (and often unexpected) successes, the community willingly acknowledges its limitations. As previously noted, the original Guide was conceived with one primary goal—inclusion of ocean science content in US science education standards. While this vision necessarily limited the scope, geography, and cultural considerations, it spawned the global ocean literacy campaign.

In addition to the challenges noted in "Challenges," significant limitations are inherent in current teaching and learning strategies. In terms of teaching people of all ages through a transdisciplinary approach, most educators (and people) are not trained to teach or learn in this manner. Such holistic education is a theoretical approach with strong merit and would require new training in teaching, learning, curriculum, and instruction.

Opportunities

Numerous publications and organizations have suggested a variety of ways to move forward. A subset of these potential opportunities, including a holistic and inclusive approach and more effective and consistent communication between networks, are discussed below.

Holistic View of Ocean Literacy

In terms of a broader way of thinking about ocean literacy, Santoro et al. (2017) suggest "a multiple-perspective approach promotes interdisciplinary and intercultural competencies as it addresses challenges to local or global sustainability" (p. 87). The strategy is outlined in Box 6 and has been "applied to issues such as climate change, disaster risk reduction, and biodiversity conservation" (p. 87). This broad approach, particularly inclusive of values, culture, and sustainability, has the potential to promote citizenship and inspire action.

Box 6 Holistic approach to ocean literacy

- Scientific perspective
- Historical perspective
- Geographic perspective
- Gender equality perspective
- Value perspective
- Cultural perspective
- Sustainability perspective

While educators, particularly those teaching at the secondary level, are not traditionally trained in transdisciplinary approaches to education, there are resources to support them to do so. For example, UNESCO (2010) developed "a multimedia teaching and learning programme" to support the UN Decade of Education for Sustainable Development from 2005 to 2014. This toolkit for professional development includes modules that lead educators through background information, activities, and modules to help them to explore what it means to teach for sustainable development. Paired with *Ocean Literacy for All: A Toolkit* (2017), which has been translated into multiple languages, could be a powerful set of resources. In support of the UN Decade of Ocean Science for Sustainable Development (2021–2030), perhaps a similar program can be created to support teaching about the ocean from multidisciplinary (or transdisciplinary) perspectives.

Unifying the Global Ocean Literacy Network

The interest in ocean literacy as a unifying concept is global in scale. Since the first virtual conference in 2002, numerous organizations have formed to represent the interests of different geographic areas. Essentially, the global community should consider a "network of networks," which has been proposed previously (ORAP 2013). Implementation will require a sustained, inclusive international effort with a strong organizational foundation. The network should include professionals and experts from multiple disciplines including education, policy, the humanities, corporate interests, and natural and social sciences (Table 4).

Table 4 Ocean literacy organizations

Geographic area	Ocean literacy organizations
Africa	African Ocean Literacy Network
Asia	International Pacific Marine Educators Network (IPMEN); Asia Marine Educators Association (AMEA); Korean Marine Educators Association
Australia	International Pacific Marine Educators Network (IPMEN)
Europe	European Marine Science Educators Association (EMSEA); EMSEA-Med (Mediterranean Sea); Ocean Literacy Italia
North America	National Marine Educators Association (NMEA); Canadian Network for Ocean Education (CaNOE); Canadian Ocean Literacy Coalition
South America	International Pacific Marine Educators Network (IPMEN), Relato océano
Global	Consortium for Ocean Science Exploration and Engagement[a] (COSEE); United Nations Educational, Scientific and Cultural Organization (UNESCO)

[a]Different from the original US NSF-funded Centers for Ocean Science Education Excellence (COSEE)

Conclusions and Recommendations

Many educators, education researchers, and scientists around the world are conducting successful programs aimed at improving ocean literacy and empowering a global ocean-literate citizenry. Moving forward, all involved should work together to effectively gather empirical data to demonstrate program and project effectiveness, and publish the results in peer-reviewed journals. In this manner, practitioners and researchers can learn from one another's successes and challenges and advance our relatively small field.

Additionally, a transdisciplinary approach to ocean literacy, as proposed by UNESCO, will move the community forward. However, explicit connections to other topics (e.g., gender equality, values, culture, systems, sustainability) should be conducted by experts in those fields. In this manner, this book will help make connections with geography.

The ocean literacy community, although global in scope, is small in terms of individuals and organizations. Open, inclusive discussions and communications via a global ocean literacy network are necessary and doable (Marrero et al. 2019), as suggested in "Unifying the Global Ocean Literacy Network." We all share the common goal of promoting ocean literacy. Let us work together to make it happen.

Resources:

- *Ocean Literacy* Guide v.3: https://static1.squarespace.com/static/5b4cecfde2ccd188cfed8026/t/5eb99cc530a3d76767dc7aea/1589222614314/OceanLiteracyGuide_V3_2020.pdf
- *Ocean Literacy Scope and Sequence for Grades K-12*: https://www.marine-ed.org/ocean-literacy/scope-and-sequence
- *Alignment of Ocean Literacy to US Next Generation Science Standards*: https://www.marine-ed.org/ocean-literacy/ngss-alignment
- *United Nations Decade of Ocean Science for Sustainable Development*: https://oceandecade.org/
- Donert K, Fauville G, Gotensparre S, Mäkitalo Å, Van Medegael L, Zwartjes L. (2015) Review of marine formal education. EU Sea Change Project. https://platform.europeanmoocs.eu/users/30446/Review_of_marine_formal_education.pdf
- Marrero ME, Payne DL, Breidahl H (2019) The Case for Collaboration to Foster Global Ocean Literacy. Frontiers in Marine Science, 6, 325. https://www.frontiersin.org/articles/10.3389/fmars.2019.00325/full

References

American Association for the Advancement of Science (1989) Science for all Americans. Washington, DC

Anderson A, Dorph R, Kwon P (2012) COSEE's influence on scientists' professional practices: findings from the COSEE scientist study. http://www.cosee.net/files/coseenet/2012_COSEE_Scientist_Study_Report_Final.pdf

Ballantyne R, Packer J (2005) Promoting environmentally sustainable attitudes and behaviour through free-choice learning experiences: what is the state of the game? Environ Educ Res 11(3): 281–295

Berkes F, Colding J, Folke C (2000) Rediscovery of traditional ecological knowledge as adaptive management. Ecolo Appl 10:1251–1262. https://www.fws.gov/nativeamerican/pdf/tek-berkes-2000.pdf

DeBoer GE (2000) Scientific literacy: Another look at its historical and contemporary meanings and its relationship to science education reform. J Res Sci Teaching 37(6): 582–601

Donert K, Fauville G, Gotensparre S, Mäkitalo Å, Van Medegael L, Zwartjes L (2015) Review of marine formal education. EU Sea Change Project https://platform.europeanmoocs.eu/users/30446/Review_of_marine_formal_education.pdf

Fauville G, Payne DL, Marrero ME, Lantz-Andersson A, Crouch, F (eds) (2019) Exemplary practices in marine science education: a resource for practitioners and researchers. Springer, Cham, Switzerland

Fauville G, McHugh P, Domegan C, Mäkitalo Å, Møller F, Papathanassiou M et al (2018a) Using collective intelligence to identify barriers to teaching 12–19 year olds about the ocean in Europe. Marine Policy 91:85–96

Fauville G, Strang C, Cannady MA, Chen YF (2018b) Development of the international ocean literacy survey: measuring knowledge across the world. Environ Educ Res 25(2):238–263

Fortner RW (1991) Abstracts of research in marine and aquatic education 1975–1990 https://www.noaa.gov/sites/default/files/atoms/files/2010_Ocean_and_Water_Education_Research_Abstracts_preface.pdf

Frederick JA, Gotensparre S, Jacobs D, Källström B, Olsson M (2019) The virtue project and the biofilms and biodiversity project: an international collaboration in marine science education. In: Fauville et al. (eds) Exemplary practices in marine science education: a resource for practitioners and researchers, Springer, Cham, Switzerland, pp 257–287

Gallagher SM, Downs RM et al (2012) Geography for life: national geography standards. National Council for Geographic Education, Washington, DC

Greely T (2008) Ocean literacy and reasoning about ocean issues: the influence of content, experience and morality. Graduate Theses and Dissertations http://scholarcommons.usf.edu/etd/271

Hoffman M, Barstow D (2007) Revolutionizing earth system science education for the 21st century, report and recommendations from a 50-state analysis of earth science education standards. TERC, Cambridge, MA

Lambert J (2006) High school marine science and scientific literacy: the promise of an integrated science course. Int J Sci Educ 28(6):633–654

Marrero ME, Payne DL, Breidahl H (2019) The case for collaboration to foster global ocean literacy. Frontiers in Mar Sci 6:325. https://www.frontiersin.org/articles/10.3389/fmars.2019.00325/full

Marrero ME, Mensah FMM (2011) Socioscientific decision making and the ocean: a case study of 7th grade life science students. Electron J Sci Educ 14(1):1–27

Marrero MEC (2009) Uncovering student conceptions of the ocean: a critical first step to improving ocean literacy. Unpublished doctoral dissertation: Teachers College, Columbia University

National Oceanic and Atmospheric Administration (NOAA) (2020) Ocean literacy: the essential principles and fundamental concepts of ocean sciences for learners of all ages. Washington, DC

National Research Council (2009) Learning science in informal environments: people, places, and pursuits. National Academies Press, Washington, DC

Niedoszytko G, Wojcieszek D, Podlesińska W, Borowiak K (2019) Implementing ocean literacy through the bond of informal and formal education. In: Fauville et al (eds) Exemplary practices in marine science education: a resource for practitioners and researchers, Springer, Cham, Switzerland, pp 123–142

Ocean Research Advisory Panel (2013) Leveraging ocean education opportunities: a report to the National Ocean Council. https://www.nopp.org/wp-content/uploads/2010/06/Leveraging-Ocean-Education-Opportunities.pdf

Payne DL, Zimmerman TD (2010) Beyond terra firma: Bringing ocean and aquatic sciences to environmental and science teacher education. In: Bodzin AM, Klein ES, Weaver S (eds) The inclusion of environmental education in science teacher education, Springer, New York, pp 81–94
Plankis BJ, Marrero ME (2010) Recent ocean literacy research in United States Public Schools: results and implications. Int Electronic J Environ Educ 1(1):21–51
Robinson A, Murray, N (2019) Evaluating ocean learning—the principles and practicalities of evaluating formal education audiences in an informal education environment. In: Fauville et al (eds) Exemplary practices in marine science education: a resource for practitioners and researchers, Springer, Cham, Switzerland, pp 143–156
Santoro F et al (eds) (2017) Ocean literacy for all: a toolkit. UNESCO, Paris
Schoedinger S, Uyen Tran L, Whitley L (2010) From the principles to the scope and sequence: a brief history of the ocean literacy campaign. Curr: The J Mar Educ, Special Edition #3: 3–7
Thornton TF, Maciejewski Scheer A (2012) Collaborative engagement of local and traditional knowledge and science in marine environments: a review. Ecology and Society 17(3):1–25. https://doi.org/10.5751/ES-04714-170308
UNESCO (2017) Strengthening indigenous knowledge and traditional resource management through schools in Vanuatu. http://www.unesco.org/new/en/natural-sciences/priority-areas/links/knowledge-transmission/projects/strengthening-indigenous-knowledge-and-traditional-resource-management-through-schools-in-vanuatu/
Weiss E, Chi B (2019) ¡ Youth & The Ocean!(¡ YO!): partnering high school and graduate students for youth-driven research experiences. In: Fauville et al (eds) Exemplary practices in marine science education: a resource for practitioners and researchers, Springer, Cham, Switzerland, pp 27–58

A Framework for the Assessment of the Effectiveness of Ocean Literacy Initiatives

Owen Molloy, Matthew Ashley, and Conor McCrossan

Abstract Central to the design of good tools and initiatives to increase ocean literacy is the problem of how to measure their effectiveness. In this paper, we introduce the different aspects or dimensions of ocean literacy and the importance of fostering a system understanding of human–ocean interaction and the use of cause-and-effect models, such as DPSIR, to capture such systems knowledge. We then proceed to show how this model can be used to actors with specific tools designed to bring about specific responses such as behaviour change, and present a framework developed to formalise the measurement of effectiveness of ocean literacy tools and initiatives based on this approach. The application of this framework to evaluating the effectiveness of specific initiatives and our initial findings are presented.

Keywords Ocean literacy · Systems thinking · Mental models · Behaviour change · Measurement of effectiveness · Theory of change

Introduction

Making informed decisions that will have a positive impact on the sustainability of our oceans and environment is not simple, and often not obvious. For example, eutrophication of the Baltic Sea is a major threat to its biodiversity. There are many causes, including run-off of agricultural nitrogen and phosphorous from fertilisers and waste. Multiple actors are involved from environmental policy makers to farmers and consumers. All must be educated on the impacts of their choices, and the efforts they could make to reduce eutrophication in the long term.

Ocean literacy tools and initiatives are used to create awareness about how humans are impacting the ocean and it is important that the effectiveness of these tools and initiatives are measured in order to ensure that people are becoming more aware

O. Molloy (✉) · C. McCrossan
National University of Ireland, Galway, Ireland
e-mail: owen.molloy@nuigalway.ie

M. Ashley
University of Plymouth, Plymouth, UK

K. C. Koutsopoulos and J. H. Stel (eds), *Ocean Literacy: Understanding the Ocean*, Key Challenges in Geography, https://doi.org/10.1007/978-3-030-70155-0_3

about how they are impacting the ocean and how the ocean impacts their day-to-day lives. The measurement of the effectiveness of the tools and initiatives can also provide insights into how the tools and initiatives can be improved to better inform and educate people on important ocean-related topics.

Defining Ocean Literacy

Ocean literacy is aligned with the objectives of environmental education as defined by UNESCO (UNESCO 1975):

- Awareness: to help social groups and individuals acquire an awareness of and sensitivity to the global environment and its allied problems.
- Attitude: to help social groups and individuals acquire a set of values and feelings of concern for the environment, as well as the motivation to actively participate in environmental improvement and protection.
- Skills: to help social groups and individuals acquire the skills for identifying and solving environmental problems.
- Participation: to provide social groups and individuals with an opportunity to be actively involved at all levels in working towards resolution of environmental problems.

These objectives align very well with a number of other initiatives in Environmental Literacy, which stress the importance of not just knowledge, but skills and attitudes. For example, the Oregon Environmental Literacy Program[1] describes Environmental Literacy as 'An individual's understanding, skills and motivation to make responsible decisions that considers his or her relationships to natural systems, communities and future generations'. The P21.org initiative, a coalition of business community, education leaders and policymakers focused on US K-12 education and the twenty-first-century skills, states that an Environmentally Literate[2] student can:

- Demonstrate knowledge and understanding of the environment and the circumstances and conditions affecting it, particularly as related to air, climate, land, food, energy, water and ecosystems.
- Demonstrate knowledge and understanding of society's impact on the natural world (e.g. population growth, population development, resource consumption rate, etc.).
- Investigate and analyse environmental issues, and make accurate conclusions about effective solutions.
- Take individual and collective action towards addressing environmental challenges (e.g. participating in global actions, designing solutions that inspire action on environmental issues).

---

[1] http://oelp.oregonstate.edu/oelp-plan/what-environmental-literacy.

[2] http://www.p21.org/about-us/p21-framework/830-environmental-literacy.

A common theme across definitions of literacy is that it is not just about knowledge, but our attitudes, our ability to understand complex issues and see alternatives, and how we communicate and act personally and within our communities. The Environmental Literacy Ladder (ELL) outlines five essential components of environmental literacy (ELL 2007). The components are arranged into the following steps: Awareness, Knowledge, Attitudes, Skills and Collective action. It is a loose hierarchy with each step increasing in complexity and building on the previous step.

The most widely used framework for Ocean literacy is the Ocean Literacy Framework, which was developed by ocean scientists and educators, primarily in the USA. It was based on previous work to define Ocean literacy, and was intended to help redress the lack of ocean-related content in educational materials (OLF 2015). The main component of the OLF is a set of seven Ocean Literacy Principles (OLP 2013). Of these, the sixth (*'The oceans and humans are inextricably interconnected'*) is the one most explicitly concerned with human–ocean interaction (Fig. 1).

Survey instruments based on the OLP such as the International Ocean Literacy Survey (IOLS) are based on questions which test an individual's understanding of the OLP. There are a number of different approaches taken to the definitions of

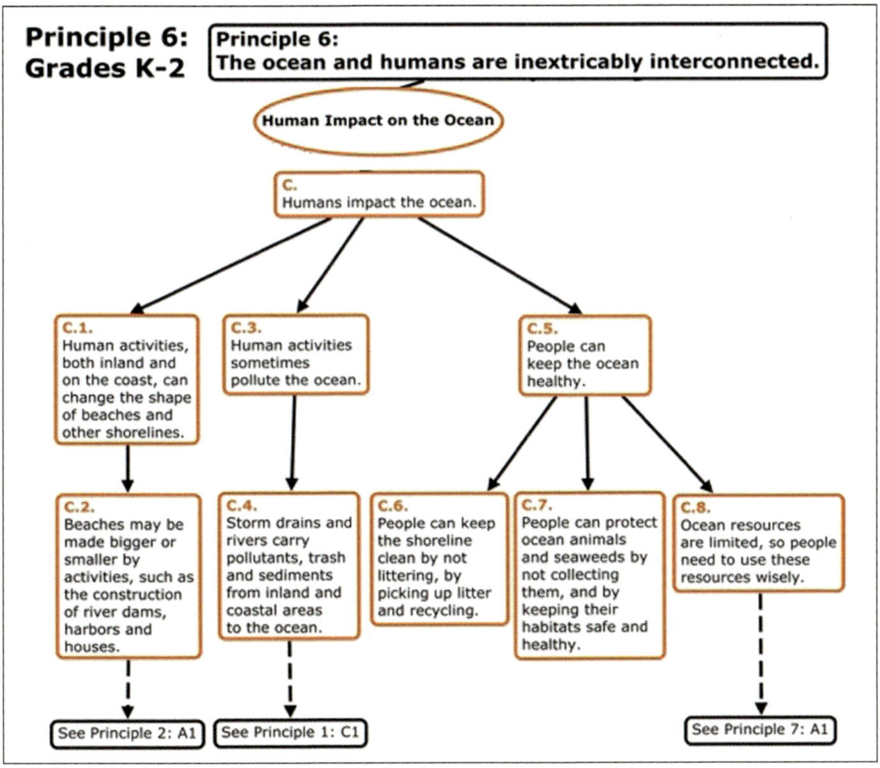

Fig. 1 Ocean Literacy Principle 6—the oceans and humans are inextricably interconnected

ocean/environmental awareness, knowledge, attitude, concern and behaviour. Readiness to take personal part in marine environment actions is included in the definition of Ocean Environmental Awareness proposed by Umuhire and Fang (2016). However, capacity for personal action is placed at the top of the ELL as a step above awareness, knowledge and attitude. It is realistic to assume that *informed* personal action would be expected to occur only after a certain level of awareness and knowledge has been achieved. Umuhire and Fang (2016) state that concern for the marine environment is a part of awareness, while the ELL states that it is related to attitude. Both Umuhire and Fang (2016) and ELL (2007) agree that understanding the relationship between human activities and the environment/ocean is part of awareness. Kikuchi-Uehara et al. (2016) assumed that behavioural intention is part of environmental awareness, whereas Schultz et al. (2004) stated that it is part of the construct of environmental attitudes. The perception that the environment is under serious threat is included in the dimensions of environmental consciousness proposed by Sánchez and Lafuente (2010). Umuhire and Fang (2016) include an ability to identify the sources of marine environmental threats as part of their definition of Ocean Environmental Awareness.

It would seem more logical to treat the various *dimensions* of environmental or Ocean literacy as independent. If we focus on behaviour change then the dimensions of ocean literacy overlap with recognised 'predictors' of behaviour change studied in social and behavioural sciences (Phal and Wyles 2017). However, within the broader concept of ocean literacy, these dimensions can be focused on independently for separate ocean literacy objectives and audiences. For example, an attitude of concern is not a pre-requisite for problem-solving skills. Likewise, one's attitudes may be formed through moral conviction or social convention, rather than based on in-depth knowledge of the subject. The importance of viewing the OL dimensions in this way is that we can immediately see the need for and benefit of studying and measuring them separately. It gives us a more fine-grained view of OL, and allows us to tailor campaigns or educational interventions targeted at specific dimensions independently.

The Ocean Literate Citizen

When we consider the notion of citizenship, as well as being defined in terms of 'membership' of a particular country, which brings certain rights and privileges, it also comes with certain duties. These duties generally involve requirement to pay taxes, to vote and in general to act in a way that is viewed as being a good and constructive member of that society. In much the same way, the notion of 'Ocean Citizenship' describes our obligation to act in a way so as to improve rather than harm our marine and coastal environment (Fletcher and Potts 2007). Fletcher and Potts state that '*individual citizens should have an understanding of the environment impact of their behaviour and an associated understanding of how to modify their behaviour to ameliorate or remove that impact*'. This is an important statement, which essentially acts to show us the way in linking the steps along the road to environmental (or ocean) literacy as embodied by the Environmental Literacy Ladder (ELL 2007), and the need for the development of an understanding of cause-and-effect, as demonstrated in the following sections relating to the development of understanding of human–ocean systems and their interactions. We can help individual 'ocean citizens' move from awareness to knowledge, understanding and ultimately taking responsibility for individual and collective action to protect and improve our ocean environment.

Understanding the Importance of Systems

Before we look at some of the recent research in assessment of Ocean and Environmental Literacy, it is useful to consider the relevance of the *systems* view of ocean literacy topics. As stated earlier, knowledge of the causal mechanisms at play greatly enhances our ability to make informed decisions regarding our behaviour. In order to help us develop and communicate our understanding of the complex cause-and-effect relationships between ourselves and the ocean, we need to develop *mental models* which map these relationships. One such modelling framework is the DPSIR framework (DPSIR 2013).

The DPSIR framework (Fig. 2) has evolved to co-exist in many variations, depending on the type of system being modelled (Elliott et al. 2017). For example, the DAPSIWR is a framework for describing interactions between society and the environment. Drivers (D) are driving forces which cause changes in society and the

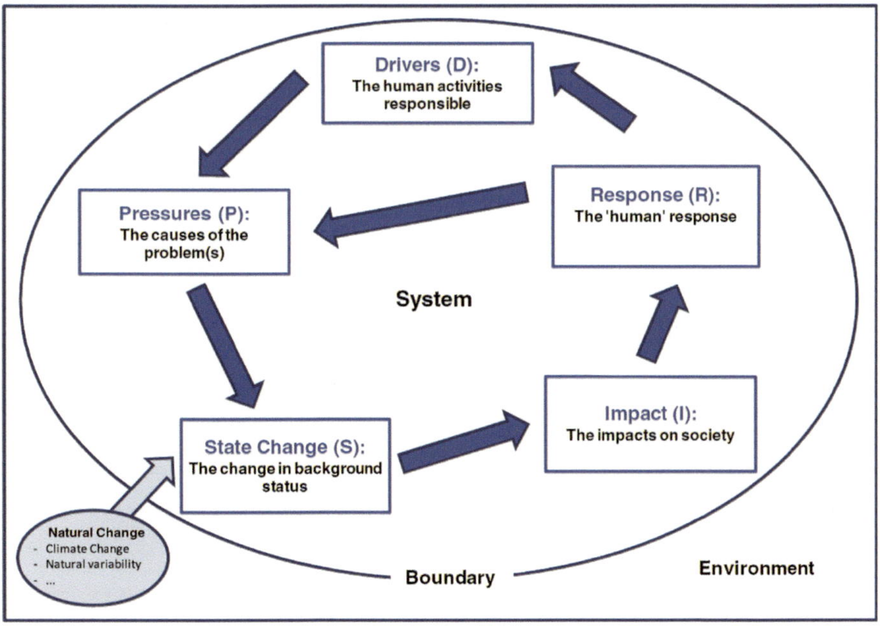

Fig. 2 The DPSIR framework (after Atkins et al. 2011)

environment, e.g. the need for food. An activity (A) is an action that is performed based on an existing driver, e.g. fishing based on the need for food. Pressures (P) are factors that originate from human activity and induce environmental change, e.g. over-fishing. State (S) can refer to a wide range of features including the characteristics of ecosystems, quantity and quality of resources, and human living conditions, e.g. the age distribution of remaining fish stocks. Impacts (I) are changes in environmental functions including resource access, and water and air quality, e.g. reduction in the resilience of fish populations. Responses (R) are policy actions which are triggered by the perception of impacts and they attempt to prevent, eliminate, compensate for or reduce the consequences of the impact, e.g. the implementation of fish quotas. Welfare (W) refers to impacts on human health and well-being, e.g. less fish available for human consumption.

A DPSIR-type model is just one way of capturing what is happening in a specific human–ocean system. The overall goal in such approaches is to gain a system-level understanding of what is going on, rather than focusing on small parts, and missing the overall dynamics and interactions. The old Indian story of the blind men and the elephant (Fig. 3) is often used to illustrate a number of different phenomena. They each touch a different part of the elephant, and have very different opinions on what it is. The story is used to illustrate things such as:

- the nature and limitations of subjective experience,
- the importance of communication,
- respect for different perspectives,
- and the importance of getting a complete picture before drawing conclusions or taking action.

Fig. 3 The blind men and the elephant

There are many different definitions of system, and they seem quite different. But in fact, they have a lot in common. Systems are defined not just by the individual parts or components, but by their connections, or relationships, and how these parts interact with each other.

Systems can be closed. By this we mean that they interact very little or not at all with other external systems. A Closed Circuit TV (CCTV) system is designed to be isolated from interference or transmission to or from cameras or monitors outside the system, making it more secure. Open systems, on the other hand, are open to exchange with the outside world. For example, a thermos vacuum flask is often used as an example of a closed system, as it doesn't allow heat exchange with the outside environment. However, if you take the lid off the flask, it immediately becomes an open system, interacting with the air around it.

Clearly, a system must be composed of elements which interact. This interaction goes to defining the behaviour of the system. While individual elements of the system may have behaviours of their own, when these are combined with the behaviours of other elements, and their interactions, the behaviour of the system can become very complex indeed. It is very difficult to see both the big picture and the smallest level of detail at the same time. That is why, when we study systems, we must decide the level of abstraction at which we can comfortably work, and which aids our understanding of the system and how it behaves. For example, when studying forests, it is better not to model down to individual cell level. If we want to understand trees, for example, we can study the tree, but to understand the forest as a system, we must understand how the various elements of that forest system interact with each other.

Systems thinking is the development of understanding of how the individual elements interact with each other, thereby influencing each other and determining the behaviour of the system as whole. We need mental models of things in order to understand them and discuss them with others. Effectively we create *representations* of reality which help us to understand them and how they behave. We can go further, and use our models to simulate various scenarios or situations, to see how the system might behave—if our model is good and accurate, it will give us useful insights!

A mental model is a kind of internal symbol or representation of external reality. So mental models are simply our own explanation or understanding of a real-life system or situation. Fixing complex problems cannot be done using simple linear thinking (Fig. 4), where for example, we treat individual symptoms, rather than understanding and treating underlying causes, or where we try to optimise the performance of individual parts of a system without optimising the performance of the system as a whole. This kind of thinking inevitably leads to some undesirable side-effects.

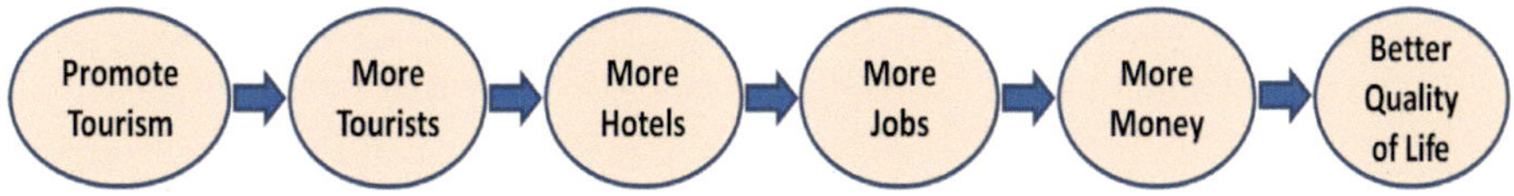

Fig. 4 Linear thinking mental model

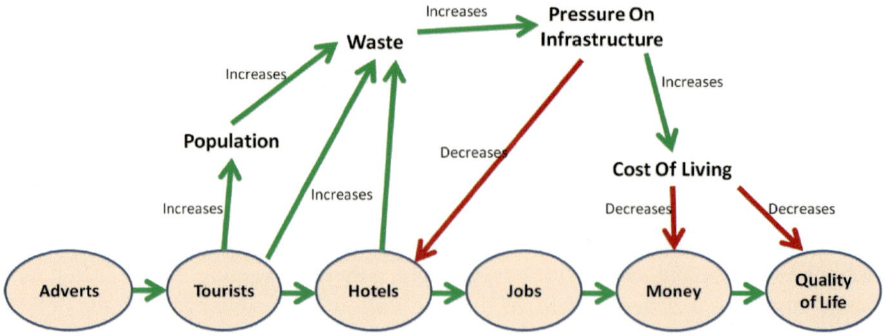

Fig. 5 Systems thinking mental model

Clearly when we are dealing with complex situations, we need to consider the consequences of our actions, and adopt a systems perspective (Fig. 5). While promoting tourism will probably increase local wealth in the short term, the long-term consequences can be quite undesirable. Increased pressure on the environment, due to waste and increased use of water resources, can have knock-on effects on existing industries and ultimately damage what made the area so attractive in the first place.

To summarise, when dealing with Ocean literacy topics, it is important to understand what we mean by knowledge and understanding. Real understanding means that we have decent mental models of the phenomena in question. Proper assessment of Ocean literacy therefore means interrogating those mental models to establish whether they really reflect reality. In the ResponSEAble project, we used 6 Key Stories to describe significant ocean issues, such as Microplastics, Sustainable Fisheries and Eutrophication. Central to the description of those stories are comprehensive DAPSIWR causal models of the interactions between the human and ocean systems. By identifying the human actors and activities and their interfaces with the ocean, we provide a better understanding of the impact of our activities and where interventions can make the most difference.

Why Measure Ocean Literacy?

It is possible to improve levels of Ocean literacy, but why bother? Surely it is how people act that matters, rather than what their mental models of various human-environmental systems are? Numerous studies (Hines et al. 2010) (Mobley et al. 2010) have shown that there is a strong link between knowledge and attitudes about the environment on the one hand, and more environmentally *responsible* behaviour on the other. Díaz-Siefer et al. (2015) developed a scale of human–environment system knowledge which focused on knowledge of environmental problems caused by humans rather than on knowledge of how the ecosystems themselves function.

The scale included the topics of climate change, pollution and resource availability because, in their opinion, these topics are the most important environmental issues that human beings are currently facing. The results of their research showed that there is a correlation between greater human–environment system knowledge and pro-environmental behaviour. Therefore numerous initiatives have been and are ongoing around the world to increase environmental and Ocean literacy. One of those initiatives, funded through the Horizon 2020 programme, was the ResponSEAble project, whose over-riding objective was to encourage Europeans to take a closer interest in their oceans and to treat them with greater respect and understanding.

The ResponSEAble Project

ResponSEAble was a European Union Horizon 2020 project focused on increasing ocean literacy. The project focused on key ocean health issues using 6 ResponSEAble Key Stories, namely eutrophication and agriculture, ballast water and invasive alien species, sustainable fisheries and aquaculture, microplastics and cosmetics, coastal tourism and marine renewable energy (ResponSEAble 2015).

Assessment of effectiveness focused on behaviour change objectives (of ocean literacy initiatives) designed for specific actors within each key story. If actors adopt behaviours and lifestyle choices that support sustainable use of the marine and coastal environment, the ocean literacy initiatives can be proven to provide tools that aid the concept of 'ocean (or marine) citizenship' (Fletcher and Potts 2007). Later in this chapter we describe a framework to assess the effectiveness of the ocean literacy initiatives, developed within the ResponSEAble project.

To identify the process required to achieve behaviour change objectives a Theory of Change logic model was adopted (Connell and Kubisch 1998). The stages, from identification of key stories, actors and behaviour change objectives through to development of OL initiatives and assessment of tools, were also summarised within the phases of a step-by-step planning and evaluation model, 'PRECEDE–PROCEED', developed by Green and Kreuter (1999). The PRECEDE–PROCEED model was originally aimed at directing change/behaviour change processes in health promotion, and has been widely adapted in environmental awareness programmes including the predictors of behaviour change, related to OL dimensions. Applying these models to guide development, assessment and evaluation of ocean literacy initiatives in ResponSEAble should enable wider application of the assessment framework to other ocean literacy programmes. The framework presented in this chapter is summarised in Fig. 6.

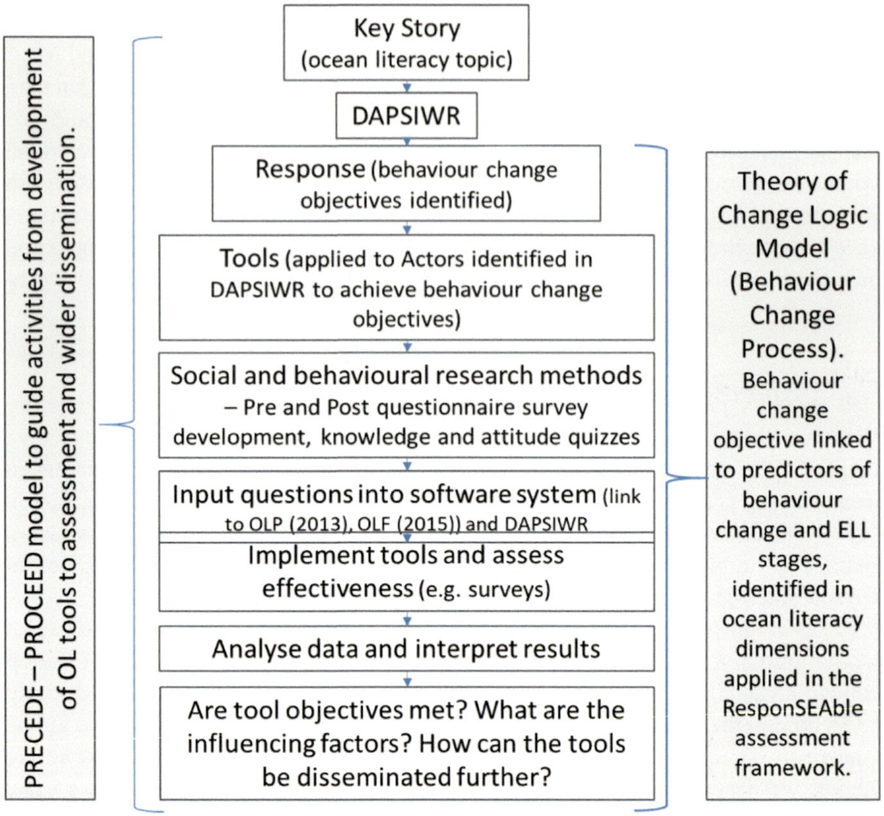

Fig. 6 Schematic of the framework applied in assessment of effectiveness of ocean literacy initiatives in ResponSEAble, including link to the identification of key stories, actors and behaviour change objectives

Our Methodology for the Application of the Behaviour Change Model

The research techniques applied in RepsonSEAble apply social and behavioural research methods, focusing on the predictors of behaviour change, and comparing actors' responses in relation to influencing factors such as age, location and pre-existing knowledge and level of activism. The assessment methods within ResponSEAble aimed to incorporate and build on previous assessment of ocean literacy.

To assess effectiveness of ocean literacy, the OL dimensions were integrated with predictors of behaviour change identified in social science and psychology literature (Klöckner 2013; Phal and Wyles 2017). We applied the 'Theory of Change' logic model (Connel and Kubisch 1998) to each tool (in relation to the key story/ocean literacy topic and the DAPSIWR for that topic) to identify the process the tool used to promote the desired response (behaviour change), and how each of the predictors of behaviour change may be met by the tool.

Application of the Theory of Change logic model allowed researchers trained in social and psychology research techniques, tool developers and key story authors to map out a path of how the desired behaviour change, promoted by the tool, would be achieved. Objectives were identified for each predictor of behaviour change and an indicator was set to assess if the objective was achieved (e.g. assessment results, such as questionnaire responses reporting an increase in knowledge about the issue, an increase in belief that it was important to act and increase in support for adopting behaviour that would reduce the impact on the marine environment).

The Theory of Change approach of Connel and Kubisch (1998) was developed to guide planning and implementation of social community initiatives (e.g. health and welfare projects). The approach has since been adopted to provide a guiding process in design, implementation and assessment of community conservation programmes, such as increasing support among fishermen to adopt sustainable fishing methods in Madagascar (Andriamalala et al. 2013) and reducing hunting pressure on the endangered Indochinese tiger population and their prey in Lao People's Democratic Republic (Saypanya et al. 2013). The benefit of the Theory of Change logic model approach was that the multi-disciplinary team working on ResponSEAble could be involved in a systematic and cumulative study of the links between ocean literacy tool activities, outcomes and the context for each ocean literacy tool. Tool developers (e.g. educators, film makers, game designers) and researchers could thereby work together to identify objectives for each OL dimension within a clear, structured approach.

The full assessment of effectiveness of ocean literacy tools applied in ResponSEAble, from a review of the DAPSIWR for each ocean literacy topic through to implementation tools with participants and assessment of effectiveness, can be summarised within the stages of the PRECEDE–PROCEED model (Gielen and Eileeen 1996; Green and Kreuter 1999). The PRECEDE–PROCEED model was originally developed in health research to guide the efforts that promoted the uptake of behaviours that supported healthy lifestyles. The model focuses on the principle that a change process should focus on the outcome, not the activity and was applied to initiatives that promoted behaviours to reduce the occurrence of leading causes of disability and death, such as heart disease, stroke, cancer and diabetes (Gielen and Eileeen 1996; Green and Kreuter 1999). Summarising the activity taken in the ResponSEAble project within these established models was intended to provide a transferable framework, to aid the assessment of effectiveness of ocean literacy tools and the initiatives at all scales in any regions they may be applied.

The role of assessment is to ensure tools are as effective as they can be at achieving ocean literacy goals, such as ensuring everyone understands about the ocean, its influence on them and their influence on the ocean. Reviewed examples of ocean literacy assessment often approach increase in knowledge or awareness about an ocean literacy topic, but not full behaviour change. In the ResponSEAble project, tools were applied to specific ocean literacy topics (key stories) and target audiences. The tools targeted specific responses or behaviour change objectives. Therefore, assessment aims to ensure tools are as effective as they can be at aiding the goals of ocean (or marine) citizenship to tackle large-scale and seemingly insurmountable international problems (Fletcher and Potts 2007), such that, participants of tools adopt behaviours and lifestyle choices that support sustainable use of the marine and coastal environment.

In this section, the adaptations to the OL dimensions, incorporating research on predictors of behaviour, are summarised in relation to the OL dimensions used in assessment in the ResponSEAble project. The relationship of assessment of effectiveness of tools to stages within the PRECEDE–PROCEED model is also discussed. Using the key story of sustainable seafaring, we provide a case study of the application of a Theory of Change logic model and the development of a pre- and post-intervention survey questionnaire, to assess the effectiveness of education courses for industry professionals in their early career stages.

Adapting the ELL to Include Predictors of Behaviour

People, their attitudes, decisions and behaviours or actions are at the heart of topics approached by ocean literacy topics (such as key stories in the ResponSEAble project). However, there is very little social research examining the problems ocean literacy considers, such as microplastics, sustainable fishing, spread of invasive species, eutrophication or impacts of coastal development (Phal and Wyles 2017).

Phal and Wyles (2017) highlight that information (informing people) is sometimes considered as the key factor for changing perceptions and behaviour by scientists outside of the behavioural sciences. Informing people (raising awareness) can be important, especially with emerging issues, but information alone is not very effective (Schultz 1999; Steg et al. 2013). Phal and Wyles (2017) conclude that understanding the influences of human thought and behaviour is just as important for understanding the human dimension and identifying the best ways of addressing human–ocean environment impacts.

The application of behavioural models and research techniques from psychology and behavioural sciences provides the opportunity for a greater level of understanding of uptake of messages from ocean literacy education tools and most importantly, assessment of behaviour change or change in intended behaviour.

Phal and Wyles (2017) present the results of an integrative model, tested by Klöckner (2013), combining data from 56 data sets targeting different environmental behaviours. Klöckner (2013) concluded that the best direct predictors of behaviour were:

- Intentions ('I will do this').
- Perceived behaviour control ('It is up to me whether I do this rather than other people or contextual factors').
- Habits (behaviours that have become automatised through repetition).

Factors shown to have indirect effects on behaviour were:

- Attitudes (favourable or unfavourable evaluations).
- Norms (what is seen as commonly done by others).
- Responsibility (ascriptions of who should deal with the problem).
- Awareness of consequences (knowledge about impacts).
- Values (general trans-situational goals such as equality or individualism).

Further studies have identified:

- Negative and positive emotions such as worry or hope (Dietriche 2013).
- Whether people see themselves as environmentalists (Whitmarsh and O'Neill 2010) may also have a role in dictating environmentally conscious behaviour.

These ten social and psychological concepts can be measured and distinguished empirically, and provide a rich toolbox for changing behaviour beyond information and knowledge provision (Phal and Wyles 2017). The stages within the ELL (2007) are closely allied with general principles and processes within psychological research to explain and predict human thought and behaviour (Fig. 7).

The predictors of behaviour change identified by Klöckner (2013) can be integrated with the ELL to enable assessment of effectiveness of an OL tool at promoting a behaviour change objective as well as increasing knowledge and awareness of a topic (Table 1).

Assessment of predictors of behaviour change and OL dimensions can be achieved through the application of research techniques such as surveys designed to be repeated pre- and post-interaction with an ocean literacy intervention (Phal and Wyles 2017). By asking the same question, pre and post interaction with tools, and over longer follow-up studies, change over time can be assessed. Comparison between different groups of people and influence of determining factors (such as initial level of environmental concern or interaction) can also be assessed (Phal and Wyles 2017).

Assessment of Effectiveness Within Stages of the PRECEDE–PROCEED Model

The PRECEDE–PROCEED model separates the planning of an initiative, such as an ocean literacy tool, and the implementation phase, such as use of the tool with participants (Green and Kreuter 1999). PRECEDE stands for Predisposing, Reinforcing and Enabling Constructs in Educational/Environmental Diagnosis and Evaluation (Green and Kreuter 1999). As the acronym implies, it represents the process that precedes, or leads up to, an intervention and is broken down into four phases (Fig. 8).

PROCEED spells out Policy, Regulatory, and Organizational Constructs in Educational and Environmental Development; this stage describes how to proceed with the intervention itself. PROCEED also has four phases that cover the actual implementation of the intervention and the careful evaluation of it, working back to the original starting point—the ultimate desired outcome of the process (Green and Kreuter 1999) (Fig. 8).

	Predictors of behaviour (Klöckner, 2013)	**Environmental Literacy Ladder (ELL).** (Text in brackets indicates an in-direct connection)	**Ocean Literacy Dimensions Applied in ResponSEAble Assessment (modified ELL)**
Best direct predictors of behaviour	**Intentions** ("I will do this.")	**Collective action:** capacity for personal and collection action and civic participation.	**Behaviour decisions, choices, actions, and habits,** with respect to ocean related issues (either intended behaviour "I will do", self reported behaviour "I do" and level of repetition of behaviour). **Activism**, the degree to which a person engages in protesting and campaigning to bring about political and social change.
	Perceived behaviour control ("It is up to me whether I do this rather than other people or contextual factors").	**Skills:** problem solving and critical thinking skills	
	Habits (behaviours that have become automatized through repetition).	(Collective action)	
Factors having an indirect effect on behaviour	**Attitudes** (favourable or unfavourable evaluations).	**Attitudes:** of appreciation and concern for the environment	**Attitude:** level of agreement with (favourable or unfavourable evaluations) or concern for a particular position, related to impacts and welfare in the DAPSIWR framework i.e. they exist, they are important, and a response is needed.
	Norms (what is seen as commonly done by others).	(Collective action)	**Communication:** extent to which a person engages with ocean related information and how much they communicate about the ocean with family and friends (influence of the social norm)
	Responsibility (ascriptions of who should deal with the problem).	(Attitude)	(Attitude)
	Awareness of consequences (knowledge about impacts).	**Knowledge,** and understanding of human and natural systems and processes	**Knowledge**, what a person knows about an ocean related topic and the links between topics (such as knowledge about impacts of human actions on the ocean environment)
		Awareness: general awareness of the relationship between the environment and human life.	**Awareness:** basic knowledge that a situation, problem or concept exists.
	Values (general trans-situational goals such as equality or individualism).	(Attitude)	(Attitude)
	Negative and positive emotions (such as worry or hope).	(Attitude)	(Attitude)
	Whether people see themselves as environmentalists.	(Attitude)	**Activism:** the degree to which a person engages in protesting and campaigning to bring about political and social change.

Fig. 7 The relationship between predictors of behaviour change, identified by Klöckner (2013) to the Environmental Literacy Ladder stages and the ocean literacy dimensions applied in ResponSEAble Assessment

The ResponSEAble Assessment Framework

The ResponSEAble framework uses a combination of the Ocean Literacy Framework (OLF), Driver-Activity-Pressure-State-Impact-Welfare-Response (DAPSIWR) framework and the Ocean literacy dimensions to expand on existing approaches to the measurement of OL. Ultimately, assessment is carried out using questionnaires, but the design of the questions is supported through the use of the framework to ensure that each question is linked not only to a specific topic in the OLF, but also to specific Drivers, Activities, etc. in the DAPSIWR framework. This allows us, if required, to develop measurement instruments that are focused on specific causal

Table 1 Sample of question/response types handled by the software system

Question types	Response types	Example question
Open response	The user inputs a text response to a question	What was the % of total energy consumed in the UK, from imported sources, in 2015?
Single response	Yes or No, true or false, Choose one option, Finish the graph and 5 and 10 point scales	Hull fouling is the undesirable accumulation of microorganisms, plants, algae and animals on submerged structures and especially on ships' hulls (Yes or No)
Single response and open response	(Single response) and text input option	To what extent do you agree with the following statement … and please explain why you have chosen that response
Multiple responses	Choose relevant items from the list	Please choose the main effects of coastal development from the list below
Single response and multiple responses	(Single response), with a list of items chosen from list	Have you heard of the problem of 'Invasive Alien Species' in relation to the ocean and if so, what were the main sources of your information on the subject?
Multiple responses and multiple responses	Choose relevant items from two-related lists	How recently have you discussed 'Invasive Alien Species' with your family or friends and if so, what communication methods did you use?

As the famous systems-thinking educator, Donella Meadows, said 'the behaviour of a system cannot be known just by knowing the elements of which the system is made' (Meadows 2008)

links between DAPSIWR elements in specific Key Stories. Finally, each question is also categorised in terms of the OL dimension (e.g. attitude) which is being measured.

The classification of the questions in relation to the OLF positions each question in the broader context of OL and will indicate the level of difficulty of the question based on which educational grade level the question matches best. Having the questions classified in the OLF will allow for the analysis of the question responses with respect to the different concepts and ideas contained in the OLF.

This approach will also create a way of visualising a person's ocean literacy, by incorporating recognised predictors of behaviour change. These can be empirically tested, through the study of the responses of participants, using ocean literacy education tools and social and behavioural science research techniques.

The OL dimensions used are based on the steps of the ELL with the OL dimension 'Communication' added. Increase in 'Communication' with friends, family and colleagues about a topic has also been identified as a predictor of a change in 'social

Phase of PRECEDE-PROCEED	Stage of Development and Assessment of OL tool in ResponSEAble
PRECEDE	
• Phase 1: Identifying the ultimate desired result.	DAPSIWR: identification of topics to be addressed and the OL tool behaviour change objective (Response).
• Phase 2: Identifying and setting priorities among health or community issues and their behavioural and environmental determinants that stand in the way of achieving that result, or conditions that have to be attained to achieve that result; and identifying the behaviours, lifestyles, and/or environmental factors that affect those issues or conditions.	DAPSIWR: identification of actors and tool participants. Reflection on audience: who the tool will be used by and who will interact with it? Including where tools will be disseminated and existing OL of participants. Development of Theory of Change.
• Phase 3: Identifying the predisposing, enabling, and reinforcing factors that can affect the behaviours, attitudes, and environmental factors given priority in Phase 2.	
• Phase 4: Identifying the administrative and policy factors that influence what can be implemented.	
PROCEED	
• Phase 5: Implementation – the design and actual conducting of the intervention.	OL tool development and testing (living lab). Implementation of tools.
• Phase 6: Process evaluation. Are you actually doing the things you planned to do?	Assessing objectives within Theory of Change, pre and post intervention surveys.
• Phase 7: Impact evaluation. Is the intervention having the desired impact on the target population?	Assessment and analysis of results. Reflection with tool developers, practitioners delivering the tool and researchers who authored the DAPSIWR review: how did the participants respond to the tool? Were the objectives met?
• Phase 8: Outcome evaluation. Is the intervention leading to the outcome (the desired result) that was envisioned in Phase 1?	Reflection on results: what worked well, what could be developed further, how could limitations be addressed, how could suite of tools be disseminated further to develop OL behaviour change objectives in Europe and globally?

Fig. 8 Relationship of PRECEDE–PROCEED model stages to development and testing and assessment of effectiveness of ocean literacy tools

norm', whereby a topic or behaviour becomes socially sanctioned, and undertaken to conform to accepted standards of behaviour (Klöckner 2013).

The analysis of the before and after questions will give a clear indication of how much the participant has moved on the OL dimension scale. Questions on ocean-related topics can be grouped into a questionnaire/survey and used to check if correlations exist between some of the OL dimensions defined in our framework. The correlations that exist between the OL dimensions of two different OL initiatives could be compared to give an indication of which OL initiative or tool is more effective, e.g. which OL tool has the better correlation for knowledge and attitude.

Where possible, questions in the pre- and post-questionnaire survey will relate to predictors of behaviour, identified in relation to the tool's behaviour change objective. By applying assessment to tools when they are used in different locations and with different groups of actors, it will be possible to investigate which tools and which messages are effective for different groups, for instance, urban and rural participants, general public and professionals entering an industry such as shipping and different nationalities.

Description of Software System

The browser-based software system associated with the framework provides tools to help the user to create the measurement items, receive the responses to the items and analyse the resulting data. When the user is creating a question, the system captures information such as the relevant OL principle (Fig. 9), Key Story, DAPSIWR nodes (Fig. 10), etc.

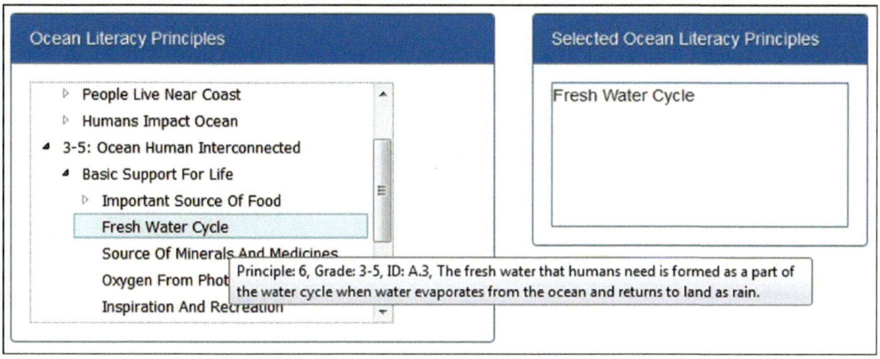

Fig. 9 Ocean literacy scope and sequence idea for question

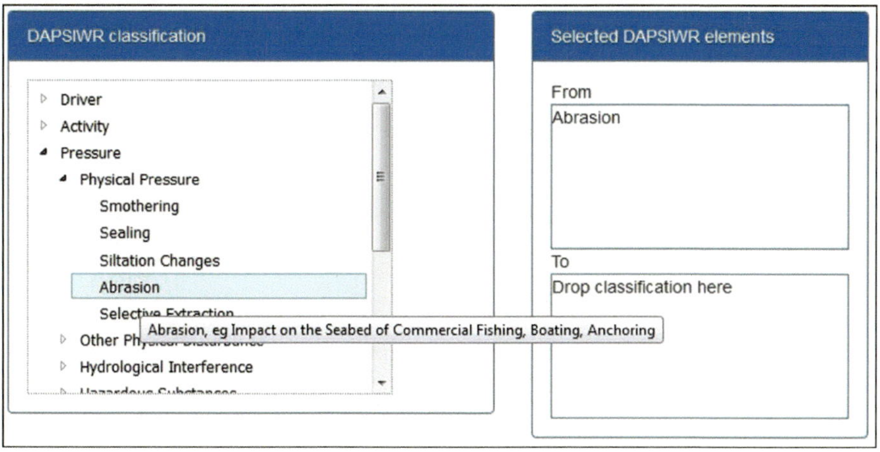

Fig. 10 DAPSIWR classification for question

It also captures which of the OL dimensions the question is measuring, e.g. a question related to awareness of eutrophication or a question related to a participant's attitude towards eutrophication. The key story indicates which of the ResponSEAble key stories the question is related to, e.g. coastal tourism.

The user can choose to link a question to a specific Activity, for example 'Coastal Tourism', or to link it to the causal link between an Activity and a Pressure. In this way, it is possible to know whether we are testing knowledge on nodes within a particular system, or knowledge on the causal mechanisms between them.

There are 12 different types of question which can be set up on the system. These question types relate to the format of the question and how the responses to the questions will be stored and analysed. A sample of the question types is shown in Table 1. Responses are sent in JSON format (JSON 2018) to a web service from where they are stored in a relational database for analysis.

Applying Existing Assessment Methods to the Framework for ResponSEAble

When analysing the data we use statistical analysis techniques to: calculate the difficulty of each of the items, obtain a score of our ocean literacy dimensions for each person, compare the scores and compare the correlations. The comparison of scores is related to before and after instruments, and comparing scores with similar scores already measured in the literature.

Predictors of behaviour change, such as those identified in social and behavioural science literature (Klöckner 2013), will also be assessed through survey instruments conducted pre- and post-participants' interaction with OL tools. The influence of variables, including age, location, existing level of interaction with the ocean environment and existing pro-environment behaviour, will be investigated in respect to change in response to scale-based questions, recorded before and after interaction with an OL tool.

Initial Experiences—Case Study: Developing Assessment of Effectiveness of Education for Professionals' Courses, 'sustainable Seafaring'

The case study summarises the initial experiences of applying the Theory of Change logic model and the development of a pre- and post-intervention survey questionnaire. The framework was applied to assess the effectiveness of education courses for shipping industry professionals (engineering officers and bridge officers) in their early career stages.

Training courses for professionals were developed by ProSea. These initiatives aimed to increase the acceptance of sustainable fishing techniques (fishermen) and the best practice in ballast water management to reduce the spread of invasive species (mariners). The courses provided knowledge and awareness raising around the topic and highlighted the benefits to people, the planet and also the profit for the industry from undertaking environmentally conscious actions.

A Theory of Change logic model was completed in collaboration with social and behavioural science researchers (University of Plymouth) and tool developers (ProSea). The Theory of Change identified the process by which the behaviour change objective of the course could be reached. The Theory of Change steps were guided by predictors of behaviour identified in the ocean literacy dimensions applied in the framework for assessment in the ResponSEAble project (Fig. 11).

Questionnaire surveys applying behavioural and social science research methods were conducted with course participants before and after attending the course. Each survey question addressed an objective within the Theory of Change framework and therefore an ocean literacy dimension within the framework for assessment of effectiveness applied in the ResponSEAble project. Scale-based questions (0–10) assessed if a change in awareness, knowledge, attitude, social and personal norms (communication), and reported or intended behaviour had occurred for participants (Fig. 12).

Initial Results of Pilot Study

ProSea's education for professionals, sustainable seafaring course, was piloted for the ResponSEAble project in Pasaia, Spain. Increases in predictors of behaviour change were observed for each dimension used in the framework. The largest increases were seen for 'how informed' participants felt about the effects of ballast water and alien invasive species on the marine environment, and their intention to 'undertake action to reduce or cope with the effects of alien invasive species'.

Participants' responses to survey questions showed a 77% increase in mean level (0 not at all—10 very well informed) of 'how informed' the participants felt they were about the effects of invasive species on the marine environment, and a 58% increase in how concerned the participants felt about the effects of invasive species on the marine environment. A 73% increase in mean level (0 not at all—10 all the time) of participants' intention to undertake actions to reduce or cope with the effects of invasive species was recorded after the course. Survey responses indicated that the participants were motivated to undertake behaviours and were provided information to enable them to carry out actions. Follow-up surveys will be important to find out if intended behaviours and actions were carried out, and what were the enabling factors or barriers. The delivery of the course, mixing lectures, videos and especially group work were popular with the participants.

Ocean Literacy - Behaviour Change: Theory of Change	Problem Awareness	Knowledge	Attitude	Interpersonal Communication / Social Norm	Barrier Removal	Behaviour Change
Theory of Change: AIM	Following the intervention participants will be aware of the issue or problem in the key story.	Following the intervention knowledge about the issue (key story) will have increased.	Following the intervention attitude towards the issue would have changed, and change in behaviour supported. Participants feel the response action will be effective.	Following the intervention participants will communicate about the issue or topic with friends, family and at work.	Barriers that prevented the behaviour change will be reduced/ removed.	Behaviour adopted or intention expressed:
Seafaring example	**i) Shipping industry professionals to recognise threat to resources that support welfare. ii) Shipping industry professionals to recognise threat to marine environment (loss of native species and native biodiversity).**	**Shipping industry professionals to recognise: (i) Maritime transport carries goods along with 3,500 million tonnes of ballast water calculated to be transported internationally, containing 4-5 000 taxa, in 480000 ship movements each year. ii) In Europe 45% of INNS are introduced through discharge of ballast water. iii) The potential negative environmental impact and impact on economic activity INNS may have. iv) Legal responses have driven technological solutions - ballast water treatment systems and ballast water free ship designs.**	**Shipping industry professionals believe/recognise that i) Ballast water contributes to spread of INNS ii) That installation of ballast water treatment systems will aid reduction of spread of INNS globally iii) That correct use of ballast water treatment systems is important and essential to reduce spread of INNS.**	**1. Shipping industry professionals will communicate with friends and family, colleagues and teachers about i) the threat of INNS to a) the health of the marine environment and b) to human wellbeing, ii) the options to reduce spread of INNS. 2. Shipping industry professionals seek information (internet, social media, books, magazines, newspapers, friends, family, expert advice) on the topic of INNS.**	**Knowledge of options or actions that reduce the spread of INNS.**	**1. Shipping industry professionals intend to undertake best practice recommendations for ballast water treatment. 2. Shipping industry professionals support industry to improve ballast water treatment and stay updated on developments in best practice recommendations. (understand the business benefits due to policy and customer need for treatment systems) 3. Shipping industry professionals maximise the opportunities they have to reduce impact of INNS.**
Measurable objective (indicator)	**Pre-post or Retrospective post survey - Analysis over time.**	**Pre-post survey- Analysis over time**	**1. Pre-post survey/Analysis over time. 2. Social media and communication analysis.**	**1. Social media and comm. analysis. 2. Pre-post survey - Analysis over time**	**1. Pre-post survey/Analysis over time.**	**Pre-post survey - Analysis over time. (Survey results of intended / reported behaviour).**
Result (from pilot study) (mean # pre – mean # post on 0 (*not at all*) -10 (*a lot / very*) scale).	77% increase in mean level of '*how informed*' participants reported being between pre and post survey. (4-7 – 8.3).	52% increase in mean level of participant's self-reported knowledge about '*how invasive species affect native marine life*.' (5.0 - 7.6) 45% increase in mean level of self-reported knowledge about '*how invasive species affect the availability of seafood*.' (5.3 – 7.7).	58% increase in mean level of participants reported '*concern about the effects of invasive species on the marine environment*.' (5.3 – 8.4). 23% increase in mean level of agreement from participants that '*it will be better for the marine environment, if ballast water is treated to reduce or eliminate the spread of invasive species*.' (7.4 – 9.1).	28% increase in mean level of how often participants reported '*talking about effective means of helping to reduce or cope with the effects of invasive species on the marine environment.*' (3.9 – 5.0).	52% increase in mean level of participant's self-reported knowledge. (5.0 - 7.6). 7% increase in mean level of participant's response that they '*felt capable they can reduce or eliminate the spread of invasive species through their everyday actions*.' (6.9 - 7.4).	73% increase in mean level of participant's intention to undertake actions to reduce or cope with the effects of invasive species. (4.5 – 7.8).

Fig. 11 Theory of Change logic model developed collaboratively between project partners to assess the effectiveness of education for professional's course

	Predictors of behaviour (Klöckner, 2013)	**Ocean Literacy Dimensions Applied in ResponSEAble Assessment.** (Text in brackets indicates an in-direct connection)	**Questions used in ResponSEAble pre and post intervention survey tool (sustainable seafaring courses)** (0 (*not at all*) - 10 (*a lot / very*) scale based responses)
Best direct predictors of behaviour	**Intentions** ("I will do this.")	• **Behaviour decisions, choices, actions, and habits.** • **Activism**	Do you currently undertake any actions to reduce or cope with the effects of invasive species on the marine environment? If you do, what actions do you currently undertake?
			In the future, do you think you will do something else to reduce or cope with the effects of invasive species on the marine environment? If yes what actions will you undertake?
	Perceived behaviour control ("It is up to me whether I do this rather than other people or contextual factors").		*'I feel capable that when I start to work as a seafarer, I can reduce the spread of invasive species through my everyday actions.'*
			In your opinion, how effective are the options below at reducing or coping with the spread of invasive species ? *(options range from actions participant can take and actions other people can take or contextual factors)*
	Habits (behaviours that have become automatized through repetition).		Do you currently undertake any actions to reduce or cope with the effects of invasive species on the marine environment? If you do, what actions do you currently undertake?
Factors having an indirect effect on behaviour	**Attitudes** (favourable or unfavourable evaluations).	• **Attitude**	*'I believe it will be better for the ocean environment and marine life, if ballast water is treated to reduce or eliminate the spread of invasive species.'*
	Norms (what is seen as commonly done by others).	• **Communication**	How often do you talk about effective means of helping to reduce or cope with the effects of invasive species on the marine environment with your family, friends, colleagues or teachers?
	Responsibility (ascriptions of who should deal with the problem).	• (Attitude)	In your opinion, how effective are the options below at reducing or coping with the spread of invasive species ? *(options range from actions participant can take and actions other people / industry or policy makers can take or contextual factors)*
	Awareness of consequences (knowledge about impacts).	• **Knowledge**	*'I have good knowledge about how invasive species affect native marine life.'*
			According to estimates in 2010, how many species are transported around the world in shipping ballast water every day? (Please circle the correct answer)
		• **Awareness**	How informed do you feel about the effects of invasive species on the marine environment?
	Values (general trans-situational goals such as equality or individualism).	• (Attitude)	Not assessed, though indirectly approached by the question: *'I always think about how my actions affect the marine environment'*
	Negative and positive emotions (such as worry or hope).	• (Attitude)	*'I am concerned about damage to the natural environment'*
			How concerned are you about the effects of invasive species on the marine environment?
	Whether people see themselves as environmentalists.	• (Attitude) • **Activism**	Have you supported a campaign or petition on marine environmental issues in the last 1-2years? (Yes / No)
			'I always think about how my actions affect the marine environment'

Fig. 12 Relationship between questions developed to assess the effectiveness of the sustainable seafaring, education for professional's course, predictors of behaviour and ocean literacy dimensions assessed in ResponSEAble

Opportunities and Limitations Identified in the Pilot Study

The results of pilot study data summarise percentage change of mean value for participant's responses between pre- and post-course surveys. The use of percentage change reduces emphasis on change where there was already a strong positive response to a question. Bar charts of mean values and error bars would better display consistent strong agreement between pre- and post-surveys. For instance, participants showed strong agreement to the attitude statement *'it will be better for the marine environment if ballast water is treated to reduce or eliminate the spread of invasive species.'* (0 *not at all*—10 *strongly agree*), prior to attending the course (mean 7.4). After the course response to this question provided the strongest agreement of any question (mean 9.1), but a comparatively moderate percentage increase of 23%. The influence of demand characteristics also need to be considered in this instance. The participants were aware they were attending a 'sustainable seafaring' course, so there may have been an influence of providing the answer that was felt to be 'expected' in pre- and post-course surveys.

The high agreement that treatment of ballast water would be better for the marine environment is of interest when considered alongside the moderate increase in agreement with the statement, *'I feel capable that when working as a seafarer, I can reduce or eliminate the spread of invasive species through my everyday actions'*, (7% increase, mean 6.9 pre–mean 7.4 post). Comments by participants in course workshop sessions suggested that, as they work or would be working in junior positions, they were required to follow orders. Participants suggested a potential barrier, that they would not have personal control over what ballast water treatment systems are installed and procedures of the shipping industry, beyond following best practice in their operation and encouraging others to do so.

Identification of barriers displays the benefit of a pilot study to identify questions where there is likely to be influence of outside factors, such as ship owners and senior officer's decisions. Influence of industry regulation and ship owners' decisions highlights the need to approach multiple actors identified in the DAPSIWR analysis, such as ship owners, port authorities and policy makers, to create an industry-wide change in behaviour.

Follow-up surveys, to identify whether participants have undertaken intended behaviours, and the enabling and inhibiting factors will provide valuable data on long-term behaviour change. The survey also collected data on course location, participants' age, job role, years working in the industry, interest in environmental activism and time spent in the marine and coastal environment in their spare time. Responses of the final surveys will be analysed in relation to these factors to investigate the level of influence they have on participant's responses and predictors of behaviour change.

Conclusions

Testing and evaluation of initiatives and indeed testing of assessment frameworks and assessment methodologies are integral in the development of ocean literacy initiatives which strive to reach common goals of responsible ocean citizenship and sustainable use of the marine environment.

In the framework presented in this chapter we have applied experience in psychology and behavioural research concepts and methods (predictors of behaviour change) to the development of a framework for the assessment of ocean literacy initiatives and tools.

This framework is being applied and tested in ResponSeable to identify strengths and weaknesses of ocean literacy initiatives. The results will also provide the opportunity to evaluate the framework itsself and the assessment techniques applied, and so inform standardised techniques and good practice in assessment of OL programmes.

In the case study, it was demonstrated that predictors of behaviour change could be assessed through a pre- and post-survey design, and the assessment framework could be applied effectively. Follow-up studies and use of control groups are recognised as being important to future assessment.

If applying the social and behavioural science survey methods (pre- and post-intervention survey tool) to future studies, it is also important to consider the balance between the time available for participants to complete surveys, how surveys will be facilitated and the complexity of the survey design. The questionnaire design in this case study was designed to be easily applied by teaching professionals independent of researchers trained in social and behavioural science research techniques. A limited number of questions were used to assess each dimension of the predictors of behaviour change. A more detailed survey could include a series of relevant questions (e.g. for the attitude dimension) that can be aggregated into a better combined score, thus, reducing the error associated with a single item/question (Phal and Wyles 2017). It is also important to consider comparing the treatment group (who have used the tool, or attended the course in this case) with control groups, who complete the same assessment but have not attended the course. Undertaking a pilot study identifies these limitations and enables them to be addressed in the future.

References

Andriamalala G, Peabody S, Gardner CJ, Westerman K (2013) Using social marketing to foster sustainable behaviour in traditional fishing communities of southwest Madagascar. Conservation Evid 10:37–41

Atkins Jonathan P, Burdon Daryl, Elliott Mike, Gregory Amanda J (2011) Management of the marine environment: integrating ecosystem services and societal benefits with the DPSIR framework in a systems approach. Mar Pollut Bull 62:215–226

Connell JP, Kubisch AC (1998) Applying a theory of change approach. In: Fulbright Anderson K, Kubisch AC, Connell JP (eds) *New approaches to evaluating community initiatives, vol 2, theory, measurement, and analysis.* Aspen Institute, Washington, DC

DPSIR (2013) Driver, pressure, state, impact, response framework [Online]. Available: http://www.ejolt.org/2013/02/dpsir/. Accessed 4 Apr 2018

Díaz-Siefer P, Neaman A, Salgado E, Celis-Diez JL, Otto S (2015) Human-environment system knowledge: a correlate of pro-environmental behavior. Sustainability (Switzerland) 7:15510–15526

ELL (2007) Environmental literacy ladder [Online]. http://www.fundee.org/facts/envlit/whatisenvlit.htm. Accessed 14 Feb 2018

Elliott M, Burdon D, Atkins JP, Borja A, Cormier R, de Jonge VN, Turner RK (2017) And DPSIR begat DAPSI(W)R(M)!" - A unifying framework for marine environmental management. Marine Pollution Bulletin 118(1–2):27–40 May 15

Fletcher S, Potts J (2007) Ocean citizenship: an emergent geographical concept. Coastal Manag 35(4):511–524

Gielen AC, Eileen MM (1996) The PRECEDE-PROCEED planning model. In: Glanz K, Lewis F, Rimer KB (eds) Health behavior and health education. Jossey-Bass, San Francisco

Green Lawrence W, Kreuter Marshall W (1999) *Health promotion and planning: an educational and environmental approach*, 4th edn. Mayfield Publishing Co, Mountain View, CA

Hines J, Hungerford H, Tomera A (2010) Analysis and synthesis of research on responsible environmental behaviour: a meta-analysis. J Environ Educ 18(2):1–8. https://doi.org/10.1080/00958964.1987.9943482

JSON (2018) JavaScript object notation [Online]. Available: https://www.json.org/. Accessed 5 Dec 2017

Kikuchi-Uehara E, Nakatani J, Hirao M (2016) Analysis of factors influencing consumers' proenvironmental behaviour based on life cycle thinking. Part I: Effect of environmental awareness and trust in environmental information on product choice. J Clean Prod 117:10–18

Klöckner CA (2013) A comprehensive model of the psychology of environmental behaviour—a meta-analysis. Global Environ Change 23(5):1028–1038

Dietriche H (2013) Role of Emotion in Environmental Decision Making. Lincoln: University of Nebraska.

Meadows D (2008) Thinking in systems: a primer (Edited by Diana Wright, Sustainability Institute). Chelsea Green Publishing Company, USA

Mobley C, Vagias A, DeWard S (2010) Exploring additional determinants of environmentally responsible behaviour: the influence of environmental literature and environmental attitudes. J Environ Behav 42(4)

OLF (2015) Ocean literacy framework [Online]. Available: http://oceanliteracy.wp2.coexploration.org/ocean-literacy-framework. Accessed 4 Apr 2018

OLP (2013) Ocean literacy, the essential principles and fundamental concepts of ocean sciences for learners of all ages [Online]. Available: http://www.coexploration.org/oceanliteracy/documents/OceanLitChart.pdf. Accessed 14 May 2018

Phal S, Wyles KJ (2017) The human dimension: how social and behavioural research methods can help address microplastics in the environment. Anal Methods 2017(9):1404

ResponSEAble (2015) ResponSEAble project [Online]. Available: https://www.responseable.eu/. Accessed 6 Mar 2018

Sánchez MJ, Lafuente R (2010) Defining and measuring environmental consciousness. Revista Internacional de Sociologia 68:731–755

Saypanya S, Hansel T, Johnson A, Bianchessi A, Sadowsky B (2013) Effectiveness of a social marketing strategy, coupled with law enforcement, to conserve tigers and their prey in Nam Et Phou Louey National Protected Area, Lao People's Democratic Republic. Conservation Evid 57–66

Schultz PW (1999) Changing behaviour with normative feedback interventions: A field experiment on curbside recycling, Basic Appl Soc Psychol 21(1):25–36

Schultz PW, Shriver C, Tabanico JJ, Khazian AM (2004) Implicit connections with nature. J Environ Psychol 24:31–42

Steg AE, den Berg Van, De Groot JIM (2013) *Environmental psychology: an introduction*. Wiley, Chichester, UK, p 2013

Umuhire ML, Fang Q (2016) Method and application of ocean environmental awareness measurement: Lessons learnt from university students of China. Mar Pollut Bull 102:289–294

UNESCO (United Nations of Education Scientific and Cultural Organisation) (1975) The international workshop on environmental education final report, Belgrade, Yugoslavia. Paris: UNESCO/UNEP

Whitmarsh I, O'Neill S (2010) Green identity, green living? The role of pro-environmental self-identity in determining consistency across diverse pro-environmental behaviours. J. Environ Psychol 30(3):305–314

Exploring and Exploiting Deep Ocean Space

Jan H. Stel

Abstract Jules Verne's life was framed by the second phase of the British Industrial Revolution. In his nineteenth-century world, a transition to steam occurred. It was a time of rapid technological developments and explorations of every corner of the Earth's surface. Then the world population clock ticked slowly and was well below 1.3 billion. Today, more than 7.8 billion people live in the fourth phase of that revolution, and the world population clock is ticking faster and faster. To solve our urgent demand for resources, we will shortly exploit the unknown treasure troves of deep ocean space. However, only some 15% of the ocean floor is mapped in detail, and less than 0.0001% of the deep-sea is explored. Since the 1990s a transition to global operational oceanography is occurring, with advanced monitoring systems, new technology like Argo floats, gliders and state-of-the-art ocean modelling. A new wave of ocean exploration is urgently needed, as is an adaptation of the prevailing international legislation, to keep up with the coming sustainable exploitation of ocean space. Blue resources discussed in this chapter are: fisheries, bioprospecting and deep-sea mining. In a low-carbon society, citizens should be aware of and be involved in this through ocean literacy.

Keywords Deep-sea resources · Pollution · British Industrial Revolution · Jules Verne · Ocean exploration

Introduction

'Oh, oh, that's not good! I have a series of error messages. We have a problem', mumbled James Cameron at a depth of 10,908 m and a hydrostatic pressure of nearly 1,100 atm. He was in the Mariana Trench, the inaccessible cellar in deep ocean space, near Guam in the Pacific. Cameron was alone in this pitch-dark, hostile and unknown environment. After seven years of construction in complete secrecy at

J. H. Stel (✉)
Maastricht Sustainability Institute, Maastricht University, Maastricht, The Netherlands
e-mail: janstel@skynet.be

Puurs-Sint-Amands, Belgium

K. C. Koutsopoulos and J. H. Stel (eds), *Ocean Literacy: Understanding the Ocean*,
Key Challenges in Geography, https://doi.org/10.1007/978-3-030-70155-0_4

an industrial site in Sydney, Australia, an innovative, high-tech, cigar-shaped, lime-green submersible was ready for this unique dive. His solo dive took place on 26 March 2012, Guam time.

During its trip through ocean space, the 7.3m-long and 2.3 by 1.7m-wide *Glomar Challenger* slowly turned on its vertical axis, making the vessel stay on course. The descent took two hours and 37 minutes. After staying for half of the projected five hours at the seafloor, malfunctions of thrusters and other equipment forced Cameron to return to the top of ocean space. He successfully released half a tonne of ballast for a speedy ascent. After surfacing he said: 'It was bleak. It looked like the moon'.

This was the second human-occupied dive to this remote spot in ocean space. The first took place on 23 January 1960 by US Navy Lt Don Walsh and Swiss engineer Jacques Piccard. They spent some twenty minutes at the ocean floor, in their bathyscaphe *Trieste*. The descent to the ocean floor took four hours and 47 minutes. There were two plexiglas windows and some floodlight, for observations, but once at the bottom their sight was hindered by clouds of silt caused by the vessel. Nevertheless, they claimed to have seen some fishes. Unfortunately, the *Trieste* was not equipped to take samples from the ocean floor nor to take pictures. The ascent to the ocean surface lasted three hours and 15 minutes. As a tough old man, Dr. Don Walsh also participated in Cameron's and Vesocovo's expeditions.

The next two solo dives into the Mariana Trench took place on 28 April and 1 May 2019 (Five Deeps Press 2019). During the first dive the multimillionaire Wall Street trader and adventurer Victor Vesocovo, reached a record depth of 10,928 metres. By staying more than four hours at this depth he also established a new time record. These dives were part of a daring exploration, *The Five Deeps Expedition*, to the deepest spots or cellars of ocean space in the five oceans we distinguish. In four of them humans have never reached the bottom at all. His two dives were followed by two scientific and one commercial dive (see websites).

The Mariana Trench is a 2,400-km-long scar in the bottom of the western Pacific Ocean near Guam, halfway Japan and Papua New Guinea. It is part of a subduction zone, where the Pacific Plate, with a speed of a few centimetres a year, moves under the Mariana Plate. This geological process creates the Mariana Trench and the arc of the Mariana Islands. The trench is 120 times larger than the Grand Canyon, and 1.6 km deeper than Mount Everest, the largest mountains on the continents, is tall.

Don Walsh and Jacques Piccard just had a peek into the perpetual darkness of this natural laboratory. James Cameron was better prepared. His sub was outfitted for scientific exploration: a hydraulic manipulator arm for taking samples and 3-D cameras for capturing images and films. LED lights, making the sub looking like a floating film studio, illuminated the ocean floor.

As stated by Cameron, the ocean floor 'looked like a parking lot covered with new fallen snow'. The only larger animals he saw, most likely was the amphipod *Hirondellea gigas*, which thrive at the bottom of the Mariana Trench. The one sample taken from the ocean was teeming with invisible microbial life. An analysis showed that it contained genomes of some 20,000 microorganisms. More than a hundred were new. In 2013, Cameron gave this innovative submarine to the Woods Hole Oceanographic Institution, WHOI, in the US.

The period between the beginning of the twentieth century and the global Great Depression in the 1930s, was the 'golden age of ocean liners'. Mainly driven by immigration to the US, British, Dutch, German and French shipping companies were building bigger, faster and fancier ships to accommodate both the wealthy and poor immigrants. RMS *Titanic*, the largest and technologically most advanced ship of its time, was the pride of the British White Star Line.

It was state-of-the-art, thought to be 'unsinkable' because of innovations like watertight compartments, and could carry 2,453 passengers. Due to a British coal strike, the number of passengers at its maiden voyage only was some 1,317. In the early hours of 15 April 1912, the doomed liner struck an iceberg and sunk within three hours. More than 1,500 people died. New research capabilities such as deep diving, manned submersibles like *Alvin*, remotely operated vehicles (ROVs) and deep-towed sonars, led to the discovery of *Titanic*'s wreck on 1 September 1985, by Bob Ballard and his team (Ballard 1987).

On 8 January 2005, UUS *San Francisco* collided at full speed at a depth of 160 m with an uncharted undersea volcano, some 675 km southeast of the naval base of Guam (Fig. 1). The ship's crew was tossed around by the forceful collision: 98 of the 137-member crew were injured; one died from head injuries. The vessel was a Los Angeles class nuclear-powered fast attack submarine. The ship's forward ballast tanks and her sonar dome were severely damaged. Despite its sophisticated technology, it still is possible that a state-of-the-art submarine, moving around in the uppermost part of the ocean, collides with a seamount. The collision demonstrates that ocean space is still an unknown territory.

Taking an airplane for travel, has become an everyday affair. In 2018, just over 4.3 billion passengers were, according to the International Civil Aviation Organization, transported on scheduled flights. But on 8 March 2014, flight MH370 of Malaysia Airlines suddenly disappeared near Phuket island in the Strait of Malacca (Fig. 1). It is the greatest aviation mystery so far. On board the plane were 227 passengers and a crew of twelve persons. Despite an intensive international search with the best technology available today, no sign of flight MH370 has been found. This is underpinning the fact that, even today, the ocean floor is not mapped in detail.

Fig. 1 The nuclear attack submarine USS *San Francisco*, ran into a mountain in January 2005 (left) and a Malaysian Airlines Boeing 777, similar to flight MH370 that went missing in March 2014(right). © Wikipedia

Exploration

An old Flemish farmer went for his first holiday to the Belgian coast. All his life he had lived in the Belgian Ardennes in the south-eastern part of the country. He was sitting on a bench at the famous boulevard of Knokke's seafront, aweing the impressive sea. 'So much water, so much water', he mumbled softly. Whereupon an old Belgian fisherman, sitting at the bench next to him, and hearing this, answered: 'and then you have only seen the top'! It is an old anecdote, indicating two absolutely different perspectives. Farmers cultivate the land and know their soils; fishermen know the ocean surface and probe its depths to catch fish.

Yet, for most of us, being landlubbers, the ocean is unfamiliar. The surface of the ocean covers almost 71% of the surface of the planet. This is an area of 361 million km^2. Interesting in this anecdote is, that the farmer, like most people today, considers the top of ocean space, as a surface (Fig. 2). The fisherman, however, has an eye for the third dimension of the ocean. After all, he knows that his catch comes from below the surface. However, during most of the historic times, we have seen the ocean as a surface to cross, as an easy way to move around, and to explore 'the other side'.

The Ocean in 2D

There is some speculation that *Homo erectus*, who started to spread throughout the Old World some two million years ago, was also using boats or rafts to explore new horizons (Davis 2018). The seafaring capabilities of our more recent ancestors

Fig. 2 The ocean as a surface to cross. © Author

are, however, better documented. Archaeological and genetic evidence indicate that Australia and New Guinea were colonized some fifty thousand years ago.

A modelling exercise showed that the colonization of Australia took place from the Indonesian islands Timor and Roti. It was achieved by well-planned, marine voyages of a couple of hundred people (Norman et al. 2018). Within 2,000 years the Aboriginals successfully colonized the entire continent. The West-European exploration and colonization by Dutch, French and British explorers, only started after the Dutch navigator Willem Janszoon (c.1570–1630), scouted the Australian west and north coasts in 1606 with the *Duyfken*.

In the early part of the fifteenth century, Chinese exploration was peaking with the famous seven voyages to countries around the Indian Ocean or the Western Sea, by Zheng He (1371–1433). He led expeditions to South and Southeast Asia, East Africa and the Middle East, every time showing the power of the newly established Ming empire under the Yongle Emperor (1402–1424), and demonstrating the supremacy of Chinese shipbuilding and navigation.

The fleets of Zheng He were overwhelming. As a rule, they consisted of dozens of nine-masted treasure ships, with a length of around 140 m, a width of up to 62 m and a displacement of over ten thousand tonnes. They carried a crew of one thousand people, had watertight compartments, and were the largest wooden ships ever. There also were specialized vessels to carry horses, grain, water and troops. The first fleet, consisting of some 250 sailing ships, left Nanjing in Southern China in the autumn of 1405. Basically, they did not come to colonize, as the West-Europeans did almost a century later. Zheng He died on the return voyage of his seventh expedition in 1433, as his fleet was sailing east from Calicut. He was buried at sea (Dreyer 2007).

The great maritime expansion of Western European countries in the sixteenth century started with the voyages of discovery organized by the Portuguese prince Henry (1394–1460). These journeys were financed by the Catholic military Order of Christ (1316–1789). Henry was especially interested in gaining access to West African gold and slaves.

With brutal force, the construction of forts, the spread of their religion, the West-European maritime powers established colonies and imperialism. This process peaked during the British Industrial Revolution, which marks a major transition in human history. The mechanization of the British cotton industry and the use of water and steam, as a power source, was the kick off. Yet, it was, according to Joel Mokyr, preceded by a dramatic change in worldview as a consequence of exploration and discovery. Moreover, there was a knowledge revolution, leading to 'useful knowledge' in Western Europe (Mokyr 2002).

The use of fossil fuels such as coal and, later, oil and gas to generate power, transformed society. It also cracked the dependency of ships on wind and ocean currents. It allowed the construction of large steamships, such as the doomed *Titanic*, as well as the present super container and giant cruise ships. Today, 90% of the world's trade is carried by sea. It is the most cost-effective way to move en masse goods and raw materials around the world. Since the Second World War technology has advanced rapidly during the third and present fourth phase of this revolution.

New innovations such as supercomputers, robots and artificial intelligence, pave the way for deep ocean space exploration.

Unlimited resources like coal fuelled a massive nineteenth-century expansion of innovative technology and infrastructure, such as rail and telegraph lines. It was a time of awe and wonder, of an unconditional belief in technology and of excitement while new scientific insights were constantly paving the way to innovation and progress. People were also convinced that marine resources like fish were inexhaustible.

But the British Industrial Revolution also was a monster leading to unprecedented local pollution in sprawling grubby cities with rivers becoming open sewers. In retrospect, it marks the beginning of an unprecedented geoengineering project, too: CO_2-pollution of the global atmosphere, leading to the present human-made climate change which in turn leads to a series of derivative environmental problems, like ocean acidification.

Today, there is a widespread scientific and political consensus that the burning of fossil fuels has led, and will lead, to global warming not seen in recent geological history. It is a shame that oil companies like Exxon and Shell knowingly concealed this effect since the 1960s. It is a crime they often supported climate change deniers to frustrate discussions and decision-making.

Jules Verne, a Great Storyteller

The United Kingdom of the nineteenth century was leading the industrial world. To underpin this position, the Great Exhibition of the Works of Industry of All Nations was organized in the Crystal Palace in London in 1851 (Fig. 3). It was an

Fig. 3 The Great Exhibition of the Works of Industry of All Nations in the Crystal Palace in London in 1851, was the first world expo. © Wikipedia

idea of Prince Albert, Queen Victoria's husband, and Sir Henry Cole (1808–1882) of the Royal Society of Arts. The Great Exhibition was an international platform where countries displayed their technological achievements. It led to a series of world exhibitions with a focus on industrialization until 1935, on cultural exchange (1939–1987), and since 1988 on nation branding. Most of the expos until 1900 were organized by the French government. One of the visitors was the famous French visionary and futurist Jules Verne (1821–1905).

He lived during the second phase, roughly 1870–1914, of the British Industrial Revolution, when it was spreading across the European continent. It was a time of new scientific ideas and technological inventions, economic progress through the adoption of steam power, urbanization in the new industrial areas and a new wave of globalization. It was the time in which knowledge collected by ocean explorers like James Cook (1728–1779), Louis Antoine de Bougainville (1729–1811), Alexander von Humboldt (1769–1859), Charles Darwin (1809–1882) and many others matured. It was the time of the *Challenger* expedition (1872–1876) with Sir John Murray (1841–1914), who was the founding father of oceanography. It was the beginning of the 'Heroic Age' of polar exploration, with names like Jules Dumont d'Urville (1790–1842), Fridtjof Nansen (1861–1930), Adrien de Gerlache (1864–1934) and many others. It was the time in which the American Matthew F. Maury (1806–1875) published in 1855 the first modern oceanographic textbook on *The Physical Geography of the Sea,* a book that Verne has read and refers to in his books.

Verne was destined to become a lawyer just like his father. In 1848 he went to Paris to finish his law study. But he also was spending much time in the Bibliothèque nationale de France, to learn about science and technology, and recent discoveries in geography. At the world expos in 1867, he saw a model of the French submersible *Plongeur*, and learned about the new possibilities of electricity. This inspired him to his bestseller *Twenty Thousand Leagues Under the Seas.*

Box 1: 20.000 Leagues under the Seas

Jules Verne's *Nautilus* was powered by electricity and explored the deep-sea. Courtesy Dutch Jules Verne Society

Verne's bestselling novel *Twenty Thousand Leagues Under the Seas* was published in two parts in 1869 and 1870. Central to the story is a giant electrically powered, submarine *Nautilus*. Sailors were reporting a large fast-moving creature showing a strange glow. Several ships, trying to approach the monster, were seriously damaged or sunk. The US equips its fastest well-armed frigate *Abraham Lincoln* to take care of the monster. The famous French professor Pierre Aronnax, from the Museum of Natural History in Paris, concluded that the monster was a huge narwhal.

Together with his assistant Conceil, and a well-known Canadian harpooner Ned Land, he was invited to join the hunt. During the first encounter they fell overboard and were taken prisoner on board the *Nautilus*. Here they meet the mysterious troubled Captain Nemo, a hero driven by hatred, as stated by Verne. He shows them the submarine, which was equipped with powerful electric lights and engines, which had large windows to view the underwater world, and diving suits to explore the ocean floor as well as to collect treasures from sunken ships.

He takes them, during a travel of some 80,000 km, to the unknown corners of ocean space. They use a secret underwater passage—the Suez Canal was opened on 17 November 1869—from the Mediterranean to the Red Sea, where they bury a sailor in the reefs. They visit the unknown continent of Atlantis, and sail under the ice towards the South Pole, where they become entrapped in the ice. They were attacked by a giant squid that nearly sinks the *Nautilus*, and go hunting sharks with electric guns, the teasers of today. Finally, the prisoners escape off the Norwegian coast, where they were rescued by Lofoten fisherman.

In his books the main characters were exploring, with new inventions, the planet's unknown territories such as the centre of the earth, the deep-sea, the African continent and the poles. This led to the famous series of *Voyages Extraordinaire* (Extraordinary Journeys), a very popular series of fifty-four, thoroughly researched adventure and educational novels. According to his publisher, Jules Verne's stories were no science fiction, but just educational entertainment. That is ocean literacy in its infancy.

Since Jules Verne's death in 1905, the world population increased from some 1.6 billion to 7.8 billion in March 2021. Today climate change is a reality. Human activities are now profoundly influencing the earth system. As a consequence, scientists have coined a new notion in earth's history: the Anthropocene. Starting with a huge nuclear explosion which, in 1945, erased the Japanese cities Hiroshima and Nagasaki, it is a new informal geological period in which human activities are a geological force.

During the Anthropocene, human activity changed the earth's land surfaces, rivers, oceans, atmosphere, flora and fauna, and humans themselves. By realizing this it also could be seen as an urgent call for sustainability and global and regional governance. Today's exploration and exploitation of deep ocean space, the last frontier of the planet, is mainly driven by green or blue economic considerations, and facilitated by the present-day technological developments. Just as in Jules Verne's life, there again is a firm trust in future technology, even if it would be applied for geoengineering projects, to fight today's climate change and to correct the problems we created.

The Ocean in 3D and 4D

We now know that the ocean inherently has a third, and actually also a fourth dimension: time. Deep ocean space is framed by the Earth's lithosphere (Fig. 4).This is the rigid outermost shell of the planet, that is composed of the crust and the upperpart of the underlying mantle, which behaves elastically on a geological timescale. The lithosphere is broken up into seven very large continental- and ocean-sized plates, and a number of smaller ones. These plates move relative to each other at a rate of 5–10 cm per year. The coming and going of continents as well as the continuously changing shape and dimensions of the ocean was coined by the Dutch geologist Jan H.F. Umbgrove (1899–1954) as the 'pulse of the earth' (Umbgrove 1942).

Through this geological process the earth's surface is constantly reconfigured by an episodic assembly and breakup of continents. This is the supercontinent cycle that lasts between 300 and 500 million years. The reconfiguration defines the earth system and, among others, the size of the ocean, which can cover up to 80% of the earth's surface, the global climate as well as the evolution of the biosphere (Nance et al. 2014). Due to this cycle the deep-sea floor shows highly diverse landscapes, like impressive undersea mountain ranges, deep canyons like the mariana Trench, seamounts, deep-water coral reefs, hot hydrothermal vents, cold seeps and endless abyssal plains. Some of them contain large amounts of mineral resources.

The ocean, with an average depth of 3.9 km, operates at a different timescale, but again a very long one when compared to humans and their societies. The pulse of the global ocean is beating at a timescale of roughly a thousand years (Fig. 4). Motion is a characteristic feature of the ocean. At the surface there are wind-driven currents and tides, caused by the gravity of the Moon and Sun. Yet, the most important motion in terms of the distribution of heat is the thermohaline circulation (THC). It is moving water around the world and operates at a timescale of hundreds of years.

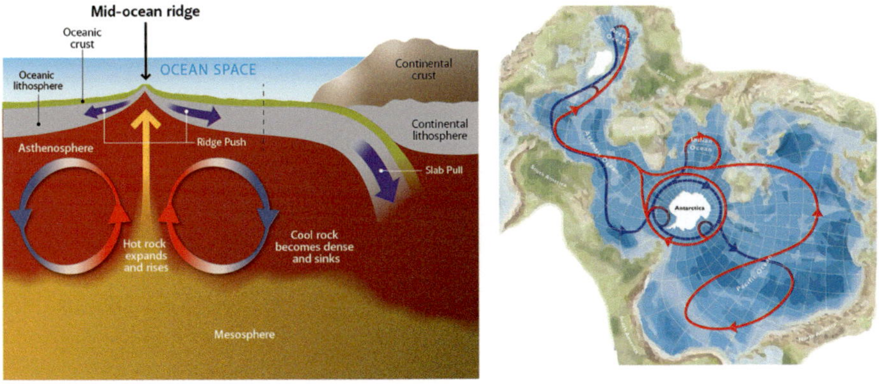

Fig. 4 The pulse of the earth (left) is measured in hundreds of million years; the one of deep ocean space (right) in hundreds of years. @ Flanders Marine Institute, Ostend, Belgium and from Meredith MP, Ocean Challenge, 2019, Vol. 23, No. 2, p. 27

The Arctic Ocean with a surface area of 15.5 million km^2 and a volume of just 1.4% of ocean space functions as a mixer and heat exchanger. Warm water is cooled, and tumbles, like a gigantic 160-km-wide and some 3,500-m-tall underwater waterfall, into the deep-sea. The Atlantic meridional overturning circulation system includes the Gulf Stream. It forms the Northern Atlantic part of the THC. Recent research (Caesar et al. 2018, 2021) shows that it is slowing down due to anthropogenic climate change. This can have severe consequences for Europe. Yet, information on the cooler part of the THC in the deep ocean space is rare and lacks monitoring.

The Southern Ocean around Antarctica is another THC key player. Here old deep water is reprocessed into less dense intermediate and denser bottom water. During this process interaction with the atmosphere and cryosphere takes place. This has profound climatic effects as heat and carbon are taken from the atmosphere and stored in ocean space (Meredith 2019).

Deep-sea exploration accelerated since World War II and during the Cold War as a result of the military situation. Innovative technology, a knowledge revolution and rapidly increasing computer power allowing the development of constantly improving complex computer models are the major drivers. The new technology includes multibeam echo-sounders; submersibles like *Alvin, Mir, Nautile* and *Jiaolong*; autonomous underwater vehicles (AUVs) and underwater drones, that can work together. ROVs, like the Japanese *Kaiko* and the US *Sentry*, are robots operated by controllers onboard a vessel. AUVs, like Argo floats and gliders, travel underwater without direct input from an operator.

New techniques include telemetry, ocean observing systems, genetics and biologging using tagged animals as 'scientists'. Jointly, these developments signal a new phase in deep ocean exploration. Deep ocean space, as such, is defined as the ocean below a depth of 200 m, which is the maximum penetration of natural light. It covers 65% of the earth's surface and makes up 95% of the total volume (EMB 2015) of 1.335 billion km^3.

Transition to Global Monitoring

Over the last half-century, new and innovative technology has dramatically changed our insights in ocean processes. This has laid the basis for a transition from basic oceanographic research by academia towards global operational oceanography by dedicated organizations often linked to national weather services. The international Global Ocean Observing System (GOOS) is an example of this transition. It was initiated by UNESCO's Intergovernmental Oceanographic Commission (IOC) in 1991, and was supported by the 1992 UN Conference on Environment and Development in Rio de Janeiro. Similar observing systems were developed for the land (GTOS), climate (GCOS) and the earth (GEO; see websites).

GOOS consists of both in situ and satellite observing platforms, and has a focus on: climate, ocean health and real-time services. It has been developed through a regional approach with organizations like EuroGOOS for Europe and NEARGOOS for North

East Asia (see websites). The former now is linked to the EU flagship programme Copernicus, Europe's eyes on Earth. Copernicus collects its data through a series of dedicated and specialized satellites, the Sentinels missions of the European Space Agency (ESA) and in situ systems operated by EuroGOOS (see websites). The US Ocean Observatories Initiative (OOI) is a modern example of the complexity, technology, science and educational and outreach aspects of such an observing system (see websites).

Argo takes the 'pulse of deep ocean space' by collecting high-quality temperature and salinity profiles with a global array of about 3,950 battery-powered autonomous floats (Fig. 5).These drifting buoys cycle to a maximum depth of 2,000 m every ten days before resurfacing and send the data by satellite to a ground station. Each float is designed to make about 150 cycles. They are, so to speak, the 'underwater satellites' that, together with dedicated satellites, map and monitor the ocean and its health. Like satellites in outer space they 'die' when the battery is too weak to pump the float to the surface, and drift around as ocean space junk.

This international project started in 2000. In June 2020, Argos collected some 2.3 million profiles. This is at least two times more than the profiles taken by all research vessels in the world during the twentieth century (Argo 2017). Through this fleet of underwater robots, we learn about the internal structure of ocean space. The next step will be the development of Deep Argo (Fig. 5), with a capability to sample the ocean up to a depth of 6,000 m. This is important as large-scale ocean circulations that extend from the sea surface to the ocean bottom play an important role in Earth's climate system. Some 1,228 floats are needed for a full coverage. In May 2020, 122 deep floats were active in the Argo network.

Another next step is the development of Biogeochemical-Argo having sensors to measure pH, oxygen, nitrate, chlorophyll and suspended particles and downwelling irradiance (Claustre et al. 2020). Such an array enables direct observations of seasonal and decadal-scale biological productivity variability, ocean acidification, nutrients upwelling, oxygen depletion and the uptake of carbon dioxide. As such, it would extend ocean colour observations by satellites into deep ocean space. To meet the defined requirements, an array of about 1,000 biochemical floats, and a deployment

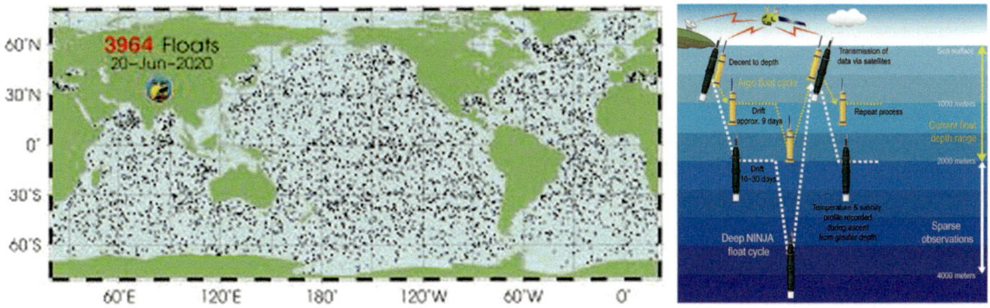

Fig. 5 Argo floats on June 20, 2020 (left), the cycles of such a float and the new deep Argo floats, © Argo and JAMSTEC

of 250 new floats per year to sustain it, is needed (see websites). In May 2020, 397 deep floats were active in the Argo network.

Henry Stommel (1920–1992) was one of the driving forces behind the development of the now very successful Slocum glider, which was technically developed by Douglas C. Webb. These gliders are underwater drones, which use either small amounts of electrical power from batteries or harvest the thermal energy of the ocean itself. They can operate for months, in the future possibly for years. A suit of forty sensors can now be applied to study a wide variety of physical, chemical and biological processes in ocean space or to measure a standard set of ocean characteristics, like temperature, salinity, depth and currents. They migrate vertically by changing ballast, and they can be steered horizontally by gliding on wings at about a 35-degree angle. Once deployed they can be controlled from a land-based operation centre. Moreover, they can operate in fleets of gliders.

Box 2: Health of the Ocean

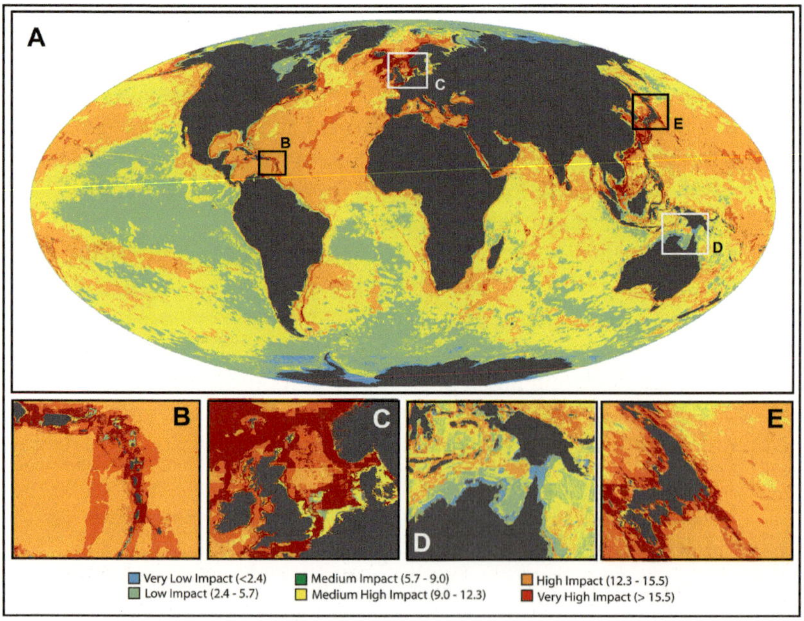

Ocean health Index 2019. The colour-coded index varies from very low (blue) to very high (red). Courtesy B. Halpern

The Ocean Health Index (OHI) is the first integrated assessment framework that combines key biological, physical, economic and social elements of the ocean's health (see websites). Overall index scores are a combination of ten components, or 'goals', of ocean health. In 2008, nineteen American scientists led by Ben Halpern published the first world map showing the effects of human activities on ocean space (Halper et al. 2008). The conclusion was that there actually is no place on earth where people have left no traces. The most affected areas include the North Sea, the Baltic Sea, the South and East China Sea, the Caribbean Sea, the American East Coast, the Mediterranean Sea, the Red Sea, the Persian Gulf, the Bering Sea and various regions in the Pacific Ocean. As we hardly know our impact on ocean ecosystems, the 2008 card only gives a minimal, but nevertheless shocking, estimate.

In 2015 they published an overview of changes in the OHI 2010–2015 (Halpern et al. 2015). The situation has not improved. About 66% of the ocean, and, coming closer ashore, 77% of the Exclusive Economic Zones (EEZs) and coastal areas, show an increased cumulative impact by human activities.

The OHI is annually updated since 2012, and is related to climate change (Halpern et al. 2017). It also is related to the UN Sustainable Development Goal 14, to conserve and sustainably use the oceans, seas and marine resources. The global scores vary between 69 and 71.

However, individual countries show notable increases or declines due to improvements in the harvest and management of wild-caught fisheries, the creation of marine protected areas (MPAs), and decreases in natural product harvest. The global OHI in 2019 was 71. The conclusion is that there are no places in ocean space without human influences (Halpern et al. 2019).

Century of Ocean Exploration

The twenty-first century is often seen as the century of ocean exploration. Innovative technology will pave the way to a better understanding and monitoring of this last frontier of the planet. This might form the basis for a better management and sustainable use of its resources. We know that deep-sea ecosystems provide vital services for both the ocean and the global biosphere. But there are many threads and uncertainties about the effect of anthropogenic global warming in deep ocean space.

In 2018, WHOI started to explore the ocean's cold and dimly lit twilight zone. This mesopelagic zone (Fig. 8) spans the world at a depth of 200–1,000 m. It teems with microscopic to very big life (Fig. 6). Part of it migrates every night to the surface to feed and return to escape predators in the morning. This creates the largest migration on earth, which transports huge amounts of carbon into deep ocean space. Moreover, life in this zone plays a largely unknown role in oceanic ecosystems and global ocean biogeochemical cycles. The biological pump plays a critical role in regulating the climate.

The twilight zone is the mostly unexplored uppermost part of deep ocean space. Recent research (Irigoien et al. 2014) learned that the biomass of mesopelagic fishes is at least ten billion tonnes. That is ten times more than expected, and 125 times above the annual average marine catch. The sheer mass of it is predictably attracting the attention of the fishery.

Fig. 6 Live in the Twilight Zone. Top, left to right: angler fish, predatory dancing siphonophore *Rhizophysa* sp., squidworm *Teuthidodrilus samae*. Lower, left to right: sea-cucumber *Enypniastes eximia*, cock-eyed squid *Stigmatoteuthis hoylei*, tan bristlemouth *Cyclothone pallida*. © WHOI

The Seabed 2030 (Mayer et al. 2018) is a joint project of the Nippon Foundation and the GEBCO (see websites). The aim is to produce a modern bathymetric map of the ocean floor by 2030, and to make it available to the public. As of June 2019, the GEBCO already increased the percentage of ocean floor that has been mapped in detail from 6 to 15%.

Exploitation

Hidden in the mists of time, our ancestors, as well as our Neanderthal nephews, were collecting shellfish at least 176,000 years ago (Cortés-Sánchez et al. 2011). More recently, marine fisheries in Europe have a relatively long historic record. One of the longest lasting is the trade of dry cod from the Norwegian Lofoten archipelago to Mediterranean countries. This trade started more than one thousand years ago, during the Viking expansion in the beginning of the Warm Medieval Period (c.950–1250). The Dutch North Sea herring fishery, which was the largest single fishery in the seventeenth century, is another example.

Since the beginning of the British Industrial Revolution, but especially since the 1950s, industrialization has taken its toll, leading to overfishing. Today, 90% of the world's fishing grounds are being harvested at or beyond their sustainable limits (Greiff 2017), while at the same time aquaculture production, having serious environmental risks, surpassing the wild catch. But, wasteful human behaviour during the catch at sea, during on-land transport and attitudes of consumers lead to a situation that almost half of the seafood supply in, e.g. the US is uneaten (Love et al. 2015).

Until now, most resources such as sand, gravel, diamonds, and oil and gas are mined from the continental shelf. The oil and gas industry, however, has been moving into deep ocean space in the 1990s in water depths of 2,000–3,000 m. The diamond industry also is moving into deeper waters off the Namibian coast. A wave of new exploitation is emerging: deep-sea mining of polymetallic nodules, polymetallic sulphides and cobalt-rich ferromanganese crusts; bioprospecting to develop new medicines, chemicals and cosmetics; deep-sea gas hydrates and deep-sea fishing. Other new emerging fields are deep-sea archaeology, geoengineering related to climate 'control', military activities and tourism (EMB 2015). Here we will discuss three new frontiers in ocean exploration: deep-sea fisheries, bioprospecting and deep-sea mining.

Legal Framework

The United Nations Convention on the Law of the Sea (UNCLOS), which was adopted in 1982, is the main legal international instrument for ocean governance. During the negotiations, the notion of the 'Common Heritage of Mankind' (CHM) was strongly advocated by the Maltese ambassador Arvid Pardo (1914–1999), to make sure that less advanced countries also shared in the revenues.

According to this notion, all resources of the seabed beyond national jurisdiction are part of the common heritage. They should be used for peaceful purposes and managed in the interest of all, including future generations. The notion was strongly opposed by the technologically more advanced countries having state-of-the-art marine capabilities. Basically, the notion is a non-ownership arrangement in which user rights, but no ownership rights of a common property resource, can be obtained (Mann-Borgese 1998). UNCLOS entered into force in 1994. Today it is signed and ratified by 168 countries. A notable exception is the US.

UNCLOS can also be seen as a traditional solution to an old problem. The old problem relates to intensive and conflicting use of common property marine resources such as fish. The solution concerns the application of the traditional mechanism of enclosure. As a result, some 149 million km^2, an area almost as large as the land surface and covering some 40% of ocean space, was brought under national jurisdiction through the institution of EEZs (Fig. 7). Jointly, they contain 90% of the known marine resources.

The introduction of an EEZ has far-reaching consequences. First of all, disputes regarding overlapping claims arose. Well-known examples are situated in the South China Sea with conflicting claims between China and its neighbours concerning the Paracel Islands, Spratly Islands and Senkaku Islands. Another area with possible conflicts is the Arctic Ocean.

In general, the maximal extend is 200 nm from the baseline (Fig. 8). In this space a country has the sovereign rights to explore and exploit, conserve and manage the living and non-living natural resources in a sustainable way. Every maritime state can claim an EEZ. Conflicting claims will be settled by the International Tribunal for the Law of the Sea in Hamburg, Germany.

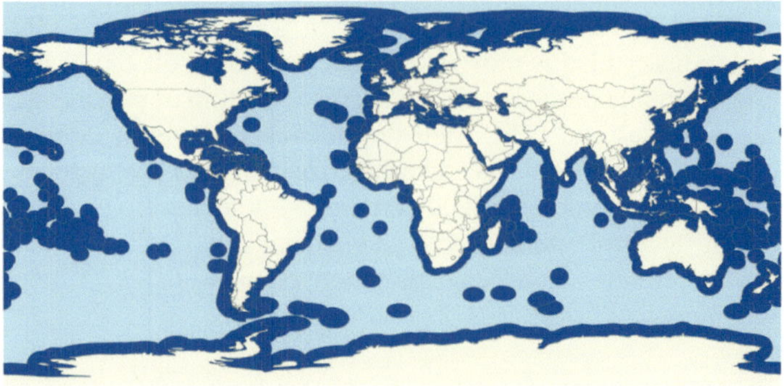

Fig. 7 Exclusive Economic Zones (dark blue) and international waters (light blue). © Wikipedia

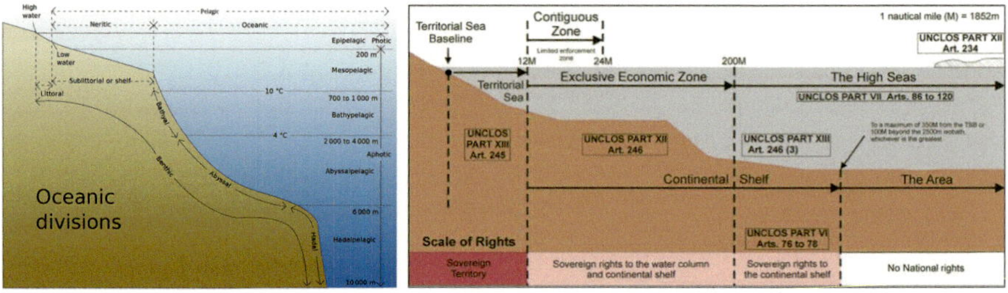

Fig. 8 The ocean is divided into: the epipelagic zone, mesopelagic or twilight zone and bathypelagic zones. UNCLOS distinguishes the following maritime zones: Internal Waters, Territorial sea, Contiguous Zone, Exclusive Economic Zone, Continental Shelf, The Area and The High Seas. These zones are based on the degree of sovereignty and start from a territorial baseline, usually the mean low-water mark of a coastal state. © Wikipedia and NOC, www.unclosuk.org/

EEZs can also lead to a fundamental change in perception: from a land-based one to an ocean-based one. Due to their EEZs, many maritime states now have sovereign rights over extensive ocean areas. These areas often are much larger than their land surfaces. From an ocean perspective the notion of 'ocean states' can be defined (Stel 2002). It might open the eyes of politicians and policymakers for the relevance of ocean space. In the European Union, the ocean–land ratio, for instance, is about five to one, making it the largest 'ocean state' in the world (Stel 2006, 2013).

The Continental Shelf is in a geological sense defined as an undersea continuation of continental crust, and is, as such, part of the land. The High Seas refer to ocean space outside the EEZs. Like the atmosphere, outer space and Antarctica, they still are a global common (Buck 1998), without any national ownership. As a result, it forms the background of a 'tragedy of the commons', in which the right of the strongest and technologically most innovative ones prevails and takes its toll in, for instance, ocean fisheries (Ostrom et al. 2000).

The Area is the ocean floor and subfloor which also is beyond national jurisdiction. It is subject to the regime of 'common heritage of mankind'. The International Seabed

Authority (ISA) was established in 1994 in Kingston, Jamaica. It organizes and controls activities in the High Seas seabed.

One of the most flagrant shortfalls of UNCLOS is the lack of regulations for, and protection of the High Seas. When UNCLOS was adopted, the High Seas were simply protected by its inaccessibility and biodiversity was not an issue at all.

Fisheries

Since the 1950s, and especially after the 1970s when coastal fisheries were overexploited, commercial fishing fleets moved, like the oil and gas industry earlier, into deeper waters with a depth varying between 200 and 2,000 metres. These activities have been taking place at continental slopes, and near oceanic seamounts and ridges, both in national EEZs as well as the High Seas.

Yet, during the mid-1990s, the reduction of fish stocks in the EEZ, as well as quota limits and new technology, led to an increase in the high sea fisheries. Deep-sea fishes live in a cold, dark and low-productivity environment, to which they are well adapted. Our ecological knowledge of their environment as well as of their life cycle is, however, limited at best. Many deep-sea fishes have long life spans, are slow-growing and have low natural mortality and late sexual maturity. This makes them extremely vulnerable to fisheries.

A well-known example is *orange roughy* or deep-sea perch, a more appetizing, branding name for the largest slimehead fish. It lives in large populations near seamounts and continental slopes. The original name relates to mucus canals that cover its head. This is not an attractive feature for consumers. *Orange roughy*, having an almost worldwide distribution, feeds at or near the bottom at a depth of more than 1,000 metres. They take some thirty years to reach sexual maturity and can live up to one century or more. It has an average length of 75 cm and weighs about seven kilos. Commercial fisheries catch *orange roughy* that are 30–80 years old. Large schools of the deep-sea perch are found over seamounts or underwater hills (Fig. 9).

Fig. 9 *Orange roughy* in its natural environment and as a favourite fried dish with a citrus cauliflower salad. © Malcolm Clark, NIWA and Deepwater Group Limited, New Zealand and Wikipedia

Deep-sea fisheries cause the collapse of benthic biodiversity and ecosystem functioning, which in turn might have a consequence for biogeochemical cycles. The footprint of global fisheries is particularly high. Industrial fishing occurs from more than 55% of the ocean surface and covers a surface more than four times the land area used for agriculture (Kroodsma et al. 2018). Even the World Bank is emphasizing that the world fisheries are in a crisis for many years (World Bank 2009, 2017). There is a need to deindustrialize these fisheries to allow fish stocks to recover. Aquaculture is seen as an answer to overfishing, and is also moving offshore, into deep waters. Today aquaculture produces about half the seafood eaten worldwide, but it is not sustainable (see websites).

Bioprospecting

Bioprospecting in deep ocean space is exploring marine biodiversity for new products that are relevant for humans and their activities. According to the Convention on Biological Diversity (CBD), biodiversity is defined as: 'the variability among living organisms from all sources including inter alia, terrestrial, marine and other aquatic ecosystems, and the ecological complexes of which they are part; this includes diversity within species, between species and of ecosystems'. Summing up, biodiversity relates to the variety of life on earth in all its forms and interactions.

Marine biodiversity, which underpins the health and the productivity of the ocean, is changing rapidly due to natural pressures and human activity, like fisheries, pollution and global warming. Yet, today's biodiversity is the result of some three billion years of evolution, which mostly occurred in the marine domain. Therefore, marine biodiversity is much larger than the biodiversity on land. But, the volume of water of the High Seas, is vastly underexplored. Hence the understanding of its biodiversity is still in its infancy. Just 5% of the ocean's diversity is known, and we use less than 1% of this (EMB 2015).

The alarming message from the 2019 Global Assessment Report on Biodiversity and Ecosystem Services (IPBES) is clear. We urgently need action to protect ocean space, and especially the High Seas. It is done and ready that marine ecosystems, from coastal to deep-sea, show the influence of human activities. The closer we get to the coast, coastal marine ecosystems increasingly show both large historical losses of extent and condition as well as rapid ongoing decline (IPBES 2019). A similar story is given by IPCC's Special Report on the Ocean and Cryosphere in a Changing Climate (IPCC 2019).

Currently marine bioprospecting is more a scientific domain than an economic one. But this will change in the near future, as the economic potential is huge. Deep ocean space is harbouring a wide variety of species well adapted to a large number of ecosystems. So far, only a few organisms like red algae, sponges, bryozoans, molluscs, soft corals and Antarctic krill could be exploited due to technological limitations. This is, however, changing rapidly, making microorganism with unique adaptations to deep-sea environments like hydrothermal vents, cold seeps and the twilight zone, accessible for exploitation.

Microbial bioprospecting is both a search for unique and valuable chemical compounds and genetic material from microorganisms as well as the study and processing of it in specialized laboratories on land. The latter form a high cost. Research is focusing on the so-called marine natural products, MNPs, which have a high economic potential for the pharmaceutical and health industry. The number of MNPs is rapidly increasing (Blunt et al. 2017). Deep-sea organisms, for instance, have properties that could revolutionize our ability to cure human diseases like arthritis, cancer and Alzheimer or to heal broken bones (Scott 2015). Yet, there still are some hurdles to be taken.

The notion of a 'blue economy' was launched at the Rio + 20 conference, which was organized in 2012. It emphasizes both the need for healthy ocean ecosystems, on which sustainable exploitation of resources can be based, and conservation measures. Presently this concept is part of the United Nations' 2030 Agenda for Sustainable Development (UN 2015). Sustainable Development Goal 14, life below water, has a focus on the conservation and sustainable use of marine resources (see websites). The transition to a blue economy is a tricky one, as our knowledge is limited,and human pressure is rapidly increasing due to population growth. That is why the time to protect the invisible world of deep ocean space, is quickly running out.

In June 2018 scientists from the Stockholm Resilience Centre and University of British Columbia published a paper on patenting marine genetic resources (University of British Columbia 2018; Blasiak et al. 2018). Out of the thirty countries surveyed, 27 are parties to UNCLOS. The authors identified 862 marine species, with a total of 12,998 genetic sequences being patented. It turned out that the German BASF, the world's largest chemical manufacturer, has patented 47% of these sequences i.e. 10% above the combined share of 220 other companies. Universities and their commercialization partners just registered 12%.

It is noteworthy that organizations located or headquartered in three countries, Germany, US and Japan, registered almost 75% of all patent sequences. It is telling that 165 countries were not represented at all. This outcome raises the question: Who owns deep-sea biodiversity?

Deep-Sea Mining

Exploiting the unknown is a slogan that fits deep-sea mining when we listen to the bulk of both scientific papers and the messages of NGOs. To take up a new challenge in the deep-sea, mostly, is the view of the industry involved. They want to provide the future world with urgently needed minerals like copper, gold, cobalt, titanium, phosphorite, iron and rare-earth elements. With a steadily rising demand for minerals and metals, and a depletion of some of the land-based resources, there now is a rapidly increasing interest in the 'promise of the deep-sea'.

Fig. 10 Deep-sea mining system consisting of a crawler collecting the nodules, a high-tech vertical transport system, and a mining vessel with a processing plant. Europe is playing a leading role in exploring both the technological challenges and environmental consequences of harvesting these resources. © IHC Merwede

Moreover, new technology allows to mine deeper and deeper. A rush on deep-sea mining licenses is ongoing. As of 28 May 2020, UNCLOSs ISA (see websites) has 30 exploration contracts, while there only were eight in 2010. The approved exploration contracts involve 22 different countries and cover more than 1.3 million km^2 of seabed, an area almost half of India. This represents 0.7% percent of the international deep seabed area and 0.3% of the world's oceans. Twelve of these contracts are sponsored by developing countries. Thirteen countries and one inter-governmental consortium currently have contracts for the exploration of polymetallic nodules, seven countries for the exploration of polymetallic sulphides and five for the exploration of cobalt-rich ferromanganese crusts.

Significant and mineable abundances of manganese nodules are found in four regions at the ocean floor. The largest area is the Clarion-Clipperton Fracture Zone in the northeast Pacific. They cover an area of some 9 million km^2, almost the size of Europe. Smaller areas in this ocean are found off the Peruvian coast and near the Cook Islands. There also is one large area covered with nodules in the central Indian Ocean.

Nodules with a size of chicken eggs or large potatoes, contain substantial amounts of iron, nickel, copper, titanium, cobalt and rare-earth minerals. They are found at flat abyssal plains in deep ocean space, at a depth between 3,500 and 6,500 m. How these nodules grow is not fully understood, but it is clear that they are formed by an interplay of precipitation from seawater, diagenetic processes in the sediment and biological processes. Countries like China, India, UK, Germany, Poland and France, as well as large companies such as Lockheed Martin, now own exploration contracts.

During the last decade, the European Union has both funded a series of research projects for a better understanding of the deep-sea ecosystems and the development of technology for mining deep-sea mineral resources (Fig. 10). It is clear that Europe will play a leading role in the upcoming mining activities at the bottom of deep ocean space (Seas at Risk 2014).

Box 3: Mining nodules in the mid-1970s

State-of-the-art deep-sea mining USNS *Hughes Glomar Explorer* was actually salvaging the wreck of the Soviet's most advanced submarines K-129. © Wikipedia

During the summer of 1974, the state-of-the-art mining vessel USNS *Hughes Glomar Explorer*, started deep-sea mining operations 1,500 miles north-west of Hawaii. The eccentric billionaire Howard Hughes (1905–1976) foresaw a new profitable market. An active PR campaign was carried out, and it was generally assumed that manganese nodules mining was feasible. An ocean mining boom and new academic studies were the consequences. The project also had its impact on the negotiations for UNCLOS. A new frontier in mining would open up. Hughes ship started the mining operations, watched by a number of Soviet spy ships. They did not trust it. Rightly so!

In February 1968, one of the Soviet's most advanced submarines K-129, left its base at Kamchatka peninsula. The vessel, armed with nuclear warheads and the best coding machines, went to its station in the North Pacific. Three months later it became clear that something was wrong: K-129 had sunk. The Russians performed a search but could not locate the wreck. Moreover, they assumed that no one would be able to find it, let alone salvage it. That is exactly what the US CIA did within the Azorian project.

Mining nodules just was a smokescreen. The whole operation was a blatant lie. The salvage was a technological highlight but went wrong. A mechanical failure of the ship's lifting grapple caused half of the submarine to break off and return to its icy grave. The project was discovered during a burglary in March 1975. Stolen documents explained the link between Hughes, the ship and the CIA. The story was made public and the project was stopped (CIA 2012).

There is, however, already a large part of deep ocean space under national authority through the EEZs. Clotting of EEZs around a number of small island states and some larger countries like Australia and Papua New Guinea (PNG) in the southwest Pacific allows companies to pass ISA. The Canadian company Nautilus Minerals negotiated a 25-year contract with the PNG government for the Solwara 1 project. Within this project the company was allowed to excavate mineral-rich sulphides, in PNG's EEZ. The sulphides are situated at a depth between 1,500 and 2,000 metres, at the ocean floor in the Bismarck Sea, off the islands of New Ireland and New Britain.

Nautilus has developed an innovative production system by using existing technologies of the offshore oil and gas sector, as well as of the mining and dredging

Fig. 11 Nautilus mining system and mining robots. Robots from left to right: collecting machine, bulk cutter and auxiliary cutter. The two robots to the right will continuously excavate material in the same destructive way as is done in coal or copper mines on land. The robot to left will collect the cuttings, which then will be transported as a seawater slurry through a flexible pipe to a surface vessel to be processed. © Nautilus Minerals Inc

industry. But the industry raised eyebrows about the blatant use of this technology. The resulting exploitation system consists of three main components: the seafloor production tools (Fig. 11), the riser and lifting system, and the production support vessel, which forms the centrepiece of their model of operation. The three huge underwater robots for the actual mining activities at the seafloor were constructed and tested in the field, but not at the required operational depth. The riser and lifting system was ready, the support vessel was under construction for several years at the Chinese shipyard Fujian Mawei Shipbuilding Ltd in Fuzhou and a preliminary economic assessment has been made.

Yet, some dark financial and environmental clouds, and a series of setbacks in the delivery of the support vessel, were already threatening the operations and the company some years ago. As a consequence, the start of the mining operation was postponed a number of times. The final blow to the company came when the contract with the Chinese shipbuilder was cancelled, and the vessel was acquired by the Indian company MDL Energy (Stutt A 2019). Finally, the company went bankrupt in November 2019, leaving the PNG government with a US$24 million debt (Deep-sea Mining Campaign 2019). This failure has spurred calls for a Pacific-wide moratorium on deep-sea mining for a decade.

Deep-sea mining is strongly opposed by many scientists and environmentalists. There rightly is a strong concern about the side effects or collateral damage, and the indelible fact that we still know far too little about deep ocean space (GEOMAR 2018; Carrington 2018; Miller et al. 2018; Van Dover 2011; Van Dover et al. 2017). That is why scientists and environmentalists plea for the application of the precautionary approach (Mengerink et al. 2014; Greenpeace 2013; Losada and Terras 2018), or a total ban for deep-sea mining (Seas at Risk 2017). On the other hand, some countries like Canada, US, Mexico and Portugal, have established protected areas to safeguard hydrothermal vents and other deep-sea features (Van Dover 2014). Damaging aspects of mining activities will, among others, be: sediment plumes, waste heaps of sediment removed and noise and light pollution.

There are, however, geostrategic factors playing a role in deep-sea mining. China is, for instance, by far the largest producer of the world's rare-earth elements. These minerals are vital for green-energy products like wind turbines, solar power, hybrid and electric cars and smartphones. In the name of fighting pollution, China temporarily closed its rare-earth industry and decreased its export in 2011. That was a wake-up call clearly showing the vulnerability of the global economy, and a nice argument for exploiting deep ocean space resources. In contrast, there also is a growing number of studies showing that recycling, applying the concept of a circular economy and, above all, changing the Western consumer lifestyle and society could be a sustainable alternative.

Knowing the Future

Jules Verne's world was one of rapid technological developments of the exploration of far away exotic places like the African continent and of an unconditional belief in technology and progress. A world with stories of the exploration of dangerous and icy polar areas by Fridtjof Nansen (1861–1930) and Jules Dumont d'Urville (1790–1842). He lived in a world with exciting stories from explorers like David Livingstone (1813–1873) and Alexander von Humboldt (1769–1859). Fed by his research in libraries and visits to exhibitions, he was enthralled by the secrets of the sea.

Verne was a man of his world. Like a top chef, he used his knowledge and imagination to produce stories that had a high appeal to the general public. This was educational outreach, ocean literacy in its infancy. Even today, he still is the second most-translated author, since 1979 (UNESCO 2018). But he also lived in a world that unknowingly and shamelessly polluted the environment and the atmosphere for economic progress. A little more than a century later, this dark side of economic growth leading to the present human-made climate emergency is well known.

Another famous Frenchman was Jacques Cousteau (1910–1997), who developed scuba diving in the mid-1950s. As a result, a wafer-thin layer just below the ocean surface became accessible to millions of scuba divers. Cousteau also developed an attractive outreach and awareness programme, in which he and his yacht RV *Calypso* played the leading role. His popular science books were well-read and his TV documentaries reached millions.

Yet, he was also living in a world that made, just to mention one, plastic pollution a severe threat for ocean space and society. Today plastic is found everywhere: from the North Pole to the South Pole, from the ocean surface to the ocean floor of the Mariana Trench (Chiba et al. 2018). Here James Cameron and Victor Vesocovo stayed just a few hours in a high-tech submersible that hardly could withstand the forces of ultradeep ocean space. Both made a documentary chronicling their expedition and airing it on Discovery Channel. Yet, the BBC documentary *Blue Planet II*, presented by naturalist Sir David Attenborough, was a blockbuster and was viewed in more than thirty countries, reaching tens of millions of viewers. That is ocean literacy in a highly connected world.

At the brink of the twenty-first century, the UN declared the 1998 International Year of the Ocean, to raise awareness about the role of the ocean environment in everybody's daily lives and to transfer knowledge through capacity building (Stel 1998). Early in 1998, the aftermath of the strongest El Niño of the last century was still making bold headlines in the news. The total cost of this natural event were some 5.5 US$ billion, but there always are winners and losers. A central rhetorical question during the *International Year of the Ocean* was: How can one protect ocean space, when one knows so little about it?

In retrospect, this UN year marks a change in our perception of the ocean's relevance for humankind, as well as the urgency to increase people's awareness and knowledge of ocean space. In 2020 there is a large difference at many levels. The ocean literacy movement is rapidly gaining momentum, outreach often is an integrated part of research efforts, and pupils and students are now exploring deep ocean space remotely, through projects like JASON with its *Nautilus* vessel (see websites) and citizen science projects. Moreover, ocean issues, although hesitantly, are addressed in science, technology, engineering and math (STEM) classes. A central theme in ocean literacy is: How do we involve teachers, the public at large, and commit policymakers and politicians?

A healthy and productive ocean provides vital services that are essential to human society. Again, there often is an unconditional belief in future technology and economic progress to solve the negative effects of an exuberant Western lifestyle. In such a technology-driven perception, sustainability has a lower priority (Stel 2016). In the wake of Rio + 20, and as a response to rapid technological, social and political changes, like the adoption of the Paris Climate Agreement in 2015, a series of promising developments are taking place at the global and European level.

International developments relevant to the exploration and exploitation of deep ocean space are, e.g.: the implementation of the Sustainable Development Goal 14, life below water, the adaptation of UNCLOS to biodiversity leading to a new framework for sustainable use of living resources in ocean space, the opportunities given by the Blue Economy and Blue Growth, the UN declared International Decade for Ocean Science (2021–2030) and the urgent need to protect the ocean, by sound management and a large number of marine reserves, also covering a substantial part of deep ocean space. At the regional level, the European Green Deal (see websites), based upon a circular economy, is a promising signal of change. Together, these developments should in turn form the framework for targeted and sustained (deep) ocean literacy activities at these and national levels.

Acknowledgements The author wants to thank Prof. Jan Mees, Director of the Flanders Marine Institute in Ostend, Belgium, for preparing Figure 4 on the pulse of the earth.

References

Argo brochure (2017) Retrieved June 1, 2018 from: file:///D:/Mijn%20documenten/Mijn%20teksten/EU%20Sea%20Change/Publikatie%202018/Argo_Brochure2017_high_res_web.pdf

Ballard RD (1987) The Discovery of the Titanic: Exploring the greatest of all lost ships. Madison Publishing Inc, Toronto, Canada, 230p

Blasiak R, Jouffray J-B, Wabnit, CCC, Sundström E, Österblom H (2018) Corporate control and global governance of marine genetic resources. Science Advances 4(6):eaar5237. https://doi.org/10.1126/sciadv.aar5237

Blunt JW, Copp BR, Keyzers RA, Munro MH, Prinsep MR (2017) Marine natural products. Nat. Prod. Rep. (Mar 17, 2017) 34(3):235–294. Retrieved June 7, 2018 from: file:///D:/Mijn%20documenten/Mijn%20teksten/EU%20Sea%20Change/Publikatie%202018/Bioprospecting/Marine%20natural%20products%20%202017.pdf

Buck SJ (1998) The global commons: an introduction. Island Press, Washington

Caesar L, McCarthy GD, Thornalley DJR, Cahill N, Rahmsdorf S (2021) Current Atlantic Meridional Overturning Circulation weakest in last millennium. Nature. Geosciences 14:118–120. https://doi.org/10.1038/s41561-021-00699-z

Caesar L, Rahmstorf S, Robinson A, Feulner G, Saba V (2018) Observed fingerprint of a weakening Atlantic Ocean overturning circulation. Nature 556:191–196. https://doi.org/10.1038/s41586-018-0006-5

Carrington D (2018) Is deep sea mining vital for a greener future—even if it destroys ecosystems? The Guardian, sun 4 Jun 2018. Retrieved June 10, 2018 from: https://www.theguardian.com/environment/2017/jun/04/is-deep-sea-mining-vital-for-greener-future-even-if-it-means-destroying-precious-ecosystems

Chiba S, Saito H, Fletcher R, Yogi T, Kayo M, Miyagi S, Ogido M, Fujikura K (2018) Human footprint in the abyss: 30 year records of deep-sea plastic debris. Marine Policy (6 April 2018). https://doi.org/10.1016/j.marpol.2018.03.022

CIA (2012) Project AZORIAN. Retrieved June 15, 2018 from: https://www.cia.gov/about-cia/cia-museum/experience-the-collection/text-version/stories/project-azorian.html

Claustre H, Johnson KS, Takeshita Y (2020) Observing the Global Ocean with Biogeochemical-Argo. Annual Review of Marine Science 12:23–48

Cortés-Sánchez M, Morales-Muñiz A, Simón-Vallejo MD, Lozano-Francisco MC, Vera-Peláez JL, Finlayson C, Rodríguez-Vidal J, Delgado-Huerta A, Jiménez-Espejo FJ, Martínez-Ruiz F (2011) Earliest known use of Marine Resources by Neanderthals. PlosOne (1 September 2011) 6(9):1–15. https://doi.org/10.1371/journal.pone.0024026

Davis N (2018) *Homo erectus* may have been a sailor—and able to speak. The Guardian (20 February 2018). Retrieved May 23, 2018 from: https://www.theguardian.com/science/2018/feb/20/homo-erectus-may-have-been-a-sailor-and-able-to-speak

Deep-sea Mining Campaign (2019) Game-over-nautilus-solwara-1-experiment-fails. Retrieved May 26, 2020 from: http://www.deepseaminingoutofourdepth.org/game-over-nautilus-solwara-1-experiment-fails/

Dreyer EL (2007) Zheng He. China and the Ocean in the Early Ming Dynasty, 1405–1433. Pearson Education Inc., New York, USA, 238p

European Marine Board (2015) Delving deeper: critical challenges for the 21st century deep-sea research. Position paper 22 of the European Marine Board, Ostend, Belgium, 223pp

Five Deeps Press (2019) Deepest Submarine Dive in History, Five Deeps Expedition Conquers Challenger Deep. Retrieved May 26, 2020 from: https://fivedeeps.com/wp-content/uploads/2019/05/FDE-Challenger-Release-FINAL-5132019.pdf

GEOMAR (2018) Deep-sea mining: serious consequences for the marine ecosystem. GEOMAR, Kiel, Germany, press release, January 5, 2018. Retrieved June 10, 2018 from: https://www.geomar.de/en/discover/articles/article/tiefseebergbau-folgen-fuer-das-marine-oekosystem-gravierend/

Greenpeace International (2013) Deep seabed mining: an urgent wake-up call to protect our oceans. Greenpeace International July 7, 2013, Amsterdam, The Netherlands, 20p

Greiff J (2017) Overfishing. Bloomberg. Retrieved June 2, 2018 from: https://www.bloomberg.com/quicktake/overfishing
Halpern BS, Frazier M, Afflerbach J, Lowndes JS, Micheli F, O'Hara C, Scarborough C, Kimberly A. Selkoe KA (2019) Recent pace of change in human impact on the world's ocean. Scientific Reports, Report 9, 11609. https://doi.org/10.1038/s41598-019-47201-9
Halpern BS, Frazier M, Afflerbach J, O'Hara C, Katona S, Stewart Lowndes JS (2017) Drivers and implications of change in global ocean health over the past five years. PLoS ONE 12(7):e0178267. https://doi.org/10.1371/journal.pone.017826
Halpern BS, Frazier M, Potapenko J, Casey KS, Koenig K, Longo C, Lowndes JS, Rockwood RC, Selig ER, Selkoe A, Walbridge S (2015) Spatial and temporal changes in cumulative human impacts on the world's ocean. Nature Communications 6(7615):7. https://doi.org/10.1038/ncomms8615
Halpern BS, McLeod KL, Rosenberg AA, Crowder LB (2008) Managing for cumulative impacts in ecosystem-based management through ocean zoning. Ocean Coast Manag 51:203–211
IPBES (2019) Summary for policymakers of the global assessment report on biodiversity and ecosystem services of the Intergovernmental Science-Policy Platform on Biodiversity and Ecosystem Services. Díaz S, Settele J, Brondízio ES, Ngo HT, Guèze M, Agard J, Arneth A, Balvanera P, Brauman KA, Butchart SHM, Chan KMA, Garibaldi LA, Ichii K, Liu L, Subramanian SM, Midgley GF, Miloslavich P, Molnár Z, Obura D, Pfaff A, Polasky S, Purvis A, Razzaque J, Reyers B, Roy Chowdhury R, Shi YJ, Visseren-Hamakers IJ, Willis KJ, Zayas CN (eds). IPBES secretariat. Bonn, Germany, 56p
IPCC (2019) Summary for policymakers. In: Pörtner H-O, Roberts DC, Masson-Delmotte V, Zhai P, Tignor M, Poloczanska E, Mintenbeck K, Alegría A, Nicolai M, Okem A, Petzold J, Rama B, Weyer NM (eds) IPCC Special Report on the Ocean and Cryosphere in a Changing Climate. Geneva, Switzerland, 35p
Irigoien X, Klevjer TA, Røstad A, Martinez U, Boyr, G, Acunã JL, Bode A, Echevarria F, Gonzalez-Gordillo JI, Hernandez-Leon S, Agusti S, Aksnes DL, Duarte CM, Kaartvedt S (2014) Large mesopelagic fishes biomass and trophic efficiency in the open ocean. Nature Communications 5(3271):1–10. https://doi.org/10.1038/ncomms4271
Kroodsma DA, Mayorga J, Hochberg T, Miller NA, Boerder K, Ferretti F, Wilson A, Bergman B, White TD, Block BA, Woods P, Sullivan B, Costello C, Worm B (2018) Tracking the global footprint of fisheries. Science 359(6378):904–908. https://doi.org/10.1126/science.aao5646
Losada S, Terras P (2018) Deep sea mining decisions: approaching the point of no return. Greenpeace International. Retrieved June 8, 2018 from: https://www.greenpeace.org/international/story/15514/deep-sea-mining-decisions-isa-jamaica-point-of-no-return/
Love DC, Fry JP, Milli MC, Neff RA (2015) Wasted seafood in the United States: Quantifying loss from production to consumption and moving toward solutions. Global Environmental Change 35:116–124. https://doi.org/10.1016/j.gloenvcha.2015.08.013Getrightsandcontent
Mann-Borgese E (1998) The oceanic circle, governing the seas as a global resource. United Nations Press, Tokyo, Japan, 240p
Mayer L, Jakobsson M, Allen G, Dorschel B, Falconer R, Ferrini V, Lamarche G, Snaith H, Weatherall P (2018). The Nippon Foundation—GEBCO Seabed 2030 Project: The Quest to See the World's Oceans Completely Mapped by 2030. Geosciences 2018 8(2):63. https://doi.org/10.3390/geosciences8020063, https://doi.org/10.3390/geosciences8020063
Mengerink KJ, Van Dover CL, Ardron J, Baker M, Escobar-Briones E, Gjerde K, Koslow JA, Ramirez-Llodra E, Lara-Lopez A, Squires D, Sutton T, Sweetman AK, Levin LA (2014) A call for deep-Ocean Stewardship. Science 344:696–698
Meredith MP (2019) The global importance of the Southern Ocean and the key role of its freshwater cycle. Ocean Challenge 23(2):27–32
Miller KA, Thompson KF, Johnston P, Santillo D (2018) An overview of seabed mining including the current state of development, environmental impacts, and knowledge gaps. Frontiers in Marine Science 4, Article 418:1–24. https://doi.org/10.3389/fmars.2017.00418

Mokyr J (2002) The Gifts of Athena: Historical Origins of the Knowledge Economy. Princeton University Press, Princeton, USA, 384p
Nance RD, Murphy JB, Santosh M (2014) The supercontinent cycle: A retrospective essay. Gondwana Res 25:4–29
Norman K, Inglis J, Clarkson C, Faith JT, Shulmeister J, Harris D (2018) An early colonisation pathway into northwest Australia 70–60,000 years ago. Quatern Sci Rev 180:229–239
Ostrom E, Dietz Th, Dolsak N, Stern PC, Stonich S, Weber EU (2000) The drama of the commons. National Academy Press, Washington DC, USA, 521p
Scott AR (2015) Polymers: Secrets from the deep sea. Nature 19:S12–S13
Seas at Risk (2014) Deep sea mining: exploring the unknowns. Seas at Risk, Multi-stakeholder conference—26th April 2016, Brussels Background paper: EU funded deep sea mining related research
Seas at Risk (2017) Deep sea mining? Stop and think. Seas at risk leaflet March 7, 2017, 4p. Retrieved June 11, 2018 from: http://www.seas-at-risk.org/24-publications/723-deep-sea-mining-stop-and-think.html
Stel JH (guest editor) (1998) Capacity Building in a changing global setting. Marine Policy, Special Issue, 22, May 1998, 175–280
Stel JH (2002) Mare Nostrum—Mare Liberum—Mare sit Aeternum, duurzaam gebruik van de oceanische ruimte. Inaugural speech, Maastricht University, the Netherlands, 47p
Stel JH (2006) Governance of Europe's Exclusive Economic Zones, a vision. In: Dahlin H, Flemming NC, Marchand P, Petersson SE (eds) European Operational Oceanography: Present and Future. Proceedings of the fourth international conference on EuroGOOS. European Communities, pp 302–311
Stel JH (2013) Ocean Space and the Anthropocene, new notions in geosciences?—An essay. Neth J Geosci 92(2/3):193–211
Stel JH (2016) Ocean space and sustainability. In: Heinrichs H, Martens P, Michelsen G, Wieks A (eds) Sustainability Science: an introduction. Springer, Dordrecht, the Netherlands, pp 193–205
Stutt A (2019) Nautilus Minerals officially sinks, shares still trading. Retrieved May 26, 2020 from: https://www.mining.com/nautilus-minerals-officially-sinks-shares-still-trading/
UN (2015) The Millennium Development Goals Report 2015 (Summary). UN, Washington, USA, pp 1–75
University of British Columbia (2018) Patenting marine genetic resources: who owns ocean biodiversity? ScienceDaily (University of British Columbia), 6 June 2018
UNESCO (2018) Index Translationum. List of most translated individual authors. Retrieved May 27, 2020 from: http://www.unesco.org/xtrans/bsstatexp.aspx?crit1L=5&nTyp=min&topN=50
Umbgrove JMF (1942) The pulse of the earth. Martinus Nijhoff, The Hague, the Netherlands, 325p
Van Dover CL (2011) Tighten regulations on deep-sea mining. Nature 470:31–33. https://doi.org/10.1038/470031a
Van Dover CL (2014) Impacts of anthropogenic disturbances at deep-sea hydrothermal vent ecosystems: A review. Marine Environmental Research 102: 59–72. https://doi.org/10.1016/j.marenvres.2014.03.008
Van Dover CL, Ardron JA, Escobar E, Gianni M, Gjerde KM, Jaeckel A, Jones DOB, Levin LA, Niner HJ, Pendleton L, Smith CR, Thiele T, Turner PJ, Watling L, Weaver PPE (2017) Biodiversity loss from deep-sea mining. Nat Geosci 10(7):464–465. https://doi.org/10.1038/ngeo2983
World Bank (2009) The Sunken Billions: The Economic Justification for Fisheries Reform. World Bank, Washington, DC 89p
World Bank (2017) The Sunken Billions Revisited: Progress and Challenges in Global Marine Fisheries. Environment and Development. World Bank, Washington, DC. https://openknowledge.worldbank.org/handle/10986/24056 License: CC BY 3.0 IGO

Websites

Argo http://www.argo.ucsd.edu/
Audacious Project https://audaciousproject.org/
Challenger glider Mission https://challenger.marine.rutgers.edu/
Copernicus http://ec.europa.eu/growth/sectors/space/copernicus/
European Green Deal https://ec.europa.eu/info/strategy/priorities-2019-2024/european-green-deal_en
EuroGOOS http://eurogoos.eu/
GEBCO https://www.gebco.net/
GEO https://www.earthobservations.org/index2.php
GCOS https://public.wmo.int/en/programmes/global-climate-observing-system
GOOS http://www.goosocean.org/
GTOS http://www.fao.org/gtos/
ISA https://www.isa.org.jm/
JASON Project https://www.jason.org/
NEARGOOS http://ds.data.jma.go.jp/gmd/goos/data/database.html
OHI http://www.oceanhealthindex.org/
OOI http://oceanobservatories.org/
Seabed 2030 https://seabed2030.gebco.net/
Sustainable Development Goals 2015 https://sustainabledevelopment.un.org/?menu=1300
Sustainable Fisheries http://sustainablefisheries-uw.org
UNCLOS http://www.un.org/depts/los/
The Five Deeps Expedition https://fivedeeps.com/

Jan H. Stel studied geology and paleontology. His thesis concerned a paleo-biological study of Silurian favositid corals of the Swedish island Gotland. As an ocean science manager, he organized ocean-going expeditions like the Dutch-Indonesian Snellius II Program in the mid-1980s, developed a European consortium of small European countries to participate in the Ocean Drilling Program, initiated capacity building programs for the IOC-UNESCO, developed the Dutch Antarctic Research Program, organized and executed a visit of the present Dutch king and queen to Antarctica in 2007 and organized projects at the interface with the European ocean industry. From 2000, he was a professor in 'Ocean Space and Human Activity' at the Maastricht University in the Netherlands. Here he coined the notion of 'ocean space' and 'ocean states'. Jan has written some 350 scientific and popular science papers and blogs to inform the public at large why ocean space is important for us.

Ocean Literacy—In the Context of Naming of Seas: Case Study: The Sea Between Korea and Japan

Rainer Dormels

Abstract Oceans and seas, which border on different states, often have different names in the respective languages. In these cases, riparian countries may have different views as to what name is most appropriate for use on international maps and in international publications. An example is the sea between Korea, Russia, and Japan. It was, until the end of the twentieth century, a common feature on most maps worldwide to name this sea "Sea of Japan." Since Korea's admission to the UN, however, the question concerning an internationally recognized name for the sea has been brought up repeatedly for discussion by Korean officials. By broadening the spectrum of the topic to non-STEM-aspects, ocean literacy can get access to curricula of social-science disciplines and can expand their reach. The discussion on the international name of the sea between Korea and Japan is connected to multifaceted matters and can, therefore, contribute to enhancing the understanding of ocean-related issues.

Keywords Naming of Seas · East Sea/Sea of Japan naming dispute · International Hydrographic Organization · Endonym · Exonym · Ocean literacy

Introduction

Sea Names and Ocean Literacy

Oceans and seas, which border on different states, often have dissimilar names in the respective languages. In these cases, riparian countries may have opposing views as to what name is most appropriate for use on international maps and in international publications.

An example of this is the sea between Korea, Russia, and Japan. It was, until the end of the twentieth century, a common feature on most maps worldwide to name this sea "Sea of Japan." Since Korea's admission to the UN, however, the question concerning

R. Dormels (✉)
University of Vienna, Vienna, Austria
e-mail: rainer.dormels@univie.ac.at

K. C. Koutsopoulos and J. H. Stel (eds), *Ocean Literacy: Understanding the Ocean*, Key Challenges in Geography, https://doi.org/10.1007/978-3-030-70155-0_5

an internationally recognized name for the sea has been brought up repeatedly for discussion by Korean officials. Korea was a colony of Japan between 1910 and 1945. The naming issue of the sea between the Korean peninsula and Japan has been going on for more than a quarter century by now. While both the Korean and the Japanese governments engage in this issue, most academics seem to avoid the topic. They may be cautious as to maintaining academic independence. On the other hand, one cannot deny that there is a certain appeal to the naming issue when it comes to interdisciplinary research, as academics must consider not only historical, political, juridical, and geographical facets but also linguistic and cultural aspects.

The giving and using of names have a social and psychological dimension (Koß 1990, vii). Toponyms in contact zones are also an indicator of a way of the coexistence of different people. They are more than etiquettes; they deal with identity. Toponyms remind us of the fact that the struggle for peaceful coexistence is a task that we cannot escape (Koß 1990, 15). Therefore, toponyms can have a historical-sociological as well as a political significance, besides their function as a basic category of geography. Knowing about naming issues regarding seas is, therefore, also a part of ocean literacy.

There is a trend toward removing the limits of ocean literacy in (STEM) education and including additional issues in the discussion regarding it. This shift broadens the whole spectrum of topics concerning ocean literacy and goes far beyond STEM disciplines. In this context, it is also relevant to mention "cultural diversity." Toponyms not only reflect the cultural identity of people or a nation but also play the role of reference point for the cultural identity of a state.

The Naming of Seas and Conflicts

Ormeling (2000, 54) distinguished categories of how sea names were named: after cardinal directions, nations, persons, places, attributes, rivers flowing into them, adjacent areas, or countries. We can see this phenomenon in East Asia where the seas initially did not have proper names. The seas were just referred to as "the Sea (海)" or the "Big Sea (大海)." According to the theory of "Four Seas," China was described as an empire in the middle of the world, surrounded by four seas. Each sea was accordingly named after their cardinal directions: East Sea (東海), South Sea (南海), West Sea (西海), and North Sea (北海). Later, European explorers and mapmakers brought up more names, such as "East China Sea," "Sea of Korea," "Sea of Japan," and so forth.

In most cases, sea names are entirely unproblematic. However, as Murphy (2005, 207) states, "when an international body of water bears the name of a nation or state, the potential for conflict arises." As a reason, he indicates "the extraordinary importance of the modern territorial state system in the perceptual and functional ordering of human affairs" (Murphy 2005, 207). Following Murphy (2005, 209–210), three sea names are under a high level of dispute between bordering countries: "Persian Gulf" (versus "Arabian Gulf"), "Japan Sea" (versus "East Sea"), and "South

China Sea" (versus "Bien Dong"). These three cases have in common that a "name commonly used for the body of water is the name of a state with a history of political or economic hegemony in the region" (Murphy 2005, 210).

Among the three above disputes, the debate over the naming of the "East Sea"/"Sea of Japan" is the liveliest. Thus, in the draft of the fourth edition, 2002 of the International Hydrographic Organization (IHO) Publication S-23 "Borders of the Oceans and Seas," the "South China Sea" and the "Persian Gulf" are listed, but not the water body between Korea and Japan.

Organization of This Book Chapter

The naming conflict of the sea between Korea and Japan is used here as an illustration of the challenges of the naming of seas. In this context, some previous studies on the naming of the sea between Korea and Japan (such as Dormels 2010, 2011a, b; Monmonier 2006; Kadmon 2007; Lee 2002) have to be taken into consideration.

In Chapters 2 and 3 both names, "Sea of Japan" and "East Sea," will be analyzed. Chapter 2 will discuss the history of the name "East Sea," as part of Korea's cultural identity. The endeavor of Chapter 3 is to show how the name "Sea of Japan" was first propagated and why it came into use instead of "Sea of Korea." Additionally, the question as to the relationship between Japanese Imperialism and the name "Sea of Japan" arises. Since when did Western mapmakers predominantly use the name "Sea of Japan"? Furthermore, the circumstances on how the name "Japan Sea" was adopted by the IHO in the "Special Publication No. 23 Limits of Oceans and Seas" will be examined.

In Chapters 4 and 5 will be analyzed: (i) the proposal made by the Korean government, to use both names "East Sea" and "Sea of Japan," (ii) to what extent a technical resolution of the IHO supports the proposal, (iii) how the discussion on the sea name between Korea and Japan contributes to understanding ocean-related issues.

The Name "East Sea"

Origins of the Naming Dispute

For a long time, the international naming of the sea between Korea and Japan was not an issue in Korea. However, as Korea opened up to the world after the chaos caused by the Korean War, greater numbers of Koreans started studying in the United States, and thus came across maps in English more often. In these maps, the sea between Korea and Japan was called "Sea of Japan." Articles that criticized that fact can be traced back as far as the 1970s (Lee 2008, 308). When the Koreans noticed that the sea between Korea and Japan had different names on historical Western maps, such as "Sea of Korea" or "Oriental Sea," it sparked their interest in the naming issue (Lee 2002, 156). Thus, the question of how the name of the sea between Korea and Japan on Western historical maps had changed in the course of time was discussed.

Raising the Naming Dispute to International Organizations

The conflict about naming the sea between Korea and Japan in the International public debate began in the early 1990s. In 1991, the Republic of Korea (South Korea) and the Democratic People's Republic of Korea (North Korea) joined the United Nations. Since 1992, the South Korean government increasingly undertakes initiatives against the sole use of the name "Sea of Japan" in international maps. South Korea has carried out different measures since this time with:

- publication of pamphlets;
- creation of internet sites on the subject;
- establishment of contact with official bodies, publishers, journalists, etc., who use the term "Sea of Japan" as the sole name of reference for the sea;
- thematization of the question with the UN and at other international organizations;
- support of research on the subject;
- organization of international symposia.

Japan's reaction to these measures was to carry out similar actions to demonstrate its position on the issue.

The Decision of the Korean Agencies Regarding the International Name

A question, which had to be decided, concerned finding a name for the sea Korea would support. The Koreans use the Korean name "Donghae (東海)"; "dong" meaning "east" and "hae" meaning "sea." Therefore, in 1992, the responsible South Korean ministries discussed three alternatives of the English-language name for the sea between Korea and Japan:

- "Tong-hae" (a Romanized form of the endonym),
- "East Sea" (a literal translation of the endonym),
- and "Sea of Korea."

They finally chose "East Sea" and presented this name to Japan and the international community.

Later, this decision faced much criticism inside and outside of Korea. The problem is—and this is no surprise—that Japan has no interest in coming to an arrangement with Korea and finding an alternative to the exclusive use of the name "Sea of Japan."

This has resulted in the fact that in many world maps two different names for the sea between Korea and Japan are used. There are maps:

- using only the name "Sea of Japan" and
- use both names: "Sea of Japan (East Sea)," "East Sea / Sea of Japan," etc.

Until reaching a final consensus with Japan, for Korea, the dual naming of "East Sea / Sea of Japan" was a temporary solution. However, such a consensus is not at hand. Hence, the dual naming became somewhat permanent in many atlases, which results in a problem for Korea. While "Sea of Japan" can be clearly associated with Japan, the name "East Sea" cannot be easily associated with Korea. During the Soccer World Cup in 2002, the international community got used to the combination of "KoreaJapan" in the 2002 FIFA World Cup Official Logo, and with that, the wish to change "East Sea" to "Sea of Korea" became more popular among Koreans as well. The ulterior motive behind this might also be a final solution of "Sea of Korea/Japan." Then why has South Korea decided to propagate the name "East Sea" as an English-language name and not "Sea of Korea"?

The Ministry of Information and Communication of Korea gives the following reasons:

1. While one speaks of "Donghae" in Korea, it would be illogical if one chose a different name internationally (besides one must consider that effectively no one uses the name "Sea of Korea" including Korea itself).
2. One has to fear that if one prefers the term "Sea of Korea" to the term "Sea of Japan," it would be seen as a possession claim and a conflict between Korea and Japan.
3. The concept "Sea of Korea" contradicts our argument that a sea, which is surrounded by several countries, should not bear the name of one of the nations (like "Sea of Japan").

Regarding the first argument of the Ministry of Information and Communication of Korea, one could indicate a counterexample from the German-speaking area: if one speaks German, one uses the term "Ostsee," and if one speaks English, one uses the term "Baltic Sea." Thus, for example, the "Ostsee-Kolleg" in Berlin is called in English "Baltic Sea School." Moreover, hardly anybody used the term "East Sea" before 1992 outside and within Korea in everyday life.

The argument according to which a sea that is surrounded by several countries should not bear the name of a nation is also weak. If one followed the third argument of the ministry, Korea would also have had to act against the names "Persian Gulf" and "Indian Ocean." However, this has not happened.[1] Moreover, not in all cases of a sea with a country name the degree of potential dispute between bordering countries is very high, as the acceptance of the "Gulf of Mexico" and the "Irish Sea" by the United States and Britain shows. As Monmonier (2006, 94) notes: "neither the United States nor the United Kingdom carries a grudge against its less powerful neighbor, and however flattering to Mexico and Ireland, directional references to

[1]The Korean school atlases call e.g. the Arab/Persian Gulf as 페르시아 만 (Persian Gulf).

neighboring states are functionally useful to the Americans and the Brits, and thus deceptively ethnocentric." So he goes on to argue that in "this sense, 'Sea of Korea' could serve both nations if the Koreans had been more assertive and the Japanese less intransigent" (Monmonier 2006, 94), and concludes that a "hyphenated toponym like Korea-Japan Sea might seem a logical compromise" (Monmonier 2006, 94–95).

The Name "Donghae" (East Sea)—A Tradition of 2000 Years

According to the Korean argument, it has already been using the term "East Sea" for 2000 years. It is a substantial component of the Korean culture. The national anthem of the Republic of Korea begins with the word "Donghae (East Sea)."

The key sources, which contain the term "Donghae," are:

- the Gwanggaeto Stele, a memorial stele for the tomb of King Gwanggaeto (r. 391–413) of Goguryeo, erected in 414,
- the Korean historical record Samguksagi (history of three empires) from 1145,
- the map of the eight provinces (Paldo Chongdo) from 1531.

Every word has many meanings, which can change over the course of time. Also, original meanings can get lost, while new ones are created.

Furthermore, geographic names can be subject to a semantic change. On this occasion, we can theoretically distinguish two kinds:

1. The geographic unit to which the geographic name refers to changes.
2. The geographic name itself goes through a semantic change over the course of time.

Both kinds of the semantic change of geographic names can proceed, admittedly, also in parallel and dependent on each other.

Regarding the use of the toponym "Donghae" (East Sea) in Korea, we can see semantical changes in three steps:

Step 1: "Donghae (East Sea)" as the eastern sea in the Sinocentric theory of "Four seas," which surround the continent of the middle (China) and to which the Korean peninsula is attached to in the east. The "Donghae" stands here in contrast to the other three seas (the North, West, and South) which surround the central continent. The sea area to the west and the south of the Korean peninsula is also in a strict sense part of this "Donghae."

This Chinese view of the world is found in hand drawings until the seventeenth century. Real and idealistic opinions interlock here. China lies in the center of an almost square continent. An example is the map "Sihai Huayi Zongtu" (四海華夷總圖); "General Map of the Chinese and Barbarian Lands within the Four Seas") from 1532 (Fig. 1).

The Korean peninsula (朝鮮) ranges in the east of the central continent. Japan (日本) is shown as an island within the "East Sea." The term "Four seas," therefore, became a synonym for the (civilized) world.

Fig. 1 Sihai Huayi Zongtu (1532) (Reproduced from Harvard Library)

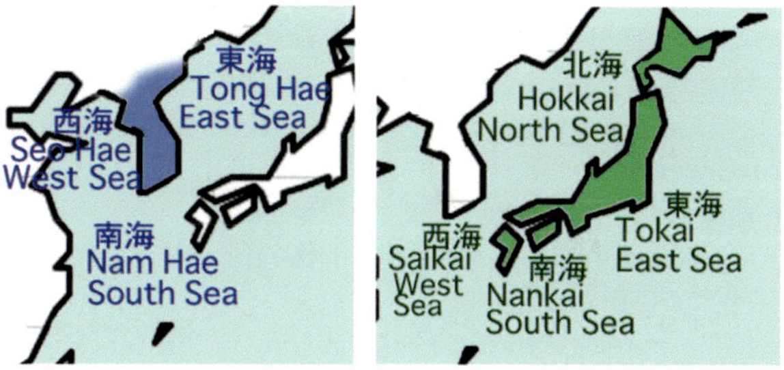

Fig. 2 Cuttings (Reproduced from Tanabe et al. 2010, 19)

Step 2: "Donghae" as the sea to the east of the Korean peninsula as a contrast to other two seas, which surround Korea: the "Seohae" (West Sea) to the west of the Korean peninsula and the "Namhae" (South Sea) to the south of the Korean peninsula (Fig. 2).

All seas pictured above of Tanabe et al. (2010, 19) are part of the "Donghae" in step 1, but now it is the sea area to the east of Korea only (here called "Tong Hae; East Sea"). Indeed, the "Donghae" has no clearly defined contour in step 2. Step 2 is practically an adaptation of step 1 on a smaller level limited to Korea. This concept was also taken over by Japan.

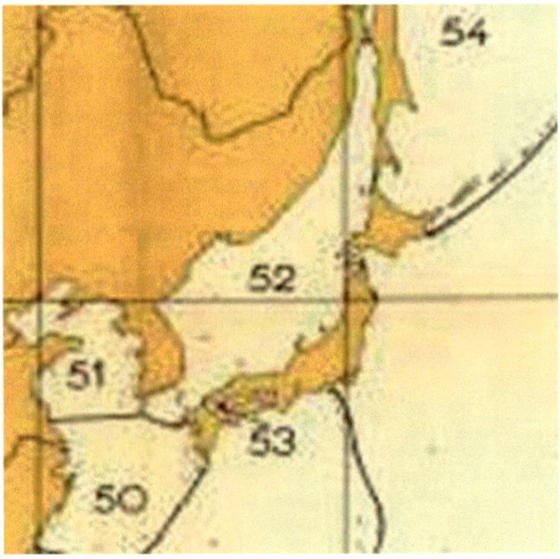

Fig. 3 Cuttings (Reproduced from "Limits of Oceans and Seas", third edition, 1953). 50. Eastern China Sea (Tung Hai). 51. Yellow Sea (Hwang Hai). 52. Japan Sea. 53. Seto Naika or Inland Sea. 54. Sea of Okhotsk

Table 1 Change in the meaning of "Donghae" (East Sea)

	Location	Meaning
Step 1	Sea to the East of China	Sea to the East of China
Step 2	Sea to the East of Korea	Sea to the East of Korea
Step 3	Sea No. 52 in the third edition of the IHO "Limits of Seas and Oceans" (1953)	Sea to the East of the Eurasian continent

Step 3: "Donghae" as a proper name for the sea between Korea and Japan in contrast to other names for seas worldwide (Fig. 3).

"Donghae" (and, respectively, "East Sea") no longer means "the sea to the East of Korea" but "the sea in the East of the Eurasian continent" and is an alternative name to "Sea of Japan" or "Japan Sea."

Schematically, the change in the meaning of "Donghae" can be classified in three steps as follows (Table 1).

The name "Donghae" has a long tradition. Even in other regions in the world, there are water bodies called "East Sea." Vietnamese publications, for example, use "East Sea" instead of "South East China Sea," without expecting the other countries to do the same. Korea, on the other hand, wants international maps to call the sea between Korea and Japan not only "Sea of Japan" but also "East Sea."

The North Sea would be an example that was named after a cardinal direction. Until the First World War, that sea was also called "German Sea" or "German Ocean." Therefore, there is no reason in principle why the sea between Korea and Japan should not be named "East Sea."

The Name "Sea of Japan"

Old Western Maps

The Japanese used to call the sea between Korea and Japan "Hokkai 北海" ("North Sea") for a long time. There are official maps of the Japanese government from the eighteenth and nineteenth centuries where the name "Sea of Korea (朝鮮海)" was used. In the second half of the nineteenth century, Japanese maps used the name "Sea of Japan."

How did ancient Western maps call the sea between Korea and Japan? The sea between Korea and Japan had a great many different names on ancient Western maps. These names were often related to Korea, such as "Sea of Korea" or "Gulf of Korea." The result is that early research regarding the naming of the mentioned sea initially concentrated on the names used in Western maps. Were there periods of time in which names related to Korea were predominant and since when are the names related to Japan dominant? Is there some connection to Japanese Imperialism?

Because of many research activities on historical maps conducted by Koreans and Japanese alike, the times, in which particular names were used more often than others can now be defined (Table 2).

We can see that:

- There were different names for the sea between Korea and Japan before the nineteenth century.
- Names referencing to Korea were most frequent in the eighteenth century.
- Influenced by European "discoverer" La Pérouse and von Krusenstern, the number of the maps which chose the name "Sea of Japan" for the sea between Korea and Japan grew at the beginning of the nineteenth century.

Koreans have a general tendency of placing the shift from using the name "Sea of Korea" to the name "Sea of Japan" as late as possible, whereas the Japanese try to set it as early as they can. Also, the Japanese strongly highlight that this change of name falls into the time of the Japanese "period of national isolation" between 1639 and 1835, while the Koreans emphasize the contact between Japan and the European countries at this time. Japan has tried to describe the name "Sea of Japan" as a purely

Table 2 Names of the sea between Korea and Japan on historical Western maps

1. Phase: till approx. 1720:	different names appear: "Sea of Japan," "Sea of China," "Sea of Korea," "Oriental Sea"
2. Phase: 1720–1740:	different names appear, among them a relative majority of maps, which carry "Korea" in the name
3. Phase: 1740–1800:	mostly names, which carry "Korea"
4. Phase: 1800–1840:	mostly names, which carry "Japan" (by the majority) or "Korea" (minority)
5. Phase since 1840:	Dominance of names, which carry "Japan"

Table 3 Shift to the name "Sea of Japan"

	Time frame	Reason
European maps	First half of nineteenth century	European perception
Japanese maps	Second half of nineteenth century	Adaptation to European names
Korean maps	about 1910	Japanese occupation

western affair without any Japanese contribution. Watanabe et al. (2008) interpreted this as follows: "The name "Japan Sea" originates from the Europeans in the process of early globalization." Therefore, the change to names such as "Sea of Japan" and "Japan Sea" can be attributed to European influence. The Japanese then adopted the name and forced it onto Korea as of 1910, when Korea was a Japanese colony until 1945 (Table 3).

Now, what sort of impact does the results of the investigation of the naming of the sea between Korea and Japan on historical western maps have? They show that, like in many other cases, too, the sea between Korea and Japan has had a great many different names in the course of history. There has been a substantial time span in which people predominantly used the name "Sea of Korea." This fact shows that sea names can change over the course of time. Monmonier (2006, 94) interprets: "Old maps seem neither relevant nor definitive, while current usage, clearly on the side of the Japanese, ignores the historical reality that toponyms, like boundaries, are political constructions, subject to change."

Determination of the Name "Sea of Japan" by the IHO

It was shown that the use of the name "Sea of Japan" was already favorable to Europeans even before the rise of Japanese Imperialism and therefore one can assume that the widespread use of the name is not directly linked to Japanese Imperialism. The Korean standpoint produced the following argument: The International Hydrographic Organization chose the name "Japan Sea," because Korea used to be a Japanese colony.

In 1997, the Republic of Korea protested the exclusive naming "Japan Sea" for the sea between Korea and Japan in the IHO Publication Limits of Oceans of Seas. While Japan was already represented since the founding of the IHO 1919 in London with three people, Korea was colonized by the Japanese from 1910 to 1945, and therefore could not participate in decisions. Only in 1957 did the Republic of Korea join the IHO, and since the eighth International Hydrographic Conference in 1962, the Koreans took part in these meetings. The DPRK is a member of the IHO since 1987.

Representatives of 24 countries took part in the meeting in London in 1919, among others, the three Japanese delegates, Sakonji, Yamaguchi, and Minato attended. The IHO determined not only the name "Japan Sea," but also the borders of the sea, so virtually the Korean "South Sea" also became a share of the Japan Sea. Korea, as an independent nation, ceased to exist in 1910. The Japanese colonialists removed the name "Sea of Korea" and other alternatives from Korean maps. Koreans had no possibility of representing their country in 1928 at the IHO. However, Japan stresses that the name "Sea of Japan" was already in frequent use and the role of Japan in the said decision was minimal as it was the sole decision of the foreign nations. Because of the original situation, the actual interest of Japan in the realization of the guideline is challenging to estimate. However, at least with the definition of the borders of the sea, it is probable that Japan might have been part of the conversation.

Whether Korea would have wanted, or would have been able, to prevent the inclusion of the name "Japan Sea" in the IHO publication, had it not been a Japanese colony in 1928, is pure speculation. The fact is that because of Japan's colonization, Korea did not have the chance to take part in the decision-making process of the IHO at that time.

The Name "Sea of Japan"—An Expression of Japanese Imperialism?

The name "Sea of Japan" originated from the west and had the original meaning of "Sea to the west of Japan." However, at the beginning of the twentieth century, the political situation in East Asia changed. In 1905, the Japanese government declared that Korea would henceforth be a Japanese protectorate and in August 1910, Korea became a formal colony of the Japanese Empire. This status remained until August 1945. The South Korean historian Young-ick Lew states that: "This was the first time in Korea's long history that the entire country and its people were subjugated under alien rule. What made this situation even more galling was the fact that historically the Korean people had always considered themselves Japan's cultural mentors" (2000, 22).

The sea between Korea and Japan became a virtual inland sea between two parts of the Empire of Japan. The name "Sea of Japan," therefore, had two meanings for Korea: Not only "Sea to the west of Japan" but also the semantic nuance "inland sea within colonialist Japan" (Table 4).

Table 4 Change of meaning of "Sea of Japan"

	Meaning
Step 1	Sea to the west of Japan
Step 2 (from 1910 on)	Meaning 1: Sea to the west of Japan Meaning 2: Inland sea within the colonialist Japan

Therefore, many Koreans associate the name "Sea of Japan" with the era of the Japanese occupation. In this respect, it may be understandable that Koreans feel uncomfortable by the sole use of "Sea of Japan."

Korea's Proposal: Double Naming Before an Arrangement with Japan

While the Koreans do not accept the name "Sea of Japan," the name "East Sea" is equally unacceptable for the Japanese. The sea lies not only to the west of Japan, but there is also no motivation for Japan to stop propagating the name "Sea of Japan." Therefore, the South Korean government proposed: "Korea neither insists on the single use of 'East Sea' nor ignores the name 'Sea of Japan' … The Korean government, therefore, calls upon the international community to use both names, 'East Sea' and 'Sea of Japan' until an agreement is reached on a name acceptable to both parties through bilateral consultations" (Shin 2010).

The South Korean government's proposal referred to the technical resolution A.4.2.6 of the IHO and the resolution III/20 of the UNCSGN. The technical resolution A.4.2.6 of the IHO from 1974 says: "It is recommended that when two or more countries share a given geographical feature (such as, for example, a bay, channel or archipelago) under a different name form, they should endeavor to reach agreement on fixing a single name for the feature concerned. If they have different official languages and cannot agree on a common name form, it is recommended that the name forms of each of the languages in question should be accepted for charts and publications unless technical reasons prevent this practice on small scale charts" (Mofat/NAHF 2007, 15). The resolution III/20 "Names of Features beyond a Single Sovereignty," which was published at the third UNSCGN in 1977, recommends, "countries sharing a given geographical feature under different names should endeavor, as far as possible, to reach agreement on fixing a single name for the feature concerned. Further recommends that when countries sharing a given geographical feature do not succeed in agreeing on a common name, it should be a general rule of international cartography that the name used by each of the countries concerned will be accepted. A policy of accepting only one or some of such names while excluding the rest would be inconsistent in principle as well as inexpedient in practice" (Mofat/NAHF 2007, 15) (Fig. 4).

However, Japan denies that both resolutions to which the ROK refers can be applied to the sea between Korea and Japan, and stresses that the sea between Korea and Japan does not fall under the sovereignty of some states, but the high sea.

Following the United Nations Convention on the Law of the Sea (UNCLOS) from 1982 in the coastal sea or the territorial waters, out to twelve nautical miles (22.2 km) from the baseline, the coastal state is free to set laws. Beyond those twelve nautical miles, there is a further twelve nautical mile zone, the contiguous zone. In this area, the state can continue to enforce laws and can check for offenses against its toll,

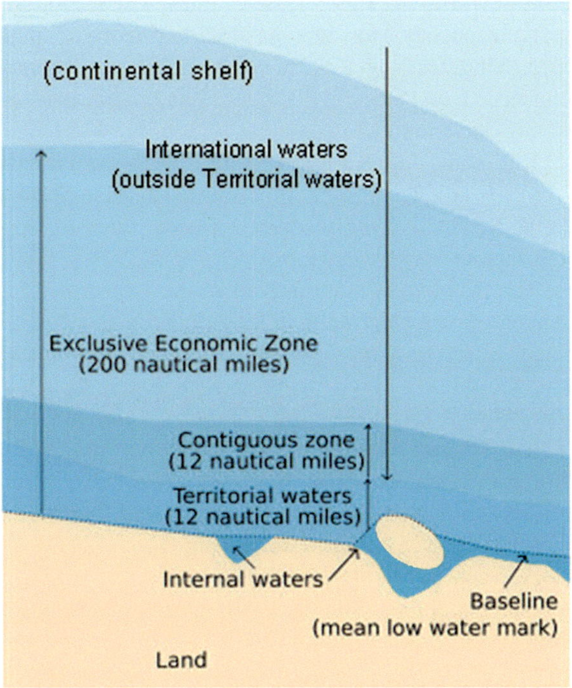

Fig. 4 The Sea and its juridical subdivision (Reproduced from UNCLOS Maritime Zones)

health, and entry regulations. Within more than 200 nautical miles (370.4 km) of an exclusive economic zone (EEZ), the state can dispose of the natural resources of the sea. However, no sovereignty can be derived from this. If the whole sea area between Korea and Japan carries the name "Sea of Japan," this would also imply that Korean territorial waters needed to bear the same name.

Most of the sea surface belongs to the "high sea." It is neither part of an exclusive economic state nor does it belong to the coastal sea or the internal waters of a state. Because it lies beyond the exercise of state power, as a matter of principle, it can carry every name, which is given to it by a language or political community of a state. Most sea areas bear different names in different languages. In such case, one speaks of allonyms.[2]

If a sea area is subordinate to one single state, this country standardizes the name of the sea. The international community then accepts the name to the same extent as other geographical names (town names, for example). Of course, not every state has to accept the standardized name, which is designated as an endonym. The state can also use another name, an exonym (Raper 2007, 9–11).

The sea between Korea and Japan partially lies within the sovereign territory of single states (Japan, ROK, DPRK, and Russia) and the remaining parts of this sea are

[2]The glossary of terms for the standardization of geographical names of the UNGEGN defines allonyms as follows: "alternative (and taxonomically undefined) names for a single geographical entity."

at least within the EEZ of these states. The Republic of Korea tries to highlight this fact particularly. However, as already illustrated, the affiliation to the EEZ does not mean that state sovereignty can be achieved. Kadmon (2007, 1) clarifies: "The sea concerned, and certainly its central and major part, lies outside the territorial waters of any one country."

Korea and Japan's interpretation of the technical resolution A.4.2.6 of the IHO from 1974 differs significantly. While Korea suggests an application of the resolution, Japan denies that the resolution Korea refers to, applies to the sea between Korea and Japan.

The problem is summarized in the interpretation and application of the phrase "share a given geographical feature (such as a bay, a strait, channel or archipelago) under different names" in the IHO resolution. Does this mean that the geographic unit should be a part of the territorial waters or is it enough if it is part of the EEZ? The issue is rendered more complicated by the fact that the recommendations of UNGEGN and IHO come from a time before the UNCLOS 1982. The sea between Korea and Japan is bigger than a bay or a strait. Whereas the map illustrates, no part of this sea is also part of the high sea and therefore cannot be compared to oceans like the Atlantic and the Pacific (Fig. 5).

As a sort of Mediterranean Sea it shows an intermediate form. If the phrase "share a given geographical feature" should be understood in such a way that a geographic unit should entirely be part of the territorial waters of the concerning states, then, indeed, the resolution mentioned above does not apply to the sea between Korea and Japan.

However, how does one apply the resolution A.4.2.6 of the IHO to other cases? What rules would apply and how strict are they? The IHO applies the resolution A.4.2.6 in 2002 draft 4th Edition of the IHO Publication S-23 three cases (English Channel/La Manche; Dover Strait/Pas de Calais and Bay of Biscay/Golfe de Gascogne).

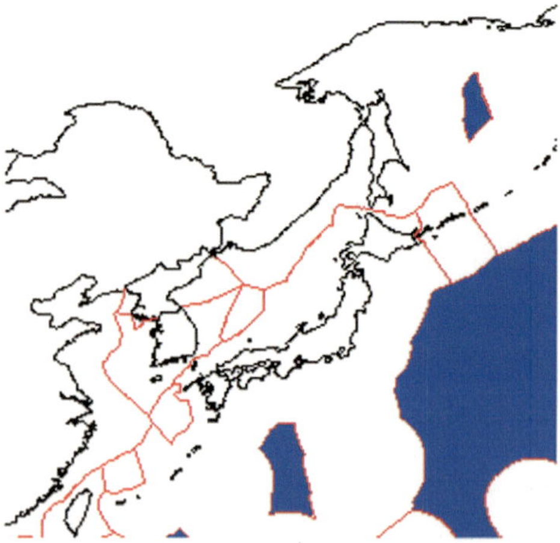

Fig. 5 Part of a map, which displays the high sea (dark blue) (Reproduced from VLIZ Maritime Boundaries)

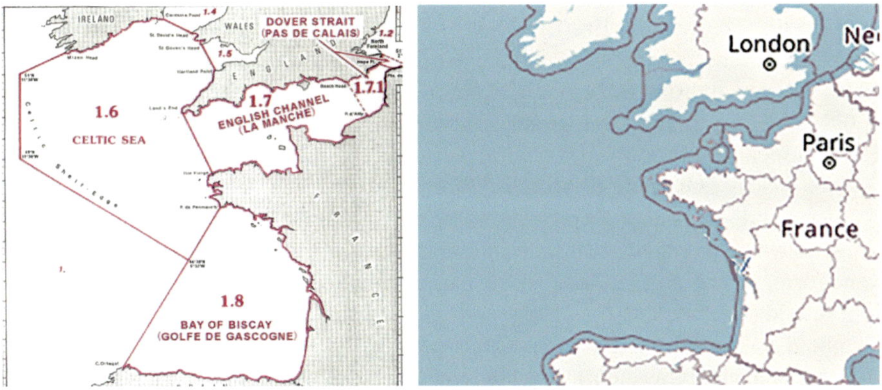

Fig. 6 English Channel/La Manche (Reproduced from IHO Publication S-23, Draft 4th Edition, 2002); Bay of Biscay/Golfe de Gascogne and the territorial sea (Reproduced from OpenStreetMap.org)

Therefore, we compare the bigger sea areas of these three with the sea between Korea and Japan and explore whether or not it is possible to apply the technical resolution A.4.2.6 of the IHO to the naming of the "East Sea / Sea of Japan" (Fig. 6).

As we can see, the English Channel/La Manche and the Bay of Biscay/Golfe de Gascogne are not entirely part of the territorial waters of the neighboring countries. Moreover, if we take a closer look, we can see that in both cases they are also not entirely part of the contiguous waters of the neighboring countries. The same applies to the case of the sea between Korea and Japan. However, the sea between Korea and Japan is part of the exclusive economic zone of the neighboring countries.

We summarize these facts in Table 5.

By using the English Channel/La Manche and the Bay of Biscay/Golfe de Gascogne as examples, we can show that the wording of the IHO resolution "two or more countries share a given geographical feature (such as, for example, a bay, strait, channel or archipelago)" does not have the same meaning as "under the sovereignty of two or more countries, such as in the case of a bay or a strait," because the IHO applies the resolution on the English Channel/La Manche and the Bay of Biscay/Golfe de Gascogne even though these sea bodies are not under the sovereignty of two or more countries.

One could argue, that the sea between Korea and Japan (1,007,300 km^2) is larger than the Bay of Biscay/Golfe de Gascogne (225,000 km^2). However, on the other

Table 5 Seas in territorial waters, contiguous waters, and EEZ

	Territorial waters	Contiguous waters	EEZ
English Channel/La Manche	No	No	Yes
Bay of Biscay/Golfe de Gascogne	No	No	Yes
East Sea/Sea of Japan	No	No	Yes

hand, one can see that the "East Sea / Sea of Japan" is a sea area that is mainly enclosed by Korea, Russia, and Japan, so that the wording "share a given geographical feature" seems entirely appropriate, especially when considering that the Bay of Biscay/Golf de Gascogne is open to the high seas in wide areas. It is therefore unlikely that the IHO technical resolution A.4.2.6 applies to the "English Channel / La Manche" and "Bay of Biscay / Golfe de Gascogne," but not to the "East Sea / Sea of Japan."

However, there is another problem. One could argue the following: the IHO resolution notes: "if they have different official languages and cannot agree on a common name form, it is recommended that the name forms of each of the languages in question should be accepted for charts and publications." In the case of the English Channel/La Manche, for example, the neighboring countries have English and French, respectively, as their official language. Therefore, the dual naming consists of a name in English ("English Channel") and a name in French ("La Manche"). But the dual naming "East Sea / Sea of Japan" uses two names in English. The South Korean, North Korean, and Japanese endonyms for the East Sea in their Romanized form are Donghae, Josŏn Tonghae, or Nihonkai. Therefore, one may think that only a notation like "Donghae / Nihonkai" would follow the example of the "English Channel / La Manche" in a strict sense. But if we once again compare the dual naming practice of IHO in the cases of "English Channel / La Manche" and "Bay of Biscay / Golfe de Gascogne" and compare it to the construction "East Sea / Sea of Japan," we will end up with a somewhat different result.

The dual naming "Bay of Biscay / Golfe de Gascogne" consists of an English name (Bay of Biscay) and a French name (Golfe de Gascogne). However, France and Spain border the sea. In case of a disagreement between France and Spain, a notation including a French and a Spain name such as GOLF DE GASCOGNE (GOLFO DE VIZCAYA) and GOLFO DE VIZCAYA (GOLF DE GASCOGNE) would certainly come closer to the IHO resolution than that of the naming in the IHO S-23 2002 manuscript quite apart from claims by the Autonomous Community of Galicia by the Basques (Table 6).

There is an IHO regulation, though, that shows that it makes sense to give the sea area between Korea and Japan a dual naming in English. The Introduction of the fourth edition of the IHO S-23 says, "For the generic naming of sea areas, with limited exceptions, English has been used for 'Oceans' and 'Seas' to conform to the title. For other areas, such as 'Straits,' 'Bays,' 'Channels' and 'Gulfs' English has

Table 6 Dual naming and the official languages of the neighboring regions

Geographical feature	Official languages of the neighboring regions	Languages of the proposed dual naming
English Channel/La Manche	English/French	English/French and French/English
Bay of Biscay/Golfe de Gascogne	French/Galician/Basque/Spain	English/French and French/English
East Sea/Sea of Japan	Korean/Russian/Japanese	English/English

been used when the areas are surrounded by more than one country and the national language has been used when only one country surrounds the area…." So, since the "East Sea / Sea of Japan" is a sea, English should be used in naming this sea body following this regulation.

However, the observations above make clear that in the case of the "Bay of Biscay / Golfe de Gascogne" the 2002 draft of the S-23 script does not follow the IHO resolution to 100% even though the dual naming "Bay of Biscay / Golfe de Gascogne" officially has been justified with the IHO technical resolution A.4.2.6.

Conclusion Even after thoroughly considering the IHO technical resolution A.4.2.6, there is no definable reason that could provide sufficient justification for the fact that the resolution has been applied in the case of the "Bay of Biscay / Golfe de Gascogne," but is not applicable in the case of "East Sea / Sea of Japan."

In November 2020 the IHO Assembly adopted a proposal to designate geographic sea areas by a system of unique numerical identifiers. It agreed to replace the S-23 "Limits of the Oceans and Seas", the current IHO standard for world map production, with the "S-130" revision, which calls for referring to all seas by numbers, rather than by specific names. In both Korea and Japan, this revision was welcomed but evaluated differently. The Japanese side emphasized the fact that the name "East Sea" did not appear in the IHO proposal and expects that that in "papers, 'Japan Sea' will remain" (The Japan Times 2020). On the other hand, Seoul's Foreign Ministry emphasized: "we laud that the IHO has made clear that it will no longer be using the S-23 as the standard, thus eliminating a major obstacle in the widespread use of the name East Sea" (Kim 2020).

How do mapmakers go about the naming of the sea between Korea and Japan? Some of them solely use the name "Sea of Japan," whereas others use both names. A study of atlases from seven Central and East European countries showed that only three of the ten called the sea "Sea of Japan," while the other seven had both names "East Sea" and "Sea of Japan" in their respective languages (Dormels 2015, 233–247). Studies of newer atlases partially showed a tendency to use endonyms for seas and oceans. In some cases, when an exonym is already widely known, the exonym was added to the endonym. When we take a closer look at Cyprus, we will see that some atlases use Turkish but also Greek exonyms. The increased use of endonyms in atlases leads to the increased use of dual naming and in return can also affect the naming of the sea between Korea and Japan. Since there are already a great many of sea bodies with dual naming, concerns surrounding the use of both names "East Sea" and "Sea of Japan" might further decrease.

Conclusion

The illustration of some aspects of the naming dispute over the sea between Korea and Japan shows that ocean literacy cannot be limited to STEM-aspects only.

Here, one can base one's thought on the idea of the Conceptual Flow Diagram, which is already connected to ocean literacy (NMEA 2000, 33): "Different parts of the world ocean have different names. ...". Based on that, one could incorporate the following ideas in a new Conceptual Flow Diagram, in connection with ocean literacy:

- Most sea names can be categorized into categories like seas named for cardinal directions, nations, persons, places, attributes, rivers flowing into them, adjacent areas, countries.
- Naming practices evolved from generic to specific with the development of maritime traditions.
- We differentiate between names "of a geographical feature in one of the languages occurring in that area where the feature is situated" (endonym) and names "used in a specific language for a geographical feature situated outside the area where that language has official status, and differing in its form from the name used in the official language or languages of the area where the feature is situated" (exonym).
- There are sea names with a high degree of dispute between bordering countries for example, "Persian Gulf" (versus "Arabian Gulf"), "Japan Sea" (versus "East Sea"), and "South China Sea" (versus "Bien Dong"). These three cases have in common that a name commonly used for the body of water is the name of a state with a history of political or economic hegemony in the region.
- Sea names are subject to change from time to time.
- Sea names are subjects of semantic changes.
- Toponyms may prove to be unpleasant to certain categories of persons.
- Following the maritime law agreement of the United Nations (UNCLOS), four principal maritime zones (the territorial sea, the contiguous sea, the exclusive economic zone, and the continent shelf) need to be differentiated.

By broadening the spectrum of the topic to non-STEM-aspects, ocean literacy can get access to curricula of social-science disciplines and can expand their reach. The discussion on the international name of the sea between Korea and Japan is connected to many complex matters and can, therefore, contribute to enhancing the understanding of ocean-related issues.

References

Dormels R (2010) Change of meaning in 'East Sea' and 'Sea of Japan'. Theories shed light on meaning of geographic names. In: Korea Herald, 14.6.2010

Dormels R (2011a) Ostmeer, Japanisches Meer, Koreanisches Meer. Zur strittigen Benennung des Meeres zwischen Korea und Japan. Praesens Verlag (=Wiener Beiträge zur Koreaforschung III), Wien

Dormels R (2011b) Dual naming of oceans and seas: theory and praxis. The 17th international seminar on Sea names Vancouver: the society for East Sea and the Northeast Asian history foundation, pp 233–250

Dormels R (2015) Some hypothesis regarding the East Sea/Sea of Japan issue: based on examinations of Sea names in maps of Central and Eastern European atlases. In: The society for East Sea.

Sea names: heritage, perception and international relations. Proceedings of the 21st international seminar on Sea names in Helsinki, Finland, 23–26 August 2015, pp 233–249
International Hydrographic Organization (1953) Limits of oceans and seas (Special Publication 23), 3th edn. Monte-Carlo
International Hydrographic Organization (2002) IHO Publication S-23, limits of oceans and seas, draft 4th edition 2002
Kadmon N (2007) Nihon Kai and Tong Hae—Sea of Japan and East Sea: Are they exonyms or allonyms, and is there a missing term? In: The 13th international seminar on the naming of Seas and East Sea, Vienna, Austria, April 26–28, pp 1–5
Kim S (2020) Korea, Japan Both Claim Victory as East Sea Is Relabeled by IHO. Korea Joongang Daily, November 17, 2020. https://koreajoongangdaily.joins.com/2020/11/17/national/diplomacy/International-Hydrographic-Organization-East-Sea-Japan/20201117183107247.html. Accessed February 9, 2021
Koß G (1990) Namenforschung: Eine Einführung in die Onomastik. Germanistische Arbeitshefte. Niemeyer Verlag, Tübingen
Lee K (2000) New trends in identification of the East Sea (Japan Sea). In The eight international seminar on the naming of Seas: Special emphasis concerning the North Pacific Ocean. Vladivostok, pp 156–166
Lee K (2008) / 이기석 (2008) ,동해 (東海)‘ 지리 명칭의 역사와 국제적 표준화를 위한 방안 (The history of the geographical naming of the East Sea and a strategy for its international standardization). In: 한국문화역사지리학회 The Society for Korean Cultural and Historical History (ed/): 지명의 지리학 The Geography of Toponyms, pp 307–334. A improved and expanded version of an article in 대한지리학회지 Korean J Geogr 33(4) (1998)
Lew Y (2000) Brief history of Korea: A bird's-eye view. The Korea Society, New York
Mofat/NAHF (The Ministry of Foreign Affairs and Trade/The Northeast Asian History Foundation) (2007) East Sea. The name EAST SEA used for two millennia
Monmonier M (2006) *From squaw tit to whorehouse meadow: how maps name, claim, and inflame.* The University of Chicago Press, Chicago and London
Murphy AB (2005) The use of national names for international bodies of water: comparative considerations. Paper presented at the 11th international seminar on the naming of Seas, October 6–8, 2005, Washington DC, USA
National Marine Educators Association (NMEA) (2000) Conceptual flow diagrams: grades K-2. In: NMEA special report #3 The ocean literacy. Campaign, p 33
Ormeling F (2000) Sea names categories and their implications. J Geogr Educ 44:54–61. 12
Raper P (2007) Extended United Nations resolution on the names of maritime features. In: NAHF (ed) Standardization of geographical names with special reference to the East Sea. S. 5–18
Shin, G (2010) One Sea, two names: the case of 'East Sea'. Understanding the naming dispute and Korea's perspective. In: The Korea Herald 16.8.2010, p 11
Tanabe H, Yaji M, Takizawa Y, Watanabe K (2010) Origin and function of geographic names. Study on the geographical name "Japan Sea". Teikyō University research group on geographical names. Tōkyō: Teikyō University
The Japan Times (2020) "IHO Approves Proposal That Maintains Exclusive Use of 'Japan Sea' | The Japan Times." November 17, 2020. https://www.japantimes.co.jp/news/2020/11/17/national/iho-proposal-japan-sea/. Accessed February 9, 2021
UN (2002) Glossary of terms for the standardization of geographical names. New York
Watanabe K, Yaji M, Takizawa Y (2008) The naming of high seas as a process of early globalization. PowerPoint Presentation of a Special lecture at the IGU Congress Tunis 2008

Education

Design-Based Implementation Research for Exploring the Ocean: A Geographical Perspective

Alfonso García de la Vega

Abstract This chapter addresses five of the Ocean Literacy Principles from the geographical perspective. An analysis and revision of several current curricula worldwide show a similar conceptual bias on sea and ocean contents. Design-based Implementation Research is the instructional method implemented in this educational proposal. The research design is developed from two methodological approaches, namely, Problem-based Learning and Project-based Learning, which apply several learning strategies, and the use of documentary, analogical, and digital resources. The learning strategies that are implemented along the different stages of both methods are collaborative learning and cooperative learning, marine fieldwork, marine geoability workshop, and spatial and analogical thinking. Main analogical and digital resources include navigation tools and systems (compass, Global Positioning System, etc.) and logs to record oceanographical campaign events (ship's logbook and fieldwork notebook). The campaign design is meant to be a pilot research for Higher Education Teachers' Training learners to enhance awareness on ocean literacy concepts and contents. The marine campaign aims at promoting ocean-committed citizenship through fostering knowledge-building and awareness-raising on the role the ocean plays in helping earth ecosystems remain in balance. Thus, Teachers' Training learners may help boost this ocean awareness-raising in their educational career in the near future.

Keywords Oceanographical campaign · Ocean geoabilities · Problem-based learning · Project-based learning · Design-based implementation research

Introduction

This chapter presents an educational proposal for classroom and oceanographical campaign implementation of pedagogical initiatives for Ocean Literacy (OL) within the curricular framework for Higher Education Teachers' Training. In section

A. García de la Vega (✉)
Universidad Autónoma de Madrid, Madrid, Spain
e-mail: alfonso.delavega@uam.es

K. C. Koutsopoulos and J. H. Stel (eds), *Ocean Literacy: Understanding the Ocean*, Key Challenges in Geography, https://doi.org/10.1007/978-3-030-70155-0_6

"Curriculum Framework for Ocean Literacy", after an analysis of OL contents and learning standards, and of ocean citizenship, a comprehensive curricular review results in a similar conceptual bias on OL worldwide. Besides, geographical ocean concepts and terms and marine geoabilities are identified for their potential inclusion into an ocean-oriented curricular framework. Section "Educational Proposal from Design-Based Implementation Research" develops the Higher Education pedagogical proposal for OL from Design-based Implementation Research. Section "Methodological Approach and Resources" focuses on Design-based Implementation Research as the instructional method implemented in this proposal on the grounds of five principles of OL from the geography perspective. The research design is based on two methodological approaches (Problem-based Learning and Project-based Learning), several learning strategies, and documentary, digital, and satellite resources. The proposed learning strategies are developed along the stages of both methodological approaches: collaborative learning and cooperative learning, marine fieldwork, marine geoability workshop, and spatial and analogical thinking. In addition to documentary resources, main analogical and digital resources include navigation tools and systems (compass, Global Positioning System, etc.), logs to record oceanographical campaign events (ship's log and fieldwork notebook, and log blog). Finally, in section "Education Initiatives for Ocean Literacy", three educational initiatives for OL are put forward, namely two classroom activities on ocean phenomena and a trans-disciplinary field trip project for an oceanographical campaign. The campaign design is meant to be a pilot project for higher education learners to enhance awareness on OL concepts and contents that are absent in the curriculum. In this respect, the campaign aims at boosting curricular development and, as a result, at wielding influence in the society to promote ocean citizenship. Thus, the campaign also pursues the goal of fostering knowledge-building and awareness-raising on the role the ocean plays in helping earth ecosystems remain in balance.

Curriculum Framework for Ocean Literacy

There prevails a great disparity among the curricula of different countries with regard to the inclusion of contents and learning standards linked to ocean literacy. However, the seven principles of OL are aligned with the contents of STEM (Science, Technology, Engineering, and Mathematics) disciplines, which offers policymakers and educators an opportunity for their adoption into the core curriculum (Ocean Project 2009; Meyer et al. 2015).

According to Beane (1995), for curriculum integration purposes, the sources of curriculum ought to be problems, issues, and concerns posed by life itself. Such concerns fall into two spheres, on the one hand, self- or personal concerns, and on the other, issues and problems posed by the larger world. Moreover, curriculum integration brings into central focus the search for self- and social meaning. This author considers that knowledge is about "knowing about, knowing how, knowing why etc.", and hence, that it may include information, skills, concepts, and processes.

A Curricular Review

A comprehensive review has been conducted by this chapter's author of the existing curriculum frameworks that have been established in different countries. This curricular revision consisted of identifying whether the national curriculum framework included areas of knowledge for OL, and if so, which of them. This analysis has not only covered several European and North American countries having a long tradition in curricular development and implementation, but also some South American, Asian, and Oceanian countries relying either on a thriving industry and a well-established curricular framework, or else, on an emergent economy and a regenerated educational boost. All curriculum framework revision leads to the same conclusion, namely, the most widely accepted curricular definitions of contents related to OL mainly describe them as "locating position of the country's surrounding seas and ocean". Actually, one of the most outstanding factors here is the fact that most of these countries are surrounded by seas and/or ocean, and that some of them even count on major fishery industries.

In the first place, the British Geography curriculum is organised on seven axes of contents: place, space, scale, interdependency, physical and human processes, environment and sustainable development, and cultural diversity (De Miguel González 2018). Thus, contents are explicitly addressed, such as "ocean and its surrounding seas" and the use of basic geographic vocabulary, with specific reference to *beach*, *cliff*, *coast*, in Key Stages 1–2 (Department for Education 2013). Other western countries with long-established sailing tradition and history, like Portugal, show similar curricular contents to countries of lesser tradition, like Poland, on the cartography of seas and the ocean (Hibszer and Szkurłat 2018). Thus, Portugal's boost to OL focuses on the curricular contents of locating position and cartographical representation of the hydrosphere (Direção Geral da Educação 2012). Alternatively, further engagement in ocean knowledge and sustainability is found in the German curriculum, regarding knowledge both of ocean systems at varying scales and of ocean pollution. Above all, there is special mention in it to the interactions between weather and ocean currents (Hermmer 2012). In fact, the German ocean-related curricular structure is better organised, so getting closer to initiatives such as framework proposals for OL (Cava et al. 2005).

Secondly, in Spain, despite it being a western peninsular country, the curriculum framework contains mere references to the position-locating of seas and ocean. Such analysis points in the Spanish curriculum to an absence of core curricular contents that may become essential subject matters for OL when compared to the aforementioned framework proposal. In particular, in the Spanish curriculum for Secondary level, ocean and seas are dealt with to learn their names and location. However, as it may happen in other countries, the curriculum omits all reference to the interaction among the universe, the earth, the atmosphere, the ocean, and the seas. The universe is addressed as a set of stars, planets, and satellites with no mention given to the ocean in the configuration of the earth. The atmospheric dynamics are reduced to a basic description of meteorological elements, dropping all references to the relevance of the ocean in climate types (Ministerio de Educación Cultura y Deporte 2014a). Besides, the ocean and seas are significantly excluded from hydrography in the Secondary level curriculum since the major interest lies on coastal and mainland fluvial dynamics and on human intervention in the continental hydrographical network. The ocean is scarcely mentioned, and the seas are "observed" from the coast. Referred landforms are *coast, beach, cliff, gulf, mouth, fiord*, etc. Again, it is undoubtedly far from relating them to the magnitude of the ocean landforms, especially the oceanic trenches, canyons, and seamounts on the oceanic plateau (Ministerio de Educación Cultura y Deporte 2014b).

Thirdly, the ocean-oriented curricular organisation differs even more markedly between Asian and South American perspectives and the Western perspective. Locating position still plays a significant role in primary- and middle-level curricula of a peninsular and fishing country like Korea (Ministry of Education 2018a, b). Geography in the Korean High School curriculum increases knowledge on coastal landforms and includes references to human exploitation of the seas and also to disputes on nearby seas (Ministry of Education 2018c). On the contrary, the ocean is poorly referenced in the social sciences curriculum of Colombia (Ministerio de Educación Nacional 2002, 2004) whereas, in Brazil there is a growing concern with raising awareness on sea pollution (Ministério da Educação 2017a, b).

According to the prior analysis, a similar cultural and academic bias on OL has been found almost worldwide. In fact, all curricula under study tangentially connect the ocean to the earth's dynamics. Coastal landforms and fisheries exploitation have been long studied, but the interactions between the ocean and the earth, and in consequence, with the climate, have been obviously neglected. Besides, all curricula lack the integrated and trans-disciplinary approach necessary to address these contents. A holistic view is needed to bring together the ocean, the seas, the human life, the wildlife, the marine flora, and the climate, because they represent altogether the crucial elements that help the earth's complex environment remain in balance.

Ocean as a Trans-Disciplinary Curricular Content

First initiatives to boost OL in the national curriculum are found in the works of different countries, like the United States (Steel et al. 2005), United Kingdom (McKinley and Fletcher 2010), and Portugal (Frazão Santos et al. 2012). Donert et al. (2015) have produced a report on "marine formal education", in the context of a Framework Programme named "Sea Change Project" where they examine and revise all contributions made on ocean issues in the sphere of education. Hence, in the United States and Canada, a deep awareness-raising campaign has been launched for incorporating OL into primary and high school education curricula, as it is the case of an educational establishment in Nova Scotia (Guest et al. 2015).

In particular, the United States, National Oceanic and Atmospheric Administration (NOAA), The Lawrence Hall of Science (LHS), National Marine Educators Association (NMEA), College of Exploration (CoE), Sea Grant, and Centers for Science Education Excellence (COSEE) have joined efforts to publish the guides *Ocean Essential Principles and Fundamental Concepts* and *Climate Essential Principles and Fundamental Concepts* (NMEA 2010; NOAA 2020). These guides are aimed at defining ocean and climate literacy and identifying the elemental principles and basic concepts of ocean and climate science. Yet, these are meant to be guides with an ocean-oriented and climate-oriented approach for learners of all ages. Other guides are conceived as practical resources for educators, since they gather the knowledge required to be considered ocean and climate literate (NMEA 2010). In the present case, for the purpose of the Design-based Implementation Research outlined in section "Educational Proposal from Design-Based Implementation Research", five Ocean Literacy Principles are selected from the geography perspective as follows: (1) The earth has one big ocean with many features; (2) The ocean is a major influence on weather and climate; (3) The ocean made earth habitable; (4) The ocean and humans are inextricably interconnected; and (5) The ocean is largely unexplored. French et al. (2015) have assigned a set of concepts to each of the seven Ocean Literacy Principles. Some of these contents are common to nearly all national curricula (e.g. 97% of the earth's water comes from the seas and ocean, and plays a part in the water cycle). In contrast, other contents correspond to major scientific achievements that have been made in the modern scientific era as a result of the research conducted on well-known oceanic events, such as El Niño–Southern Oscillation (ENSO).

In the European sphere, in the European Landscape Convention (Council of Europe 2000), the parties thereto undertake to improve knowledge, to encourage increased participation and international collaboration, to enhance awareness-raising, and to promote training and education on landscape. This Convention sustains an anthropocentric and terrestrial view of landscape, as a result of an active human intervention and influence on the landscape. However, it remains open to question whether the ocean actually plays a part in such a conception of landscape, since the Convention drops all reference to marine or ocean landscape when defining the concept of landscape. On the contrary, in the international arena, the ocean was incorporated into the study of climate. In fact, the recent report on climate change mentions the interaction between the ocean and the climate (Oppenheimer et al. 2019).

On a national scale, the Spanish *Plan Nacional de Adaptación al Cambio Climático* is a proposal to promote coordinated action against the impact of climate change in Spain (draft under ongoing public consultation by the ministerial advisory committee). The draft includes a guideline axes on coasts and marine environment. However, the author of the present chapter considers that this environmental policy initiative may become a great opportunity to promote OL in relation to climate change, by putting forward education-oriented suggestions for the inclusion of specific action on OL (Ministerio de Transición Ecológica, 2020).

One of the most innovative and emergent concepts nowadays refers to the creation of ocean citizenship, in allusion to informed citizens fully committed to ocean sustainability (Fletcher and Potts 2007). These authors advocate for fostering individual and social attitudes of respect and commitment to the ocean by forging links between citizenship and governance. Such positive attitudes may lead to ocean sustainable governance, active participation in the decision-making process, personal and social responsibility, and protection of fragile marine ecosystems due to intensive ocean resource exploitation, among other sustainable actions. Orion (2002, 2007) develops a holistic approach to the sciences, attaching utmost significance to both the relationship between learning and emotion and the pursuit of applying learning to everyday life. The approach to the sea and the ocean from science education appears established (Payne and Zimmerman 2010; Reid and Breidahl 2019), the challenge comes from geographic education.

In this respect, citizens that understand the complex world in which they live and act for positive change with regard to the ocean can be called marine citizens (Rieckmann 2017). McKinley and Fletcher (2012) advocate that "…marine citizenship requires an enhanced awareness of marine environmental issues, an understanding of the role of personal behaviour in creating and resolving marine environmental issues, and a shift in values to promote marine pro-environmental behavioural choices". Besides, building critical thinking is the key element in fostering a proactive attitude towards the present and the future of the great ocean. It is essential to achieve the socio-emotional and behavioural learning objectives and contents that are associated with the universal interpersonal values (e.g. solidarity, empathy, team spirit, generosity, etc.) being inherent to human condition in order to cope with adverse situations encountered in the ocean. In this sense, historical and non-fiction literary documents may provide educators with a set of teaching resources that facilitate addressing universal human values upon sailing the sea.

Geographical Concepts and Geoabilities Linked to Ocean Literacy

Ocean literacy allows a trans-disciplinary perspective and fosters the development of cognitive processes associated with spatial thinking, reasoning, critical thinking, and analogical thinking, among other processes. Comprehensive geographical knowledge acquisition involves understanding oceanographic terms related to deep-sea landforms, as they are major physical features to be found on the seafloor of ocean basins. In addition to oceanic trenches, there exist landforms being distinctive features of continental platforms, including *pockmarks* and *seamounts.* A pockmark, a type of cold seep, is a concave crater-like depression on mud seafloor, where the fluid seepage causes ocean floor subsiding. Pockmarks are well known by bottom trawling fishers as "holes", and they become crustacean fisheries in the Mediterranean Sea (Acosta et al. 2014). A seamount is a deep-sea mountain or peak that rises from the seafloor to 1,000–4,000 m in height. After wave erosion, they may turn flat-top seamounts, and are called *guyots* or *tablemounts* (Visser 1980). Other landforms are found, like *seavalleys* or *seacanyons,* along continental platform boundaries. All these are

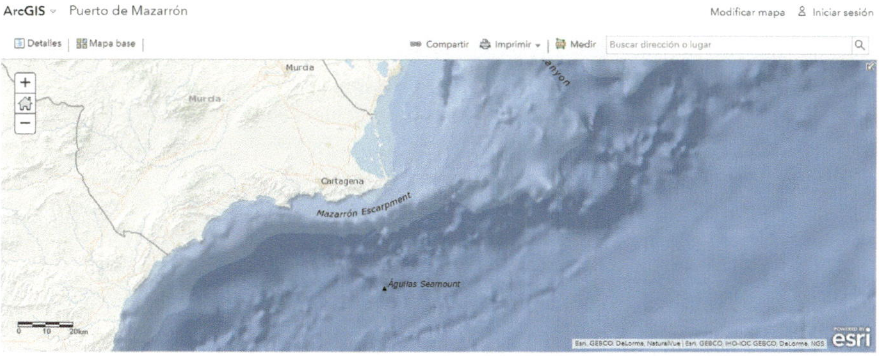

Fig. 1 Seacanyons and pockmarks on the Mediterranean seafloor, nearby Aguilas seamount and Mazarron escarpment (*Source* ESRI 2018)

significant landforms owing to the fact that they are associated with ocean flora and wildlife (seacanyons), or with regional seismicity (pockmarks) (Fig. 1).

As previously mentioned, position-locating of seas and ocean basins is a reiterative content in the curriculum of many countries. However, this content makes reference to sea and ocean position in the earth, but it does not refer to the position of physical features within the sea and/or the ocean. Thus, identifying position, distance, course, and heading are geospatial skills to acquire while navigating the ocean. In addition to these marine geoskills, other basic spatial abilities that have been formerly acquired in the terrestrial environment may be incorporated into the marine spatial-thinking skill set, including scale-comparing and route-mapping. If acquiring geoskills and geoabilities is fraught with difficulty on earth settings, the challenging question to consider here is how learners may be able to measure a distance on the ocean, or else, which spatial-thinking abilities indeed are involved in the task of veering course and taking a heading for the open sea.

The acquisition of these marine geoabilities poses an undoubted geographical challenge, since these geoabilities are an essential element in experience-based learning of seas and the ocean. That is to say, to attain these geoabilities, learners need to become familiar with observing the ocean and estimating proportions, distance, course and heading, and with marking benchmarks on the nautical chart. In addition, they may need to track evidences of ocean weather (wind direction and temperature) and seawater (wind waves, tidal surges, and rollers) conditions through meteorological observations.

Wayfinding and navigation constitute the quintessential marine geoability in the ocean. Both tasks require good performance of terrestrial geoabilities. But, above all, these tasks demand mastering reaction in the face of adverse situations in the open sea so as to be able to make an estimate of distance to the coast, and draw analogies between distances. Everyday routines and professional activities foster the experiential and experimental development of scale. Taylor and Jones (2009) point to the development of "cognitive anchor points" from kinetic activities. Therefore, these activities allow a geospatial wealth of knowledge that are essential to navigate the open sea. Having said that, the experiential and experimental acquisition of scale has nothing to do with the ability to increase and reduce a cartographical scale. The educational initiatives presented below foster marine geoabilities for OL, namely, position-locating, course and heading estimation, and navigation, which are also existing geoskills in the Geography core curriculum.

Educational Proposal from Design-Based Implementation Research

Methodological Approach and Resources

The methodological framework in this paper is based on Design-based Implementation Research. McKenney and Reeves (2018) underline that educational design research merges scientific investigation with systematic development and implementation of solutions to educational problems. The principles behind this research method, according to Fishman et al. (2013, 136), are as follows: (1) A focus on jointly defined problems of teaching and learning practice; (2) A commitment to iterative, collaborative design; (3) A concern with developing knowledge and theory through disciplined enquiry; and (4) A goal of developing capacity for sustaining change in systems. Among this method's features, this chapter pursues the goal of orienting Design-based Implementation Research towards building up a curriculum that incorporates some geographical contents to be aligned with OL principles. Choppin et al. (2013) highlight that "the model of design-based implementation research (DBIR) offers insights into complexities and challenges of enacting productive curriculum adaptations". DeBarger et al. (2013) assert that "Making productive adaptations is a kind of advanced professional practice that requires different instructional knowledge and skills, and so new models of professional development are likely needed to support teachers in this practice". Ben-Zvi Assaraf and Orion (2009) suggest Earth Systems-based Environmental curriculum on the basis of the five elements advanced by Collins (2004): (1) Goals and elements of design; (2) Setting where implemented; (3) Description of each phase; (4) Outcomes found; and (5) Learnt lesson.

Geographical and Pedagogical Approaches to Ocean Literacy

The proposal in the present methodological design on OL blends together the geographical perspective and the pedagogical approaches. On the one hand, the geographical perspective, according to UNESCO's Multiple-perspective Approach to Ocean Literacy (Santoro et al. 2018), seeks to address a problem by identifying its origin and putting forward potential solutions that may apply to different geographical contexts worldwide. In such a report, the authors acknowledge the geographical perspective as one of the seven trans-disciplinary perspectives in the study of the ocean. In geography teaching, geographical-event observation and database analysis may help to elaborate an interpretation of real-world phenomena. Puttick (2013) emphasises the difference between "looking at and looking for" in geography teaching. This author believes that looking *at* relates to abstract, external, impersonal, uninvolved knowledge, whereas looking *along* concerns participant, inhabited, personal, committed knowledge. In the present proposal, learners may develop both processes of proximity to real-world phenomena.

The present proposal for ocean teaching embodies a constructivist learning environment, where contributions from Piaget (1964), Vygotsky (1986), and Bruner (1960 and 1966) are specially considered. Bruner (1960, vii) recognised that "Its view of children as active problem-solvers who are ready to explore 'difficult' subjects while being out of step with the dominant view in education at that time, struck a chord with many". In this kind of learning environment, both discovery and social links boost students' significant learning. The essential features that distinguish the constructivist learning environment are classroom dynamics triggered by cooperative and collaborative learning situations, the student's self-regulated learning, and the teacher's role as a learning tutor facilitator. This constructivist learning environment offers a set of teaching strategies that may be applied to OL.

Donert et al. (2015) identify some pedagogical approaches linked to OL. Among them, there stands out the trans-disciplinary nature of ocean science that requires integration in an enquiry-based trans-curricular programme that brings science closely together with the arts, citizenship, and ethics. These authors advocate an increased emphasis on real-life-world problem-solving that triggers enquiry-based learning and thinking. Stepath (2006) upholds experiential approaches, and points to adopting awareness, attitude, and action as three goals of the learning enquiry-based project. In brief, OL methods prioritise hands-on, scientific enquiry-based approaches, and "real-world real-life experiences", including museum visits, contact with scientists, and experimental activities or fieldwork as effective pedagogical approaches to teaching and learning.

The present Design-based Implementation Research rests on educational foundations as follows: pedagogical methods (Problem-based Learning and Project-based Learning), learning strategies (analogy, enquiry learning, fieldwork, collaborative learning, and cooperative learning), and a wide array of digital and analogical resources. The reading of the journey diaries and the oceanic campaigns based on their protagonists' first-hand vivid experiences may be adopted as exemplary models for producing a ship's logbook and a fieldwork notebook. Some of these accounts may lay the groundwork for understanding OL-related geographical concepts in a real-life marine context. A reliable record of data readings on atmospheric phenomena may be obtained by using both analogical and digital navigation tools, nautical charts, and digital cartography. From all these data and findings, learners may keep a detailed log of their voyage in the case of embarking on an oceanographical campaign.

Regarding the method of Problem-based Learning (PBL), Barrows (1986, 1996) underlines that new knowledge is acquired and integrated from a trans-disciplinary approach to problems. Barrows (1996) and Hmelo-Silver and Barrows (2008) proposed that the teacher should be a mere learning tutor facilitator. Therefore, no intervention on the teacher's side in the learner's process of knowledge acquisition becomes a key factor in the development of such a method. Hmelo-Silver (2004) lays down five basic principles to develop the PBL method. These basic principles are as follows: (1) building up a vast and flexible body of disciplinary knowledge; (2) developing effective skills in real-life problem-solving; (3) developing self-directed, lifelong learning skills; (4) collaborating actively and efficiently with the group; and (5) becoming intrinsically motivated to learn.

Savery and Duffy (2001), Spronken-Smith (2005), Savery (2006), and Hmelo-Silver and Barrows (2008) consider knowledge scaffold building is triggered by the constructivist learning environment, in a similar way to PBL. Barkley et al. (2007) suggest some techniques for implementing collaborative learning: graphic information organisers, reciprocal peer teaching and discussion, think-aloud pair problem-solving, and structured problem-solving. The methodological strategy of learning by analogy facilitates implementing PBL (Holyoak 1995; Holyoak and Thagard 1997) by comparing resources of a different nature (satellite images, digital cartography, historical maps, and literature) on a particular topic.

Secondly, the method named Project-based Learning was defined by W. Kilpatrick at the beginning of last century. In this method, the students are protagonists of their own learning since the project arises from their interests in proportion to the learner's autonomy level granted by the instructor. This method encompasses varying learning types: discovery learning, significant learning, and cooperative learning. Cooperative learning may be the most appropriate learning type for knowledge-building on OL. Cooperative learning requires to implement two strategies: workshop and fieldwork. In the workshop, activities are carried out for accomplishing a final product, for developing ocean spatial thinking, and for acquiring the navigational routines to be recorded on a fieldwork notebook. The final cooperative product in the proposed project campaign takes shape in the form of a log blog, to which all students contribute. The fieldwork represents the development of the campaign itself where the coordinated activity of all participants plays a prominent role in cooperative learning.

The learning strategies to implement in the methodological design of a project campaign are enquiry learning, analogy, and fieldwork. Besides, they stand both aforementioned collaborative learning, associated with the PBL method, and cooperative learning, linked to the Project-based Learning method. Enquiry learning plays an intrinsic part in both methods. Analogy may become an effective learning strategy for acquiring marine geographical concepts. Finally, Bradbeer (1996) states that fieldwork provides to be an excellent place to implement PBL. Moreover, when adopting the Project-based Learning method, fieldwork applied to a project campaign has proven to be a powerful learning strategy. Even the geographical scenario being defined as the minimum conceptual unit of geographical knowledge has shown to work as a successful learning strategy (García de the Vega 2012, 2018). As Beane affirms (1995, 616) "That is, knowledge is called forth in the context of problems, interests, issues, and concerns at hand. And since life itself does not know the boundaries or compartments of what we call disciplines of knowledge, such a context uses knowledge in ways that are integrated".

Ocean Literacy Resources

This chapter intends to associate scientific documentary sources with cartographical sources, historical sources, and non-fiction literary sources, since they are all valuable primary sources for building knowledge on the ocean and the earth's dynamics. Over the nineteenth and twentieth centuries, travel books, accounts of scientific expeditions, and journey logs have been the expression and record of the close interaction between oceanic and weather dynamics. Principle 3 of OL is "The ocean is the major influence on weather and climate". Thus, countless geographical concepts may be drawn from scientific knowledge records that give evidence of oceanic features and atmospheric phenomena. Among these, water temperature and salinity in ocean currents, atmospheric pressure, and coastal relief are involved in atmospheric processes, such as El Niño in the northwest area of South America. In fact, these evidences are gathered by direct observation, measured by means of varying instruments, and captured in the writings of Humboldt, Steinbeck, and FitzRoy, amid other authors. In these books, the experience of sailing the ocean inevitably leads to identifying the signs that anticipate a storm, to daily recording the accurate data measurements from reading the navigational instruments, and to defining a strategy to face the peril aboard. As a consequence, the reading of such accounts fosters learners' acquisition of geographical concepts related to other disciplines such as Science, Technology, Engineering, Arts, and Mathematics (STEAM). These literary non-fiction works allow for comparing their data with other evidences collected from historical cartography, and modern Information and Communications Technologies (ICT), and Geographical Information Systems (GIS). By using online cartography and weather forecast summaries, learners may grasp the interaction between the ocean and the weather. This use pursues a twofold goal: understanding the earth's dynamics and raising learners' awareness on human activities across the ocean and exploitation of ocean resources.

Over the centuries, maps have offered exact measurements on the earth's representation. The cartographic realm is associated with the human interest in recognising and reaching territories that might be inhabited or entail a sailing benefit. Thus, the portolan charts symbolise the human need for shortening those huge nautical distances between ports. Since the fifteenth century the first navigation maps, or nautical charts, were essentially designed to plot the course between ports on the sailing chart by highlighting reefs, shoals, the mouth of a harbour, and wind speed and direction. Therefore, mankind's approaching of the ocean comes from the demand to apprehend it.

Since the Middle Ages, countless earth discoveries were subsequently made as a result of the conquerors' voyages. In 1513, Núñez de Balboa reached the South

Sea. Ferdinand Magellan and Juan Sebastián Elcano first circumnavigated the globe in 1519–1522 sharing, on their return, their newly acquired detailed knowledge and cartographic representations of the African and Asian continents, such as a globe of the entire earth, a sailing chart, and a world map. The discovery of Australia reflects the evident difficulty in sailing the seas and manning the vessels in the ocean, since it was only landed on by James Cook's expedition after the second half of the eighteenth century:

> *[One day of final November, 1520]*
>
> *After having entered inside this strait we found that there were two mouths, of which one trended to the Sirocco (S.E.), and the other to the Garbin (S.W.). On that account the captain again sent the two ships, St. Anthony and Gonception, to see if the mouth which was towards Sirocco had an outlet beyond into the said peaceful sea.*
>
> *The land of this strait on the left hand side looked towards the Sirocco wind, which is the wind collateral to the Levant and South; we called this strait Pathagonico. In it we found at every half league a good port and place for anchoring, good waters, wood all of cedar, and fish like sardines, missiglioni, and a very sweet herb named appio (celery). There is also some of the same kind which is bitter. This herb grows near the springs, and from not finding anything else we ate of it for several days. I think that there is not in the world a more beautiful country, or better strait than this one.* (Antonio Pigafetta 1520)

The historical documentary sources mainly evolve from a written record of events kept by the person participating in the narrated facts. Alternatively, there exist other creditable sources where the historic facts are narrated on the most accurate documentary basis. Whichever the case is, the teaching exploitation of ocean-related historic sources is a worthy method of promoting OL in the classroom. In this respect, the daily records of voyages that were written by the adventure protagonists themselves, namely, Christopher Columbus' and FitzRoy's, have proven an extremely useful tool when defining a teaching strategy based on ocean-oriented contents and processes.

Robert FitzRoy, a hydrographer and the commander and surveyor of the *Beagle* between 1831 and 1836, was accompanied by Charles Darwin, a naturalist and geologist, in his surveying voyages to South America. As a detailed account of such expedition, the "Narrative of the surveying voyages of his Majesty's ships Adventure and Beagle between the years 1826 and 1836" was written by FitzRoy. The contribution of this narrative to the meteorology and cartography of South America is worldwide acknowledged and valued. Still, FitzRoy's ability to grasp the interactions between the ocean and the weather is most significant, as it is the case when, by his own account, he vividly describes the gales and storms he was often met with on sailing the Beagle.

> *Maldonado, 30 january 1829.*
>
> *On the 30th of January, after some intensely hot and sultry weather, we experienced a very severe ' Pampero.' It was preceded by the barometer falling to 29'50, and by a strong N.W. wind, which suddenly veered round to S.W., when the pampero burst upon us. Our ship and boats fortunately escaped any bad effects from the violence of the squall, which was so strong as to lay the former, at anchor, upon her broadside; but on shore our tent was blown down, and a boat that had been lately built, and fresh painted, on the Island Gorriti, was completely destroyed. [...] The violence of this pampero, during the twenty minutes it lasted, was terrific. Old inhabitants of Maldonado declared, that they had experienced nothing like*

> *it for the last twenty years. The spray was carried up by whirlwinds, threatening complete destruction to everything that opposed them. In less than half an hour it had diminished to a strong S.W. gale, which lasted during; the night.* (Fitzroy 1829)

Non-fiction literary sources may provide detailed accounts of first-hand personal experiences and factual events. In this respect, it is advisable to hand the texts that were written by the protagonists themselves of the true-life story, which bear authentic testimony to the observed phenomena. In *Cosmos*, Von Humboldt devotes a substantial number of pages to describe the Mediterranean Sea, which is depicted as a space that is shared by contemporary cultures in subsequent times. Hence, Humboldt evokes countless significant events of ancient history that have been formerly reported by classical literary sources in Greek and Roman times.

> *Diodorus Siculus, when speaking of lovely islands which may be supposed to be the Canaries, allude to the storms which may have occasioned their accidental discovery. Phoenian and Carthaginian ships, it is said, sailing to the settlements already existing of the coast of Lybia were driven out to sea; the event is placed at the early period of the Tyrrhenian naval power, during the strife between Tyrrhenian Pelasgians and the Phoenicians.* (von Humboldt 1849, 131)

On the other hand, digital resources of both cartographical and imaging nature have proved to be invaluable tools for acquiring OL. Solem (2001, 22) highlights that "as the theoretical referent for enquiry and PBL, constructivism has also emerged as a guiding theory for teaching with GIS and the internet". Wilson (1996) considers that learning with ICTs requires collaborative interaction embedded in a technology-enhanced constructivist learning environment. He underlines that "technology-enhanced constructivist learning currently focuses on how representations and applications can mediate interactions among learners and natural or social phenomena".

Through direct empirical observation of meteorological phenomena, learners may obtain significant data and evidence for understanding ocean dynamics. Using both ICTs and GIS cartography applied to navigational routines facilitates knowledge acquisition on the marine world and fosters marine geoabilities and geoskills. Transdisciplinarity encourages yet spatial concept acquisition from varying scientific disciplines, including STEAM, according to Sinton et al. (2013). In the present case, STEAM is prioritised, owing to the inclusion of archaeology and ancient history in the proposed oceanographical campaign, with a view to adopt a holistic and transdisciplinary approach to the campaign project. Navigating the seas and the ocean requires making first-hand quantitative estimations and recording accurate data on distance and magnitude that may be obtained only through direct observation and by means of navigation-related GIS.

In brief, these documentary sources supply the ocean educator with a mine of information on ocean topics and experiences. Therefore, teaching proposals targeting OL principles, content, and values may be brought forward in the classroom and thus, some projects may be conceived and implemented to launch marine and ocean awareness campaigns at both school and educational community level (. 2).

Fig. 2 Stormy Mediterranean Sea

Education Initiatives for Ocean Literacy

This section suggests three marine education initiatives for OL that are designed to be relevant for learners of higher education and to find their application in the Spanish curriculum for Teachers' Training and Education. These educational activities are adapted to the national or regional Mediterranean context. They are also conceived to be implemented in different learning settings through a geographic perspective. The first initiative addresses the appearance of "Meddies" at the confluence of the Mediterranean Sea with the Atlantic Ocean from PBL methodology. Secondly, the Project-based Learning method is applied to tackle the issue of the floating lava islands. The third initiative, a trans-disciplinary project to be implemented through a marine field trip, is aimed at planning an oceanographic campaign for the exploration of the ancient Phoenician and Greek civilisations spreading throughout the Mediterranean Sea. For each activity, a lesson plan is outlined with specific learning standards, disciplinary cognitive contents, background information, or content necessary for teachers to conduct the activity as a case study; a precise methodological approach, classroom dynamics and procedure, and teacher's role; teaching resources and learner materials, and evaluation with assessment strategies and learning outcomes.

Classroom Activity on Deep-Water Salty Whirlpools

Case Study The origin of ocean streams has been associated with the combination of three defining features: temperature, salinity, and the earth's motion (Zhen et al. 2018). Humboldt (1846) kept a measurement record of temperature and salinity variation in the ocean water density along the western coastal border of South America, and thus proved the existence of a cold low-salinity ocean stream. He verified that the combination of the earth's rotation and the centrifugal force drives north the ocean waters towards the equator. Coming from the cold Antarctic polar waters, the Humboldt Current flows north along the western boundary of the South American continent in the direction of the equator, where it turns west to join the Equatorial Current. The study of this stream is due to the direct observations made by the naturalist to whom it owes its name.

In the Atlantic Ocean, near the area between southern Europe and northern Africa, NASA has located deep-water salty whirlpools. Deep-water dynamics derives from the circulation system of ocean waters, since these deep-water whirlpools help drive the ocean currents that moderate the earth's climate. Warm water usually resides at the ocean surface, but the outflow of warm water from the Mediterranean Sea is so salty and dense that when it enters the Atlantic Ocean at the Strait of Gibraltar, it descends to depths of more than 1,000 m (one-half mile) in the North Atlantic Ocean, forming the uppermost layer of North Atlantic Deep Water (Gordon 1986). This warm and salty underwater flow then splits into clockwise-flowing whirlpools that may continue to spin westward for more than two years. These whirlpools, or eddies flowing out of the Mediterranean Sea, are so named "Meddies", which may have an impact in the increase of sea level.

Lesson Plan	
Learning content standards	(1) Identify salt proportion and temperature variation in the ocean waters as the two key elements that help generating ocean streams (2) Relate the South Pacific Ocean current with the climate along the South-American western coast boundary (3) Search for the warm-water salty whirlpools in the circulation system of ocean waters
Disciplinary cognitive contents	Ocean streams, in particular, the Humboldt (Peru) Current of South America, the ocean thermohaline circulation and the global conveyor belt, eddies, and the effect of ocean streams on climate regional patterns

(continued)

(continued)

Methodological approach and learning strategy	PBL and geographical scenario (García de la Vega 2018). The geographical scenario shows the defining features that contribute to formulate the problem and to elaborate on a solution NASA's Figs. 3 and 4 illustrate the geographical scenario. This marks out to the right the south-western Mediterranean area where the mass of warm salty water forms within the Mediterranean Sea and where it exits as a deeper stream through the Strait of Gibraltar. To the left, the scenario encloses the largest area in the Atlantic Ocean, next to Cape St. Vincent in southern Portugal where Meddies may be found
Classroom dynamics and procedure	(1) Classroom discussion on the proposed geographical scenario to identify key features defining the problem: formation of Meddies and their dynamics in the ocean (2) Research the topic on the Internet in working groups. Position, salinity, temperature, circulation system, and increase in water level may all turn key elements to search the Internet for and retrieve information on Meddies. Results may be achieved to identify the origin of the highest salt concentration, to link the increase of salt in the seawater to its highest density, and to determine the effect of salt distribution on the ocean thermohaline circulation (3) In the final classroom session, all pieces of information obtained are being gathered, shared, and collated. Each working group puts forward its solution to the problem, and the optimal one is reached by consensus
Teacher's role	Mediator of significant learning by: (1) moderating situations of cooperative learning during classroom discussion sessions (2) monitoring situations of collaborative learning during working group sessions
Teaching resources and learner materials	• NASA satellite photograph tracking warm Meddies, https://www.nasa.gov/images/content/144990main_meddes-20060320-browse.jpg • Currents Tutorial on Thermohaline Circulation and The Global Conveyor Belt by NOAA, https://oceanservice.noaa.gov/education/tutorial_currents/05conveyor1.html • World Ocean Thermohaline Circulation (Döös et al. 2012) • Excerpts from *Cosmos: A Sketch of a Physical Description of the Universe* by Alexander von Humboldt
Evaluation Assessment method Learning outcomes	Focus on learners' ability to pose the problem, express themselves, and debate Based on the quality of individual's Internet searching results, and of his/her contribution to the final product: selection of Internet image representing an ocean geographical scenario of any oceanic circulation-related problem with well-defined features (1) Understand both the oceanic circulation and the effect of temperature and salinity on ocean currents (2) Interpret and draw conclusions on the global-scale system of ocean currents, the presence of Meddies and their influence on the climate (3) Become aware of warm eddies impact on increase in water level

Fig. 3 (Left) The Strait of Gibraltar has 9 miles of distance and conforms a semi-closed Mediterranean Sea (NASA 2013)

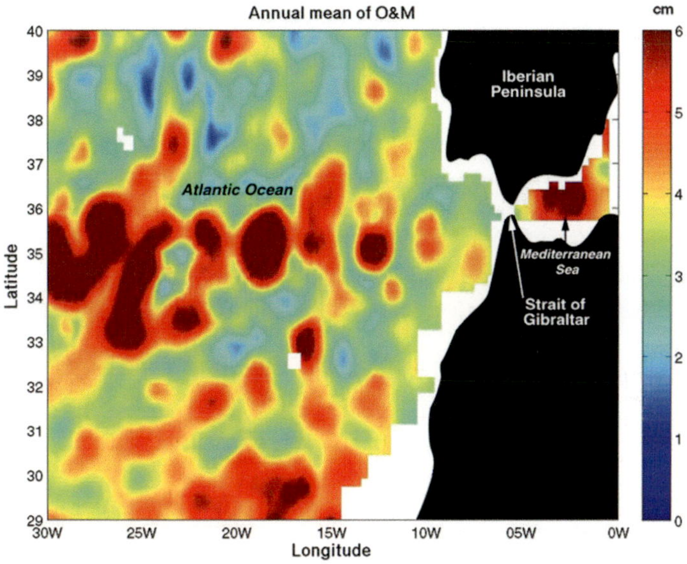

Fig. 4 (Right) Meddies are mapped in red by The Advanced Very High-Resolution Radiometer, an infrared spectrometer that shows the increase in temperature from a warm Meddy coming from the Mediterranean Sea into the Atlantic Ocean (*Source* NASA 2006)

Classroom Activity on Floating Lava Islands in the Ocean

Case Study Since antiquity, the appearance of new volcanic islands has been explicitly referenced. Santorini archipelago in the Greek Cyclades, located in the southern Aegean Sea, including Thera (now called Santorini), Therasia, Aspronisi, and the Kameni islands, encircles the current Santorini caldera (Fig. 5). This caldera was formed during the Minoan volcanic eruption in 1645 BC (Hammer et al. 1987) of Thera (Santorini) volcano, the most active volcanic complex of the South Aegean Volcanic Arc. In particular, Palea (old) Kameni and Nea (new) Kameni were formed by repeated, initially submarine eruptions of dacite lava and ash at the centre of the caldera. Pliny the Elder records the emergence of the new Kameni island in his *Natural History* (4.12.23). In 2006, submarine volcanic eruptions occurred next to Vava'u island near Tonga archipelago in the South Pacific Ocean (Vaughan et al. 2007). In 2011, in the North Atlantic Ocean, submarine volcanic activity was detected in the proximity of El Hierro, an island of the Canaries archipelago, off the north-western coast of Africa (Fig. 6). In the first case, the floating banks of pumice stone were the evidence of volcanic eruptions. In the second, the lava emissions were accompanied by gas releases that did not reach the water surface. In both cases, the largest fault systems, linked to oceanic trenches, helped drive seismicity on the earth. Bonnett (2014) highlights the floating banks of pumice stone that have been derived from submarine volcanic eruptions. Identifying the presence of floating banks of pumice stone on seawater, as well as locating their position and heading, may be particularly relevant to find their origin. Both position and heading of floating banks of pumice stone allow for detecting the appearance of a new island. Bonnett points up that a sailor caught sight of a huge group of new islands resulting from a volcanic eruption near Tonga archipelago in the South Pacific Ocean (Figs. 7, 8, and 9).

Lesson Plan	
Learning content standards	(1) Identify submarine volcanic eruptions in oceans (2) Relate lava emissions to fault systems and oceanic trenches (3) Draw analogies among different submarine eruptions
Disciplinary cognitive contents	Different submarine volcanic materials, identification of fault systems and oceanic trenches in the cartography, and location of submarine volcanic eruptive processes over history
Methodological approach and learning strategy	PBL and learning by analogy. By using both the satellite and blog imaging of volcanic eruptions in the Pacific and Atlantic oceans, and the storytelling on the yacht's blog, learners may be able to suggest analogies. These analogies are drawn from the body of evidence found in this set of resources. Therefore, comparing resources, including ship's logbooks or yacht's blogs, and satellite and blog imaging, facilitate implementing analogy in PBL. Thus, analogy-oriented learning here is targeted at the overall classroom group to find similarity of location, typology, and resulting land formation between these submarine volcanic eruptions

(continued)

(continued)

Classroom dynamics	(1) Classroom brainstorming session to help formulate the hypothesis on the origin of these volcanic eruptions (2) The teacher suggests the similarity of the two situations for learners to put forward an analogy (3) Learners search for the common and differing features in both locations
Procedure	The seismic events occurring in the ocean may account for the endogenic dynamics of landforms at the earth's surface. In fact, our witnessing these unique situations may lead us to understand the earth's dynamics. Volcanic eruptions may be recorded anywhere in the earth, with varying intensity and volcanic system development, and result in either subaerial or submarine volcanoes. In fact, the submarine volcanic activity shows the large fault systems and their relation to the oceanic trenches and in consequence, to the earth's seismicity. If the volcanic activity or the submarine seismicity are the key elements for making appropriate analogies, specific questions should be posed that help learners elaborate hypothesis and further a solution. After the formulation of such questions, learners may search for information and try to find evidences and records to work out solutions to the hypothesis. The location, together with the increase in seismic activity and temperature, may be the cause of changes in the marine ecosystems. Learners becoming aware of these causes fosters their comprehension and acquisition of a global insight into the earth's general dynamics
Teacher's role	Facilitator of the learning process in a way that he/she brings enthusiasm and dynamism onto the searching task in working groups
Teaching resources and learner materials	• ICT and GIS resources: NASA's satellite images of 2006 (Vava'u island) and 2011 (El Hierro) eruptions, and digital cartography of these geographical areas. NASA's satellite imaging before and after the eruption enables learners to (1) locate deep-water position of submarine volcanic eruptions, (2) show the relation borne to the oceanic trenches, and (3) identify the extension of the floating banks of pumice stone and to track their heading (Figs. 8 and 9) • Blog-diary *Fredrik and Crew on Maiken* by Fredrik Fransson, who discovered the volcanic emission in 2006 (http://yacht-maiken.blogspot.com). This blog bears a resemblance to FitzRoy's *Narrative*: "This is the chronicle of the adventures on the yacht Maiken", and enables the learners to recognise a ship's logbook by using modern digital resources available. Fransson posted on 12 August 2006 a sailing record titled *an amazing last few days—weird but amazing,* where he narrates his witnessing "the birth of an island". The blog post *Stone sea and volcano* includes a large imaging collection, under the titles "Maiken cruising through a sea of stones" and "This volcanic island rose from the sea in front of us", illustrating the presence of subaerial volcanic emissions and floating banks of pumice stone on the water surface in the Pacific Ocean

(continued)

(continued)

Evaluation Assessment method Learning outcomes	Focus on learners' ability to formulate hypotheses and reach conclusions from the given evidence, as well as their ability to make analogies worldwide between volcanic islands whose formation is driven by oceanic trenches Based on the quality of the individual's building up of hypotheses and pertinent solutions, as well as his/her mastery of drawing an accurate world map by using Google Earth and where to mark position of volcanoes next to the oceanic fault systems and trenches as the final product (1) Identify oceanic trenches on the basis of volcanic islands' location (2) Understand the relationship between the dynamics of the ocean and the earth

Fig. 5 Santorini islands (*Source* NASA 2020)

Fig. 6 Hierro (Spain) (*Source* NASA 2020)

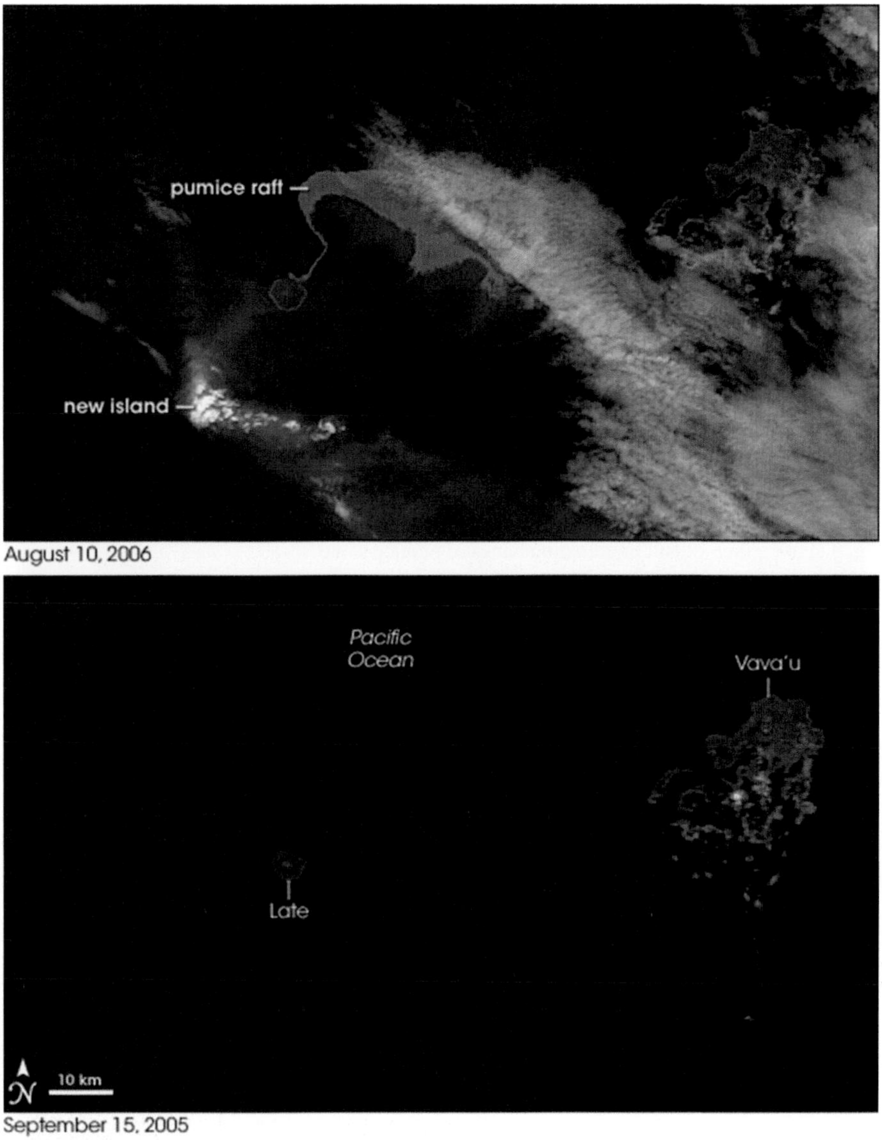

Fig. 7 In the above satellite images, in the South Pacific Ocean, the position of islands in 2005, and both the appearance of a new island and the trail of pumice stones in 2006, are observed (*Source* NASA 2006)

Fig. 8 Photograph illustrating floating banks of pumice stone

Fig. 9 The emerging islands near Vava'u (Fiji). Below, an extract is included from the blog-diary that was written by Fredrik Fransson in Maiken Yatch in 12 August 2006, where he narrates the appearance of a new island and the presence of pumices (*Source* Fredick Fransson 2020, http://yacht-maiken.blogspot.com)

Trans-Disciplinary Oceanographical Campaign on Ancient Mediterranean Civilisations

Travel books reproduce sailing expeditions in real marine spaces that may have changed over a long-time span. The option of launching campaigns by inspiring learners with lectures on the ocean trips courageously undertaken by writers or scientists in the past may be an exciting educational challenge. *The Log from the Sea of Cortez* is an outstanding literary non-fiction resource, which narrates the expedition and crossing to the Sea of Cortez (Gulf of California) in the Pacific Ocean that was undertaken on the *Western Flyer* by the novelist John Steinbeck and the marine biologist Ed Ricketts. This book contains a mine of information on the characteristics of a last-century authentic marine campaign. The sailing expedition around the Gulf of California, the collection of marine specimens, and the interviews conducted with the inhabitants of mainland coasts and islands, together with Steinbeck's direct observations and written reflections in his book, all constitute a reliable primary source of documentation and a fieldwork model for setting up an ocean awareness campaign to promote OL.

The expedition dates back to the 1940s when the writer describes the magnificent splendour of El Pulmo Reef's marine ecosystem on the east coast of Mexico's Baja California Peninsula: "The complexity of the life pattern on Pulmo Reef was even greater than at Cabo San Lucas. Clinging to the coral, growing on it, burrowing into it, was a teeming fauna. Every piece of the soft material broken off, skittered and pulsed with life, little crabs and worms and snails. One small piece of coral might conceal 30 or 40 species, and the colors on the reef were electric" (Steinbeck 1951). Since then up to the 90s, this coral reef fell into an absolute state of decay, because it had been heavily damaged and thoroughly overfished. In 1995, Cabo Pulmo was designated a national marine park, with local communities' action stepping up coral reef protection and conservation. These initiatives have led in recent days to the recovery of El Pulmo Reef's marine biodiversity to its former healthy condition. Many webpages may be found to provide information on coral biodiversity and Cabo Pulmo's historical development and recovery campaigns conducted along the Sea of Cortez (http://areefreborn3d.com/explore-the-reef/cabo-pulmo-history/ y http://ocean.si.edu/ecosystems/coral-reefs/).

In the present case, the marine campaign project is intended to be launched in close partnership with a Spanish oceanographical institute and local sponsors. The project is targeting higher education learners of Teachers' Training and Education faculties of Spanish universities. The oceanographical campaign is designed to be a sailing trip, by crossing the Mediterranean Sea, leaving from ancient Phoenician harbour cities in Spain to approaching Greek harbour cities in Turkish Armenia, or Asia Minor. This campaign field trip pursues three goals: (1) tracing archaeological evidence of the presence of these cultures in the ancient settlements; (2) the colonised cities' contribution to the cultural development of their respective civilisation; and (3) understanding how population movements correspond to the natural processes of mankind. The first part of the trip is to be made inversely to the Phoenicians' routes (from Spain to Turkey), who were the first ancient settlers in landing on the Iberian

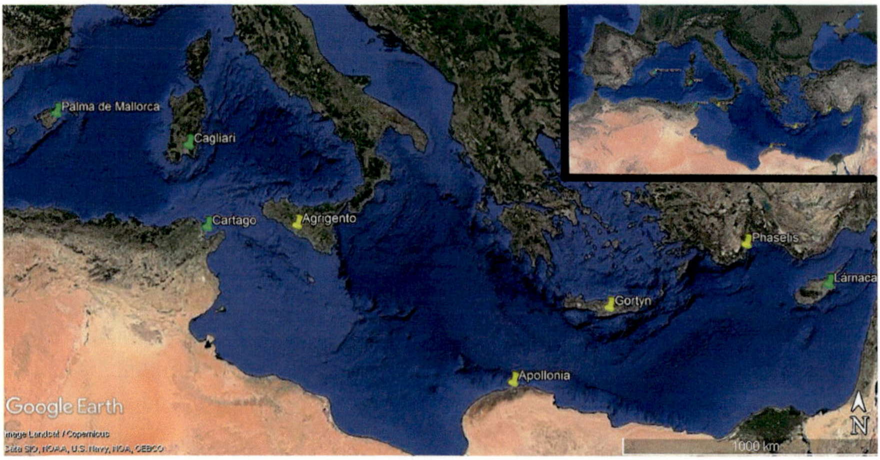

Fig. 10 Project for a Mediterranean campaign on ancient Phoenician (green) and Greek (yellow) civilisations (*Source* Palma de Mallorca-Cagliari-Carthague-Agrigento-Apollonia-Gortys-Phaselis and Larnaca. Google Earth [2020])

Peninsula coasts. Cadiz, Malaga, and Carthague were three flourishing harbour cities in the western Mediterranean that were founded by the Phoenicians, whose apogee as a great power in the ancient Mediterranean world would last till the expansion of the Roman civilisation. The first half of the sailing field trip covers the ancient Phoenician harbour cities of Palma de Mallorca (Spain), Cagliari (Sardinia, Italy), and Carthague (Tunisia). During the second half of the sailing trip, they approached ancient Greek harbour cities across the middle and eastern Mediterranean Sea: Agrigento (Sicily, Italy), Apollonia (Cyrenaica, Libya), Gortys (Crete, Greece), and Phaselis (Turkey). The last ancient harbour city to visit is Larnaca, a Phoenician settlement located in modern Cyprus. The eight harbour cities and their respective archaeological sites are expected to be visited in a 12-day sailing crossing (Fig. 3.10).

This oceanographical campaign aims to identify the ancient sites giving origin to the spreading of Phoenician and Greek civilisations across the Mediterranean Sea, from Cyprus to the Iberian Peninsula. The campaign is focused on raising learner awareness about the contents on the seas and ocean that are missing in their academic education and in the national core curriculum. Therefore, the project may encourage learners to discover the relationship of such marine contents with the ancient civilisations crossing the seas westward to colonise coastal territories on the Iberian Peninsula. Moreover, the project mainly fosters understanding on current global migration as a part of the universal dynamics in population movements since time immemorial. Learners grasping that migrations occur as a logical consequence of a myriad of determining factors may lead them to understand the driving factors forcing both Phoenician and Greek cultures to occupy new territories throughout the Mediterranean Sea.

Furthermore, this campaign across the Mediterranean Sea is aimed at reconsidering spatial thinking in an environment we are quite unfamiliar with. This revision is meant to examine marine geoabilities and operations, including egolocation, landmark location, and line of side location, network location, enclosure location, and coordination location, according to Gersmehl (1991). Marine wayfinding and navigation encourage unusual spatial capacities that are strange to terrestrial environments. Using portable devices, and introducing marine wayfinding and navigation systems, leads to decision-making on setting a course or maintaining a heading in a particular direction that differs greatly from planning a terrestrial route.

CIVILIZATIONS SPREADING PUZZLE CLUE

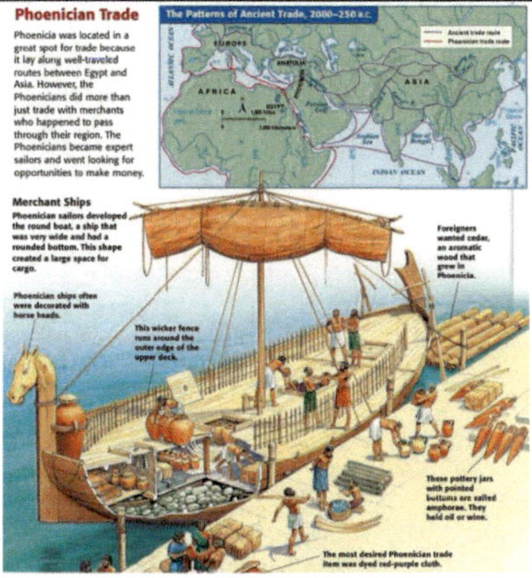

From the ninth century BC, the Phoenician and Greek civilisations were both in a position of increasing colonial dominance over the western Mediterranean territories. Different reasons may be argued in each case that motivated both cultures to spread westward. The Phoenicians were in principle ousted by the economic, political, and military pressure of the Neo-Assyrian Empire, which dominated the whole eastern region from the Persian Gulf to the Mediterranean Sea. The Greeks moved westward owing to their reduced and mountainous territory.

From the ninth century BC, the Phoenician and Greek civilisations were both in a position of increasing colonial dominance over the western Mediterranean territories. Different reasons may be argued in each case that motivated both cultures to spread westward. The Phoenicians were in principle ousted by the economic, political, and military pressure of the Neo-Assyrian Empire, which dominated the whole eastern region from the Persian Gulf to the Mediterranean Sea. The Greeks moved westward owing to their reduced and mountainous territory. According to Aubet Semmler (2001), Phoenician diaspora to the west has been traditionally attributed to both Assyrian political and military pressure on the cities of the coast (especially on Tyre), and to Assyrian demand for raw materials, which supposedly forced large masses of population in the east to flee to the west. Yet, classical authors pointed to overpopulation in Phoenicia just before the beginning of the colonisation period and to the subsequent demographic pressure on resources around the tenth century BC. This is an explanation that, jointly with a spirit of conquest and not commercial objectives, accounts for the Phoenicians' arrival in North Africa (Carthage) and the colonisation of the west. Actually, Greek colonisation seems to respond to the imbalance between population and resources, but the Phoenician diaspora has long been associated to a unique variable, that of trade. However, the Phoenician commercial and colonial expansion westward was driven by a combination of intertwined factors of an internal and/or external nature over a long-time span, until an external stimulus or one determining factor serves to trigger it off or to threaten the system as a whole. On the one hand, as far as internal factors are concerned, undertaking a commercial or colonial enterprise towards faraway lands becomes a necessity under special circumstances of shortages or political crisis. Or else, it is feasible in a situation of stability and prosperity, where an expansionist policy is motivated by the need to export excess production—by the circulation of manufactured articles—, and consequently to seek for sources of raw materials, agricultural land or trade routes. On the other, regarding external factors, Phoenician cities' prosperity relied largely on three axes: their role as intermediaries between the great powers of the east, their specialist production of luxury goods targeting foreign clients, and their concern with becoming the main supplier of precious metals to the Asian empires. Thus, the role played by Assyria was important but secondary, since the Assyrian empire could contribute to the commercial restraint of the Phoenician cities, particularly Tyre, or, on the contrary serve as a stimulus to a large naval and commercial initiative. Therefore, the ultimate causes of the expansion westwards must be sought fundamentally in the internal dynamics of Phoenician society in the east and the inexhaustible demands for raw materials—basically metals. During the ninth century BC, the Phoenician cities became the only suppliers of manufactured goods to the neighbouring states, like Assyria. In brief, the problem of overpopulation and agricultural deficit, a shortfall in food supplies—Tyre imported huge quantities of oil and cereals from abroad, explains, among other things, the Phoenician's concern with extending their territory in the tenth to eighth centuries BC.

Campaign lesson plan	
Learning content standards	(1) Acquire knowledge on two ancient cultures spreading throughout the Mediterranean Sea, namely Phoenician and Greek cultures (2) Identify archaeological sites as the historical evidence of formerly existing Phoenician and Greek civilisations (3) Acquire the geoabilities related to the marine environment (4) Understand migration movements across the Mediterranean Sea throughout the ages (5) Appreciate the heritage of sailing journeys as a valuable ocean legacy from the antiquity (6) Raise social awareness about the cultural contributions from migration movements
Discipline cognitive contents	Archaeological sites from Phoenician and Greek cultures, ocean-related geoabilities/skills, and migration movements in the marine environment

(continued)

(continued)

Methodological approach and learning strategy	Project-based Learning. The project involves the formulation of a challenging question, the subsequent pursuit of an enquiry, and coming up with a feasible solution. The project may arise from a variety of questions. When the learners show keen interest in a particular project, they proceed to pose the question themselves, even though this question may not either be properly formulated or be the optimal one for such a project. For instance: Why do current population movements occur? How can we orient ourselves in the sea? Why did the Phoenicians and the Greeks cross the Mediterranean Sea and arrive at the Iberian Peninsula? Then, upon reconsidering the core curriculum, the teacher brings up the most instructional and appropriate challenging question in relation to the curricular contents involved: Which is the historical legacy that the Phoenicians and the Greeks have left us in the Iberian Peninsula since antiquity?
Campaign project dynamics and procedure in three stages	**(A) Campaign preparative work**. This stage involves four working sessions in the classroom, prior to setting out on the Mediterranean journey, to fulfil the following purposes: (1) Formulating the challenging question, which takes place in the first whole-group classroom session. The solution is expected to be cooperatively worked out along the overall campaign process by enquiry learning, and jointly provided with at the end of the campaign (2) Elaborating secondary questions leading to solve the major question, which is carried out during a second classroom session. Small specialised work groups of 5–6 people are formed to cover a particular topic: the Phoenicians, the Greeks, their economic activities, the Mediterranean Sea, and the migrations (3) Searching for and gathering thematic and factual information and documentation of archaeological, cartographical, geographical, historical, literary, and technical nature, and preparing topic-oriented interview questions, all that serve to support their enquiry on their respective topics, is distributed within work groups in a third classroom session. Each team member is assigned a precise area of study under the topic, and individual research is conducted at home before embarking (4) In relation to navigation, exploring various literary texts, travel books, and journey diaries in the last fourth session, with a view for learners to become familiar with the contents and inner structure of ship's logbooks and fieldwork notebooks (including Steinbeck's, Cook's, and Fitzroy's), since these texts may be considered and used as campaign design models **(B) Campaign development**, during the sailing expedition across the Mediterranean Sea (1) Each specialised work group contributes to the project by sharing the pieces of information that are being collected and looked through during the days of navigation (2) The whole-group discussion will take place in plenary sessions devoted to specific topics. Learners may have the chance to collate and complement the data that was previously prepared in the classroom with their own in situ observations and accounts, and interview results (3) Specific training in issues arising from the sailing journeys, regarding marine wayfinding and navigation, and marine geoabilities and geoskills, is to be arranged in small working groups

(continued)

(continued)

	(4) Learners' findings at this stage of the project are to take shape in the form of an individual fieldwork notebook (5) Project developments become tangible in the form of an ongoing cooperative blog that will be enriched by everyday posting of experiences learnt and knowledge gained throughout the sailing trip. The vivid impressions and experiences on the sailing journeys that were individually reported are gathered in a collective on-board diary. Site accounts are narrated, and evidences are collected for understanding and fitting back together the various pieces of the civilisations spreading puzzle **(C) Campaign closure**. The three major campaign issues are solved out on the last sailing journey (1) A common viable solution is reached by consensus in response to the project's challenging question on the basis of the significant thematic, documentary, and expert findings presented and reflected upon (2) The issues concerning weather and navigation in the Mediterranean Sea are commented upon and thoroughly examined. In line with the evidence found on every topic, blog contents have been previously posted on a daily basis by each working group (3) These cooperative blog posts are revised and enhanced accordingly
Teacher's role	(1) Coordinator of project activities along the three campaign stages (2) Facilitator of the learning process in a way that he/she brings enthusiasm and dynamism onto the work group task development, and supervisor of ongoing collaborative and cooperative learning assessment work and final product during the campaign
Teaching resources and learner materials	(1) Nautical charts, apps, and GIS on portable devices for information on weather forecast and cartography, and navigation tools (compass, barometer, anemometer, etc.) for decision-making on course and heading (2) Topic-oriented question sheet for expert interviews. In order to guide their research on their respective subject matters, team members may conduct expert interviews with tourist guides and people in charge of the archaeological sites on their visits and other scientists of the oceanographical institute or other marine institutions. The need for setting up interviews with experts on each particular topic lies in the fact that gaining the evidence they require to back their findings may be collected from interview answers (3) Learner's fieldwork notebook. It serves a twofold purpose as: (a) a ship's logbook to register nautical information and routines, including date, hour, temperature, atmospheric pressure, weather, wind direction, sailing course, and heading; (b) a notebook of archaeological records on site visiting, where data is written down that enable learners to rebuild the ancient settlements and to trace the commercial or colonial enterprise. Thus, learners should pay special attention on site to any potential belongings, homeware, manufactured articles, luxury goods, and jewellery made of precious metals, containers for raw materials, as well as shipping goods and stock-in-trade, equipment of maritime transport, and navigation tools (4) A cooperative blog. It is created by the learners for their sharing out and elaborating on an open-access project document as a collaborative log of the sailing journeys. Figures, images, and photographs are selected and entered onto the blog to illustrate written information

(continued)

(continued)

Evaluation assessment method Learning outcomes	Focus on learners' ability to formulate questions and to draw conclusions from findings and their ability to discriminate information Based on the quality of the individual's note taking and blog posting. The teacher undertakes the method of formative assessment by offering feedback and constructive criticism on each individual's logbook compositions. Then, learners reconsider their essays, and revise and make amendments to their writings. By intergroup cooperative assessment, learners are also given the opportunity to offer feedback and constructive criticism on other groups' working and to develop critical thinking. Each group may collect the commentaries made by journey mates on both navigational records and thematic research and enhance the wording of their fragment before entering it onto the blog. The final public product is a sailing log blog. It will include a section of References from the mentioned sources, and proper citations. Note taking, thematic and documentary research, and interview results are to be embedded in the final product, in the form of collaborative posting on the cooperative blog Aimed at accurately solving out the challenging question based on evidence and findings

Conclusion

An oceanographical campaign sustains a set of OL-related geographical contents, and encourages knowledge acquisition of ocean concepts, abstract terms, and sailing expeditions' experiences related to oceanographic topography. The sailing trip project fosters the development of marine geoabilities and navigational routines and allows for keeping a record of navigation events and routines on the ship's log and taking notes on the fieldwork notebook. The oceanographical campaign proves to establish a relation between the sea and the earth, the ocean and the climate, and to show the ocean-defining features (e.g. seawater temperature, salinity, density), on which the ocean ecological balance depends to ensure the global habitability of the earth. In the oceanographical campaign, learners handle logs and historical accounts of sailing journeys by distinguished scientists. By using these documentary resources, together with living a sea experience themselves, learners are able to understand abstract concepts (e.g. storm formation and surge). Learners may grasp the usefulness of navigation tools to estimate and keep a record of ocean and climate measurement data. The campaign is the fulfilment of a trans-disciplinary higher education scholarship on the basis of a fieldwork project proposal that enables the participant to acquire knowledge on Mediterranean Sea civilisations. In conclusion, Design-based Implementation Research proves to be a potential learning environment for the inclusion of OL principles and concepts into the higher education curricular framework.

Acknowledgements Special thanks to Diana Payne for her review of this chapter, and suggestion of pertinent remarks for chapter improvement. Special thanks to Fredrik Fransson and the Maiken's crew for photograph and logblog courtesy.

References

Acosta J, Rivera J, Muñoz A, León C, Elvira E (2014) Mapa topobatimétrico en relieve del margen continental del sureste español. Instituto Oceanográfico Español, Madrid

Aubet Semmler ME (2001) The phoenicians and the West: politics, colonies and trade. Cambridge University Press, Cambridge

Barkley EF, Cross KP, Major CH (2007) Collaborative learning: a handbook for college faculty. Wiley, San Francisco

Barrows HS (1986) A taxonomy of problem-based learning methods. Med Educ 20:481–486

Barrows HS (1996) Problem-based learning in medicine and beyond: a brief overview. In: Wilkerson L, Gijselaers WH (eds) Bringing problem-based learning to higher education: theory and practice. Jossey-Bass Pub., San Francisco, pp 3–12

Beane JA (1995) Integration and the disciplines of knowledge. Phi, Delta, Kappan 76(8):616–622

Ben-Ziv Assaraf O, Orion N (2009) A design based research of an earth systems based environmental curriculum. Eurasia J Math, Sci Technol Educ 5(1):47–62

Bonnett A (2014) Off the map: Lost spaces, invisible cities, forgotten Islands, feral places and what they tell us about the world. Aurum Press, London

Bradbeer J (1996) Problem-based learning and fieldwork: a better method of preparation. J Geogr High Educ 20(1):11–18

Bruner J (1960) The process of education. Harvard University Press, Cambridge, MA. 97 + xxvi pages

Bruner JS (1966) Toward a theory of instruction. Belkapp Press, Cambridge, MA. 176 + x pages

Cava F, Schoedinger S, Strang C, Tuddenham P (2005) Science content and standards for ocean literacy: a report on ocean literacy. Available at: http://www.coexploration.org/oceanliteracy/

Choppin J, Debarger A, Beauvineau Y, Moorthy S (2013) Designing for productive adaptations of curriculum interventions. In: Penuel W, Cheng BH, Fishman BJ, Allen AR (eds) Design-based implementation research: theories, methods, and exemplars. Yearb Nat Soc Study Educ 112(2):298–319. Teachers College, New York

Collins A, Joseph D, Bielaczyc K (2004) Design research: Theoretical and methodological issues. J. Learn. Sci. 13:15–42

Council of Europe (2000) *European landscape convention*, of 20 October 2000

DeBarger AH, Choppin J, Beauvineau Y, Moorthy S (2013) Designing for productive adaptations of curriculum interventions. Nat Soc Study Educ Yearb 112(2):443–463

De Miguel González R (2018) El currículo de geografía en España: evolución y tendencias actuales. En: García de la Vega A (ed) *Reflexiones sobre edu-cación geográfica. Revisión disciplinar e innovación didáctica.* Madrid: Ediciones de la Universidad Autónoma de Madrid y Ediciones Lagim de la Universidad de Brasilia, pp 191–215

Department for Education (2013) *The national curriculum in England*, framework document *Secundary curriculum.* London

Direção Geral da Educação (2012) Curriculum Geral do Ensino Básico, DL n.º 139/2012, de 5 de julho, Lisboa

Donert K, Fauville G, Gotensparre S, Mätkitalo Å, Van Medegael L, Zwartjes L (2015) Review of marine formal education. EU Sea Change Project

Döös K, Nilsson J, Nycander J, Brodeau L, Ballarotta M (2012) The world ocean thermohaline circulation. J Phys Ocenogr 42:1445–1460

Fishman B, Penuel WR, Allen A, Cheng BH, Sabelli N (2013) Design-based implementation research: an emerging model for transforming the relationship of researchand practice. In: Fishman BJ, Penuel WR, Allen A, Cheng BH (eds) Design-based implementation research: theories, methods, and exemplars. Nat Soc Study of Educ Yearb 112(2):136–156. Teachers College Record, New York

Fletcher S, Potts J (2007) Ocean citizenship: an emergent geographical concept. Coast Manag 35(4):511–524

Frazão Santos C, Teixeira ZG, Janeiro J, Gonçalves RS, Bjorkland R, Orbach M (2012) The European marine strategy: contribution and challenges from a Portuguese perspective. Marine Policy 36:963–996

French V, Chu N-C, Santoro F, Sousa Pinto I, Borges D, McDonough N (2015) Review of ocean literacy in European maritime policy. EU Sea Change Project

García de la Vega A (2012) El aprendizaje basado en problemas en los itinerarios didácticos vinculados al patrimonio. Educación y futuro: revista de investigación aplicada y experiencias educativas 27:155–175

García de la Vega A (2018) El escenario geográfico en el estudio del paisaje. Propuestas metodológicas en la enseñanza de la geografía. In: García de la Vega A (ed) Reflexiones sobre educación geográfica. Revisión disciplinar e innovación didáctica. Ediciones UAM y Edições Lagim, Madrid, pp 129–152

Gersmehl PJ (1991) The language of the maps. Indiana University of Pennsylvania and National Council for Geographic Education, Indiana

Gordon AL (1986) Interocean exchange of thermocline water. J Geophys Res 91(4):5037–5046

Guest H, Lotze HK, Wallace D (2015) Youth and the sea: Ocean literacy in Nova Scotia, Canada. Marine Policy 58:98–107

Hermmer I (ed) (2012) Educational standards in geography for the intermediate school certificate with sample assignments. German Geographical Society, Bonn

Hibszer A, Szkurłat E (eds) (2018) Geography teacher in the face of the challenges of the reformed school. Works of the geographical education commission of the polish geographical society, no. 8. PTG Geographic Education Commission: Sosnowiec

Hmelo-Silver C (2004) Problem-based learning: what and how do students learn? Educ Psycol Rev 16(3):235–266

Hmelo-Silver C, Barrows HS (2008) Facilitating collaborative knowledge building. Cognit Instr 26:48–94

Holyoak KJ (1995) Problem solving. In: Smith EE, Osherson DN (eds) An invitation on cognitive science, 2nd edn, vol. 3. Cambridge, MA, MIT Press, pp 267–296

Holyoak KJ, Thagard P (1997) The analogical mind. Am Psycolog 52(1):35–44

McKenney S, Reeves TC (2018) Conducting educational design research. Routledge, London

McKinley E, Fletcher S (2010) Individual responsibility for the oceans? An evaluation of marine citizenship by UK marine practitioners. Ocean Coast Manag 53(7):379–384

McKinley E, Fletcher S (2012) Improving marine environmental health through marine citizenship: a call for debate. Marine Policy 36(3):839–843

Meyer D, Isakower A, Mott B (2015) An ocean of opportunities. Available at: http://theoceanproject.org/wp-content/uploads/2017/08/OceanOfOpportunities-ExecutiveSummary2015.pdf

Ministério da Educação (2017a) Base Nacional Común Curricular. Ensino Fundamental, Brasilia

Ministério da Educação (2017b) Base Nacional Común Curricular. Ensino Medio, Brasilia

Ministerio de Educación Cultura y Deporte (2014a) Real Decreto 126/2014, de 28 de febrero, por el que se establece el currículo básico de la Educación Primaria. Madrid

Ministerio de Educación Cultura y Deporte (2014b) Real Decreto 1105/2014, de 26 de diciembre, por el que se establece el currículo básico de la Educación Secundaria Obligatoria y del Bachillerato. Madrid

Ministerio de Educación Nacional (2002) Lineamientos curriculares para el área de ciencias sociales. Bogotá

Ministerio de Educación Nacional (2004) Estándares Básicos de Competencias en Ciencias Naturales y en Ciencias Sociales. Serie Guías Nº 7, Bogotá

Ministerio de Transición Ecológica (2020) Plan Nacional para la Adaptación al Cambio Climático. Madrid

Ministry of Education (2018a, July 27) [초등학교 교육과정] (Primary school curriculum). Seoul. Available at: http://ncic.go.kr/mobile.kri.org4.inventoryList.do

Ministry of Education (2018b, July 27). [고등학교 교육과정] (High school curriculum). Seoul. Available at: http://ncic.go.kr/mobile.kri.org4.inventoryList.do

Ministry of Education (2018c, July 27) [중학교 교육과정] (Middle school curriculum). Seoul. Available at: http://ncic.go.kr/mobile.kri.org4.inventoryList.do
National Marine Educators Association (NMEA) (2010) Ocean literacy scope and sequence for grades K-12. The Ocean Literacy Campaign, Special Report n. 3th, Ocean Spring, MS. Available at: www.marine-ed.org
National Oceanic and Atmospheric Administration (NOAA) (2020) Ocean literacy: the essential principles and fundamental concepts of ocean sciences for learners of all ages
Ocean Project (2009) America, the ocean, and climate change: key findings. Available at: http://theoceanproject.org/
Oppenheimer M, Glavovic BC, Hinkel J, van de Wal R, Magnan AK, Abd-Elgawad A, Cai R, Cifuentes-Jara M, DeConto RM, Ghosh T, Hay J, Isla F, Marzeion B, Meyssignac B, Sebesvari Z (2019) Sea level rise and implications for low-lying Islands, coasts and communities. In: IPCC special report on the ocean and cryosphere in a changing climate [Pörtner H-O, Roberts DC, Masson-Delmotte V, Zhai P, Tignor M, Poloczanska E, Mintenbeck K, Alegría A, Nicolai M, Okem A, Petzold J, Rama B, Weyer NM (eds)]. In press
Orion N (2002) An Earth systems curriculum development model. In Global science literacy (pp 159–168). Springer Netherlands
Orion N (2007) A holistic approach for science education for all. Eurasia J Math Sci Technol Educ 3(2):111–118
Payne DL, Zimmerman TD (2010) Beyond terrafirma: bringing ocean and aquatic sciences to environmental and science teacher education. In Bodzin AM, Klein ES, Weaver S (eds) The inclusion of environmental education in science teacher education. Springer, New York, NY, pp 81–94
Piaget J (1964) The early growth of logic in the child. Routledge & Kegan Paul, London
Pigafetta A (1874) The first voyage round the world by Magellan. Translated from the Accounts of Pigafetta and Other Contemporary Writers. Antonio Pigafetta. Ed. Henry Edward John Stanley. Cambridge University Press, New York
Puttick S (2013) Looking at and looking along: a conceptual framework for teaching different perspectives in geography. Int Res Geogr Environ Educ 22(4):353–366
Reid A, Breidahl H (2019) Going with/against the flow: the challenges of an authentic marine science education. In: Fauville G, Payne DL, Marrero ME, Andersson AL, Crouch F (eds) Exemplary practices in marine science education a resource for practitioners and researchers. Springer, Cham
Rieckmann M (2017) Education for sustainable development goals: learning objectives. UNESCO, Paris
Santoro F, Selvaggia S, Scowcroft G, Fauville G, Tuddenham P (2018) Ocean literacy for all: a toolkit. UNESCO, Paris
Savery JR (2006) Overview of problem-based learning: definitions and distinctions. Interdiscip J Problem-based Learn 1(1):9–21
Savery JR, Duffy TM (2001) Problem based learning: an instructional model and its constructivist framework. CRLT Technical Report No. 16-01. Bloomington, IN, W.W. Wright Education Building, ED 2201
Sinton DS, Kolvoord B, Gersmehl P, Bednarz S, Uttal D (2013) The people's guide to spatial thinking. National Council for Geographic Education, Washington, DC
Solem M (2001) Using geographic information systems and the Internetto support problem-based learning. Planet 4(1):22–24
Spronken-Smith R (2005) Implementing a problem-based learning approach to teaching research methods in geography. J Geogr High Educ 29(2):203–221
Steel BS, Smith C, Opsommer L, Curiel S, Steel RW (2005) Public ocean literacy in the United States. Ocean Coast Manag 48:97–114
Steinbeck J (1951) The log from the Sea of Cortez, 1st edn. Penguin, New York (1941)
Stepath CM (2006) Coral reefs as sites for experiential environmental education: Learning with Australian students—a foundational study. Unpublished PhD, James Cook University, Townsville, Queensland (Australia)

Taylor AR, Jones MG (2009) Proportional reasoning ability and concepts of scale: surface area to volume relationships in science. Int J Sci Educ 31(9):1231–1247
Vaughan RG, Abrams MJ, Hook SJ, Pieri DC (2007) Satellite observations of new volcanic island in Tonga. Eos 88(4):23
Visser WA (ed) (1980) Geological nomenclature. Martinus Nijhoff, The Hague
von Humboldt A (1849) Cosmos: a sketch of the physical description of the universe, 7th edn, vol. II (trans: Sabine E). Longman, London
Vygotsky L (1986) Language and thought, 1st edn. Cambridge, MA, MIT Press (1934)
Wilson BG (ed) (1996) Constructivist learning environments: case studies in instructional design. Educational Technology Publications, Englewood Cliffs, NJ
Zheng C, Xiao Z, Zhou W, Chen X, Chen X (2018) Ocean current characteristics. 21st century maritime silk road: a peaceful way forward. Springer Oceanography. Singapore, Springer

Innovative Education Strategies to Advance Ocean Literacy

Veronica McCauley, Kevin Davison, Patricia McHugh, Christine Domegan, and Anthony Grehan

Abstract The ocean covers more than 70% of the planet and yet knowledge of what occurs in the ocean among the public is poorly developed. There is an urgent need to raise levels of ocean literacy so that the public can mandate politicians and policymakers to manage the ocean sustainably. Education of the next generation is key, requiring a 'bottom up' grassroots approach. Consultation and collaboration with a broad range of stakeholders locally, nationally and internationally is essential. While interest has been recognised as an important condition for learning, educators continue to be challenged by academically unmotivated students. Education competes with instantaneous technology in grabbing attention and maintaining interest, with the vast majority of students being coined as digital residents. Therefore, new media and methodologies need to be considered in providing an engaging current education platform to increase ocean literacy. This chapter recounts the design of two educational initiatives that address the advancement of ocean literacy: educational e-book design and gaming pedagogy. A participation and social inclusion model that champions practice *with* and not *on* the stakeholder group was considered during the design process, to create resources with an improved opportunity to promote change.

Keywords Ocean literacy · Stakeholder participation · e-book · Gaming pedagogy · Behavioural change · Science education · Informal learning · Social inclusion

Introduction

The ocean plays a key regulating role in ocean climate, is a potential source of economic blue growth and is increasingly impacted by human activities (e.g. increasing CO_2 levels; eutrophication of ecosystems, alteration of marine habitats due to population increase without due consideration of climate impact) (Pörtner et al. 2014). Understanding the ocean is an essential step in protecting the world in

V. McCauley (✉) · K. Davison · P. McHugh · C. Domegan · A. Grehan
National University of Ireland, Galway, Ireland
e-mail: veronica.mccauley@nuigalway.ie

K. C. Koutsopoulos and J. H. Stel (eds), *Ocean Literacy: Understanding the Ocean*, Key Challenges in Geography, https://doi.org/10.1007/978-3-030-70155-0_7

which we live (Cava et al. 2005). The ocean, the largest habitat on our planet, covers more than 70% of the world and yet knowledge of what occurs in the ocean among the public is poorly developed (Tran et al. 2010; Hynes et al. 2014; Guest et al. 2015). Given that the ocean "binds together much of the Earth science systems content, one cannot be science or environmentally literate without being literate in ocean and aquatic concepts" (Markos et al. 2017, 231). There is now an urgent need to raise levels of ocean literacy so that the public can mandate politicians and policymakers to manage the ocean sustainably (Fauville et al. 2015).

Ocean literacy is defined as an awareness of the ocean's influence on us and our influence on the ocean (Cava et al. 2005). The movement towards a consensus position on ocean sciences education began in 2002 (Santoro et al. 2017). This research led to the development of the seven principles of ocean literacy (Ocean Literacy Network 2015). These principles continue to guide ocean literacy research, education and outreach activities to influence human behaviour and the choices citizens make in taking direct and sustainable action towards healthy seas and ocean, healthy communities and ultimately, a healthy planet. Fauville et al. (2018) contend that ocean literacy leads to more informed and responsible decisions regarding the ocean and its resources. Ocean literacy goes beyond knowledge and just knowing about the state of the ocean. It fosters a deeper understanding of the individual and collective behaviours and responsibilities needed to take care of the ocean (Fielding et al. 2019).

Engagement of young citizens at a crucial stage in the development of their value system is "likely to lead to better informed stewards of the marine environment and development of a lasting 'Marine Citizenship.' A shared knowledge and identity can instil behavioural change at the level of the individual, as well as a sense of care and responsibility within the general public, and, in turn, can empower them to act" (EMB 2017, 3). The importance of providing a marine and aquatic education has been promoted since the 1970s (Charlier and Charlier 1971; McFadden 1973; Picker 1980; Fortner and Lyon 1985), yet ocean and aquatic concepts are infrequently taught and rarely appear in school curricular materials (Schoedinger et al. 2010; Markos et al. 2017) and are not always included in STEM engagement initiatives. There are exceptions of course, "pockets of excellence, where passionate educators and innovative programs managed to bring marine science content and experiences to some students" (Schoedinger et al. 2010, 3) yet, unfortunately, these are few. Further, educational research has focused less attention to marine issues (Markos et al. 2017), in comparison to other science disciplines. Thus, education of the next generation is key, coupled with investment in research, in advancing a movement towards improved ocean literacy.

Considering this dearth in formal and informal marine education, and the potential role that a more confident ocean literate public could play in contributing to the conversation on global warming, and climate change (Britton et al. 2020), this chapter reports on two ocean education strategies. Each strategy is framed within a social inclusion framework, in terms of building citizen capacity in ocean education and empowerment. This chapter begins by briefly recounting the authors' experience over ten years, designing innovative approaches to science communication. A short discussion follows on the value and relevance of considering a social inclusion model in intervention design, followed by a particular focus on two recent educational initiatives—informal and formal—that address the advancement of ocean literacy: educational e-book design and gaming pedagogy.

Innovative Approaches to Science Outreach and Education

Over the past decade, the authors have been significantly active in terms of the design of innovative and novel strategies in science education (McCauley et al. 2015, 2018; McHugh and McCauley 2016, 2017), in particular in the realm of science outreach (Davison et al. 2008; Domegan et al. 2010, 2019; Martins Gomes and McCauley 2012, 2013a, b, 2016; Garcia-Soto et al. 2017; Fauville et al. 2018; McCauley et al. 2019; McHugh et al. 2020). Four of the authors led the first All-Island Conference in science outreach in Ireland (2007), providing an opportunity for active collaboration, coordination and communication among key stakeholders. Several other conferences followed, igniting greater dialogue between outreach providers, practitioners, policy-makers, teachers, teacher educators and scientists regarding science communication, best practices and evaluation strategies (McCauley and Davison 2015). One of the key outcomes from the conference research (Davison et al. 2008) revealed the need for a shift away from simply providing information about science and science activities to the public to instead advance an approach that influences social behaviour towards science *engagement* including multiple strategic stakeholders.

Several research projects emerged that champion these ideas. A public engagement/informal learning example is the Guerrilla Science initiative. This project drew together scientists and artists to create public art to ensure science was put in the path of the general public and not hidden away in university laboratories. This fostered new hybrid pedagogical partnerships in science education and outreach. Furthermore, a formal learning example is the creation of a curricular 'hook book.' Lesson hooks are instructional techniques that stimulate student attention, interest and engagement in the lesson that follows (McHugh and McCauley 2016). Two of the authors collaborated with student teachers and scientists to design lesson hooks for teachers to provide tools to make science teaching innovative and engaging. This work resulted in the creation of several science hook e-books as resources for teachers, for example, a Physics Resource Hook Book can be found here: https://books.apple.com/us/book/physics-resource-hooks/id549475578 and consists of a collection of eleven lesson hooks.

These approaches drew in teachers, students and the general public into a wider conversation about understanding science. The core lesson here is to work with the community and engage multiple stakeholders to fully appreciate the value added and challenges identified from all perspectives.

A resource design strategy that has the capacity to positively influence social behaviour towards improved ocean literacy requires stakeholder engagement (McHugh et al. 2018). The *participation agenda and social inclusion* is an emerging preoccupation in the science communication field (Massarani and Merzagora 2014). Socially inclusive change-making initiatives encompass a particular set of traits, outlined below:

1. Diverse audience participation is essential in all stages of the design process.
2. Recognition that social inclusion is not reducible to an issue of access, and that unintentional exclusion mechanisms may be present.
3. Consideration must be given to local social changes (Massarani and Merzagora 2014).

These traits were considered as design principles within both interventions that follow, and as such, are discussed in line with each initiative. The first step is to consider all relevant stakeholders relative to the intended output (e.g. curricular marine iBook, ocean literacy game). To be inclusive, the diversity of the audience should be considered in all phases of the initiative from design to governance to implementation. A core component of this step is the need to do more than just reach the community, but ensure their involvement and participation throughout the process. For example, participation is about speaking and listening to people on their own terms. Participation goes significantly beyond asking people for their opinions or what might be called "participation by consultation" (McHugh et al. 2018, 4). It gives a voice about the barriers to change and ownership and responsibility for solutions to influence their welfare. In short, it's an interactive process; it's *with* and not *on* the stakeholder group (McCauley et al. 2019). The success of the activities relies on the capacity to identify and interpret specific local social changes and consider them in the design strategy.

Sea Change Case Study

The case studies reported here emerged from an intra-European research project, Sea Change. Sea Change was a Horizon 2020 funded project with seventeen partners across nine European Countries, designed to bring a fundamental transformation, a 'Sea Change' in the way European citizens experience their relationship with the sea. The overarching aim of the campaign was "to empower ocean literate citizens to take direct and sustainable action towards healthy seas and ocean, healthy communities and ultimately a healthy planet" (Sea Change Consortium 2015, 3). The production of two tangible artefacts arising from the research will be described in turn below: gaming pedagogy and e-book design.

Gaming Design Initiative

Education today has to compete with contemporary and instantaneous technology in capturing student attention and maintaining their interest (Clifton and Mann 2011).

Contemporary students, often referred to as "digital residents" (White and Le Cornu 2011; Connaway et al. 2011) or millennials (Steffes and Duverger 2012), command a wide range of digital resources to manage their social lives (Kim 2014), but often suffer from lower attention spans as they are habituated with a wide range of activities (Prensky 2001). According to Clifton and Mann (2011), it is imperative that both formal and informal education find new ways of engaging students. Considering this, educational technology is presented as a means to not only improve learning (ASTI 2013; Mumtaz 2000), but also to stimulate student motivation (Heemskerk et al. 2012) and interest (Bingimlas 2009) in learning. CoderDojo is a global movement of free, volunteer-led programming clubs that embody the concept of using technology to motivate and interest students in learning. These clubs encourage children age 7–17 to learn how to code, develop websites, apps, programs, games, and to explore technology in an informal and creative environment. The authors identified this learning scenario as a seductive setting to engage students in enhancing their ocean literacy.

Initiative Design

A Future Ocean competition was organised with youth groups, inviting Dojos in two European countries (Sweden and Ireland), to use their gaming design talents in producing ocean literacy content. Data from Ireland only is reported here.

In line with our social inclusion participation strategy, the research involved diverse stakeholders relevant to the production of a student-led, ocean literacy game. Several planning meetings were organised by the coordinating design team (a marine scientist, science teacher educator, local chair of CoderDojo, game mentor and parent from the CoderDojo group). The student voice, a key stakeholder, was incorporated throughout the design process through regular advisory meetings with the game mentor, who acted as a link between the design team and student game designers. Select members from the teams across both countries received training to develop their understanding of the conceptual foundation of participation and social inclusion protocols (McHugh et al. 2015). The project was scheduled over a five-month period during the school academic year. The 2017 timeline is illustrated in Fig. 1.

A game design challenge was agreed upon by all stakeholders and a launch (see Fig. 2) was organised where student designers in unison with their mentors were tasked to design and make a game that would raise its players' awareness of marine issues. The launch was attended by approximately 70 people (two-thirds children, one-third adults) and guest talks were given by marine scientists on relevant topics

Fig. 1 Future ocean competition timeline, 2017

Fig. 2 Future ocean competition flyer

(e.g. corals, plastics in the ocean) and by game designers on tips for game conception and development. The CoderDojo community met every Saturday throughout the project (February–May) to support and inspire game development. The design team held three meetings over this period to provide updates on student game design progress, award ceremony, competition judging process and prize allocation.

Initiative Output

Approximately 100 people engaged with the Future Ocean CoderDojo competition (60 game designers across 37 teams and 37 mentors/parents), submitting a wide range of online gaming content on ocean literacy. Final submissions included 11 team members across fivesenior entrant (age 13–18) teams and 49 members across 32 junior entrant (age 7–12) teams. The final awards ceremony was attended by 60 people (approximately two-third children, one-third adults). Mentor/Guardian and Child/Game Designer surveys were distributed at the ceremony. Nine mentors and twelve Game Designers completed short surveys on the process.

Results revealed that one-third (33%) of mentors felt that the Future Ocean game design process helped them broaden their understanding of the importance of the sea and its impact on us 'a little,' and two-thirds (67%) felt that its impact was 'a lot.' When asked to expand on this, the following comments were returned (Fig. 3).

This data suggests that awareness was raised for adults who had not previously considered the importance of the ocean. This corroborates the National Marine Educators Association's first principle of ocean science—The Earth has one big ocean with many features (NMEA 2020). This new appreciation was shared by parent and child given the mentoring relationship that is a part of the coding design process. In line with The Stages of Change Model (Boston University School of Public Health 2016), developed by Prochaska in the late 1970s, participants began at the precontemplation stage (where they did not intend to take action in the immediate future, and where they are unaware that their behaviour is problematic or may produce negative consequences). Their feedback clearly indicates a move to the contemplation stage (where people are intending to start the healthy behaviour, they recognise that their behaviour may be problematic, and a more thoughtful and practical consideration of the consequences considered). Of the six stages of change, an encouraging start is certainly evident.

Results revealed (see Fig. 4) that prior to the game design campaign, 8% of game designers knew nothing about the ocean, 67% 'a little' and 25% 'a lot.' When asked how much they learned about the ocean when creating the game, 33% said 'a little' and 67% 'a lot,' revealing that the game design process in itself had a significant impact on improving ocean literacy. While this is admittedly a small sample that cannot be generalised, it offers an example of an approach through informal education that targets students where there is significant potential to catalyse interest towards behavioural change.

The young game designers were also asked to state what they felt the players of their games would learn. Their main response (50%) was that they hoped the players would realise the detrimental impact of polluting the ocean (Fig. 5).

" I never heard of the term 'micro-plastics' before. Great focus for the kids to work in a team"

"... knew we need to respect the ocean but not how bad the situation was"

"I found it very enlightening. It highlighted the importance of the sea in relation to medical research"

"Because of this project I went online and investigated the impact we are having on the sea... we live in a coastal area and [should] know how important the ocean is economically, socially and for our leisure!"

"This Sea Change project was an excellent concept especially important to emphasise how pollution of the seas impact our sea change. [It] Increased my awareness of how the small plastic particles in exfoliating shower gels breakdown and end up getting back to our food chain in shell fish etc."

"I have learned about the importance of the sea and its relevance and how important it is to take care of it. I have enjoyed it. Great learning curve for child and adult! Great engagement of a child from relevant environmental issue to be developed on a gaming platform"

Fig. 3 The impact of the mentoring the design process on mentor's ocean literacy

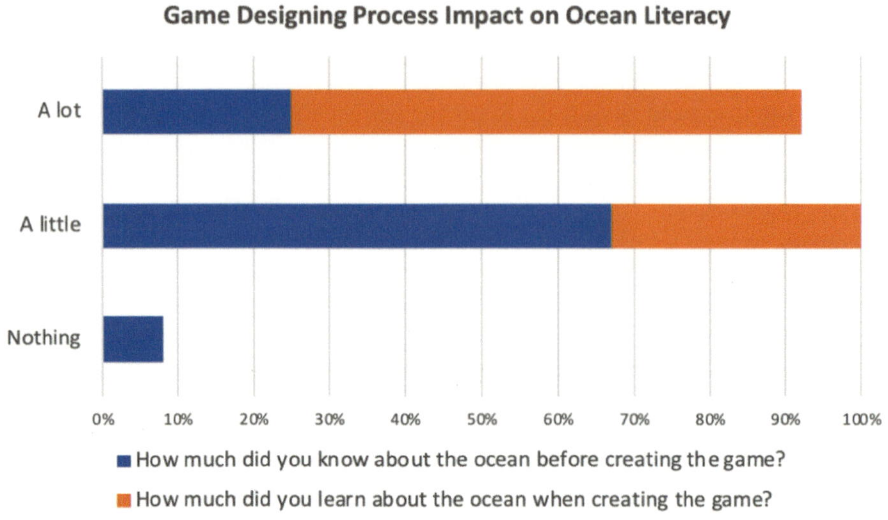

Fig. 4 The impact of the game designing process on game designer's ocean literacy

> Game Designer 2: "Not to litter. It will kill fish" [Junior Entrant (Age 7-12)]
> Game Designer 6: "I hope they will learn not to pollute and will see what happens to our animals if we do so." [Junior Entrant (Age 7-12)]
> Game Designer 8: "I hope they will learn not to carelessly dump rubbish in the sea as this is very dangerous to life there" [Senior Entrant (Age 13-18)]

Fig. 5 Game designers' reflection on what the players of their games would learn

> Game Designer 1: "I hope players of my game will learn to recycle as much as they can" [Junior Entrant (Age 7-12)]
> Game Designer 9: "I hope that people will learn to respect the ocean and the earth itself" [Senior Entrant (Age 13-18)]

Fig. 6 Game designers' reflection on what the players of their games would learn

The student responses here suggest an awareness of the need for both increased education and behavioural change. In addition to an anti-litter campaign of the ocean, they also intended that their games would educate the players on the impact of this pollution, for example, the wildlife, as indicated by Game Designers 2 and 6 above, and activities to promote the movement of healthy us, healthy ocean were also highlighted (Fig. 6).

They focus here on a hope for an ongoing commitment and a shift towards respecting the environment again suggests intentional change was built into the game

Game Designer 7: "[players of my game should learn the following:] The Mariana Trench is the most unexplored place on Earth. The sea shouldn't be abused. What and what isn't harmful to the sea" [Junior Entrant (Age 7-12)]

Game Designer 12: "That people will learn how important the sea is to us, from food, to having fun" [Junior Entrant (Age 7-12)]

Fig. 7 Game designers' reflection on what the players of their games would learn

design. Further, students outlined some benefits that they felt are drawn from marine contexts, as described below (Fig. 7).

These reflections reveal the students learning from the process, and their appreciation of the need for this information to be passed along in their games. Both mentors and students reported (in surveys distributed at the awards ceremony) an increase in their own ocean literacy as a result of participating in the gaming design process, which is a very favourable output.

As part of the dissemination process, the games are held on the CoderDojo website (https://scratch.mit.edu/users/SeaChange/), on the Sea Change website (www.seachangeproject.eu/ouroceanourhealth/competitions/241-seachange/1547-future-ocean) and also linked through the e-book (detailed below). The project was promoted via partner websites, project websites and social media channels. While the games produced were modest and not designed for marketed for mass consumption, broader dissemination may promote further interaction with the games and thus making further inroads towards increasing the public's ocean literacy.

Reflection on Social Inclusion Process and Initiative Outcome

In terms of diverse audience participation, the work with CoderDojo in Galway involved engaging with parents/mentors as well as young people through game design to open up new conversations about the marine environment. A science teacher educator with expertise in mobile learning and a marine scientist facilitated these diverse conversations. All stakeholders contributed from concept to completion. Meetings provoked discussions around barriers, success, strategies for progression (e.g. how to support the game designer's marine science knowledge? Game design knowledge? How to raise prize funds?), less about surface opinions and more about valuing contribution as part of a symbiotic team. The competition around game design for ocean literacy provided an incentive for critical thinking through the application of coding skills and in turn, introduced marine education into the space of informal learning.

With regard to an unintentional exclusion mechanism—time available to implement adequately across both countries was an issue. The restricted timeline resulted in a very poor uptake in the partner country. Further to this, prize funding was unavailable for the Swedish competition, which may have lowered the incentive to participate. Were time and funding not an issue we would anticipate that there may have been greater take up by students, and in turn, a greater rise in ocean literacy. The Irish data suggests that there would be significant value in tapping into informal learning such as coding clubs to raise greater awareness with young people about the marine environment. Also, there is significant value in securing modest external funding for prize money to drive initiatives with the potential to change behaviour and increase ocean literacy.

In terms of the consideration of local social changes, game designers were encouraged to use as little text as possible, so that the games would not be overly hindered by language differences internationally. The students and mentors were aware of the international audience so did not design their games in culturally specific ways, but rather focused on shared knowledge and care for the ocean.

Educational e-Book Design Initiative

A current global trend is the employment of educational technology tools in schools and an assessment of their impact on pedagogy (Szeto and Cheng 2014). Furthermore, there is an emerging consensus on the need to integrate mobile devices into a broader learning ecology (Pegrum et al. 2013). One of the challenges with educational technology, however, and one of the core features of this research is the complete integration of technology into naturalistic contexts (McGarr 2009). According to Pegrum et al. (2013), appropriate technology-based resources, steeped in pedagogical theory, have only begun to appear and more of these resources are sorely needed. Presently, the transition from traditional pedagogy to one based on mobile learning is in an experimental stage given that teachers find themselves as co-learners with their students as they become accustomed to the new technology (Pegrum et al. 2013).

Fernández-López et al. (2013, 77) states, "the use of electronic devices and multimedia contents increases their (students) interest and attention." Students across multiple studies are positive towards tablets and see them as essential for modern education (Clark and Luckin 2013; Pegrum et al. 2013). Tablet use allows for personalised usage that is beneficial to individual learning and motivation. Collaborative work is enhanced through the tablet's 'flicking screen.' As the device is rotated, the screen can flip from vertical to horizontal. Teachers commented on its versatility and ubiquitous nature (Burden et al. 2012) when compared to laptops and PCs that require shuffling and moving of large machines for group learning (Clark and Luckin 2013). Drawing further from Clark and Luckin (2013), parents indicated that children showed increased engagement with their schooling when using tablets. More time was spent on homework and they described their learning as more relevant and authentic (Clark and Luckin 2013).

As with any new technology, it harbours a plethora of benefits and pitfalls, the complexity of which has been subject to empirical investigation. Simply placing various technologies into classrooms does not guarantee successful learning (Liu et al. 2009) as evidence suggests that the full potential of technology is not being realised in formal educational settings (Bocconi et al. 2013). Successful technological design and incorporation require complete implementation in context by working with all stakeholders (Reeves 2006; Alghamdi and Li 2013).

Initiative Design

The e-book was co-created with a community of educators, scientists, teachers and students, in an effort to design a resource challenged with raising ocean literacy in second-level education where it was almost absent from the curriculum. More than 20,000 students in over 100 of Ireland's 730 secondary schools have adopted tablet devices for digital learning. The vast majority have chosen iPads with others generally opting for Android software (Weckler 2013), thus the iOS platform was chosen for e-book design. The *Harmful Algal Blooms* e-book was designed in collaboration with second level pupils, teachers, scientists, marine educators, second-level teacher educators (science education, sociology, geography education and technology disciplines) in higher education and research institutes, and with input from European project partners. A summary overview of the steps in the design phase is illustrated in Fig. 8.

As the e-book was intended for a multi-country EU audience for extended impact, it was essential to have input from partner countries to produce a product that would align with various instructional contexts. A cross mapping analysis of school science curricula (age 12–15) was carried out in Belgium, Sweden and Ireland and it emerged that there was little if any ocean content explicitly taught. A strategic design evolved, where a set of regular curricular topics were identified (e.g. chemistry: elements, compounds, mixtures), and a 'vignette approach' was adopted. For example, the e-book could be used to teach science topics through ocean examples. Teachers may choose to use a marine video or image as a hook to engage learners in a specific lesson, or perhaps encourage students to carry out a marine experiment, a research project or review quizzes which might help to draw out core scientific concepts in innovative and memorable ways.

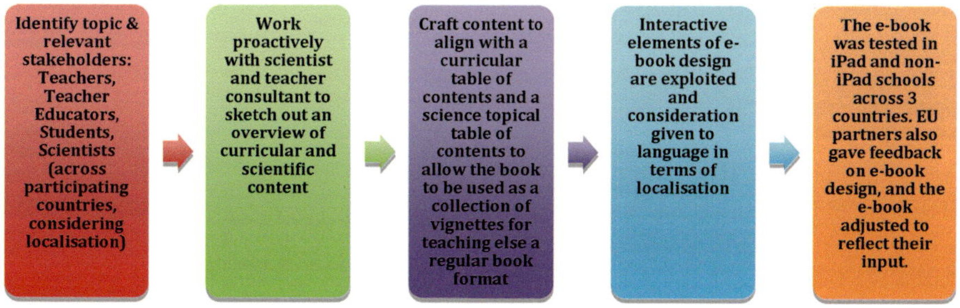

Fig. 8 An overview of the steps in the design phase

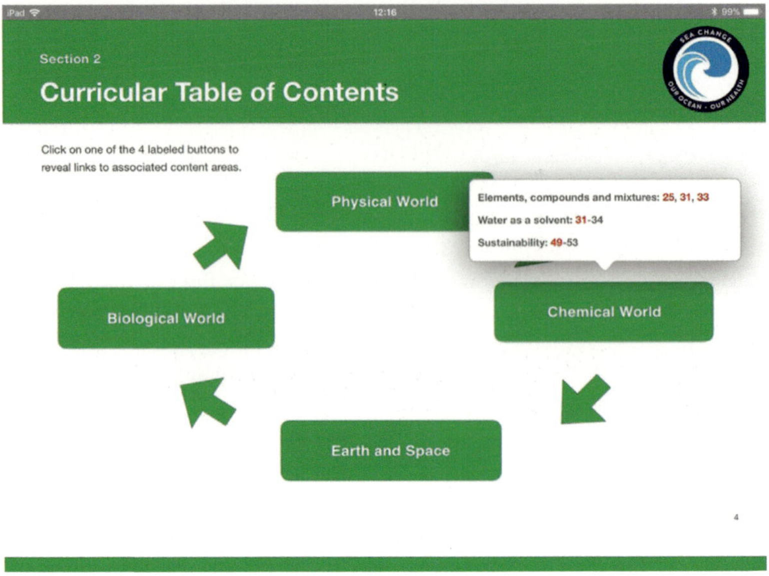

Fig. 9 e-book curricular table of contents

The design team worked with a teacher consultant who mapped the content of *Harmful Algal Blooms* across the current Junior Cycle science curriculum. A curricular table of contents (see Fig. 9) was incorporated into the e-book for this purpose. Additionally, a topical table of contents (Chapter 1 illustrated in Fig. 10) was incorporated, whereby the book could be approached as a marine topic, in this instance the linear narrative of harmful algal blooms and its incredible influence on ocean health.

The e-book represents a nascent technology that claims an adaptable and bespoke learning tool, concomitant with student engagement. The e-book includes keynote presentations, interactive images, interactive galleries, scrolling sidebars, pop-over functionality, videos, study cards, scientific glossary and curiosity facts and quizzes. It also contains a link to the winning CoderDojo Sea Change Ocean literacy games. The design considered relevant science education methodologies and although developed in English, it was created in a format that minimises the cost of future translation.

In the design phase, four teachers and eighty-four students (across Sweden, Belgium and Ireland) gave feedback via a formal research process, on the e-book as a teaching and learning resource. Sea Change task partners (UGOT, EUROGEO, AquaTT, Climar, DTU, MBA and Ecsite) were also invited to feed into the e-book design. Changes based on this input from diverse stakeholders were incorporated. Partner consultation and teacher trials began simultaneously in October 2016. Research in schools is challenging, in terms of access (negotiating teacher timetables and an unexpected teacher strike) and therefore some of the pilot design

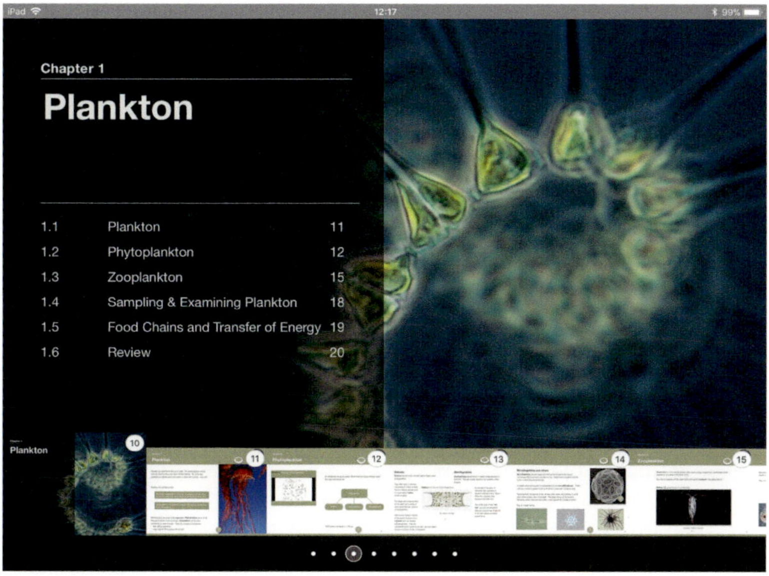

Fig. 10 Topical table of contents—Chapter 1 and sub-sections

testing of the e-book had to be delayed until January 2017 in order to accommodate quality feedback from schools.

A multiple case study methodology was employed (McHugh et al. 2020). A teacher and their respective class were treated as a case, a singular unit of inquiry, to help inform and improve where necessary the design of the e-books. Both a tablet and non-tablet school were used during this study. Data was collected from one school in Belgium ($N = 25$ students), one in Sweden ($N = 19$ students) and two schools in Ireland ($N = 15$ and $N = 25$ students) with a total of four teachers and eighty-four students participating.

Teachers were asked to incorporate the e-book into their lessons during a three-month period (November 2016–January 2017) with their respective science class. Teachers were then interviewed about the process, having taught with the e-book, to get a sense of positive and impeding design features. Interviews were audio recorded. A lesson reflection sheet detailing their use of the e-book was also completed by each teacher. Student perception was gathered through a short two-question qualitative survey which they filled out at the end of any lesson in which the e-book was used. The research process and outcomes are detailed further by McHugh et al. (2020). Proposed changes from both groups were folded into the design of the e-book in December 2016/January 2017 and the e-book was launched in March 2017. The inclusion of education stakeholders in the design of similar instructional resources such as the e-book described here is not only beneficial in term of fine tuning the architecture, but may potentially have a catalytic effect on ongoing and growing conversations about the value of our ocean, which in many instances is muted in contemporary science curricula (McHugh et al. 2020).

Initiative Output

The e-book is an introduction to phytoplankton. The story is told through the phenomenon known as harmful algal blooms (HABs), where a few of these organisms can produce toxins or else grow to such high levels they affect water quality. Chapters include plankton, HABs, eutrophication, climate change and observing the ocean through satellites. Because the e-book includes a curricular table of contents linking to topics science teachers would ordinarily teach across the curriculum, it was designed to infuse the diverse and engaging story of *Harmful Algal Blooms* into teaching across the sciences and to increase ocean literacy (Fig. 11).

In terms of reach, the HAB e-book was published on the iBookstore (http://www.apple.com/ibooks/) and was advertised on the Sea Change website (http://seachangeproject.eu/), the School of Education, NUI Galway, Ireland website (http://www.nuigalway.ie/colleges-and-schools/arts-social-sciences-and-celtic-studies/education/news/) and the College of Arts, Social Science and Celtic Studies, NUI

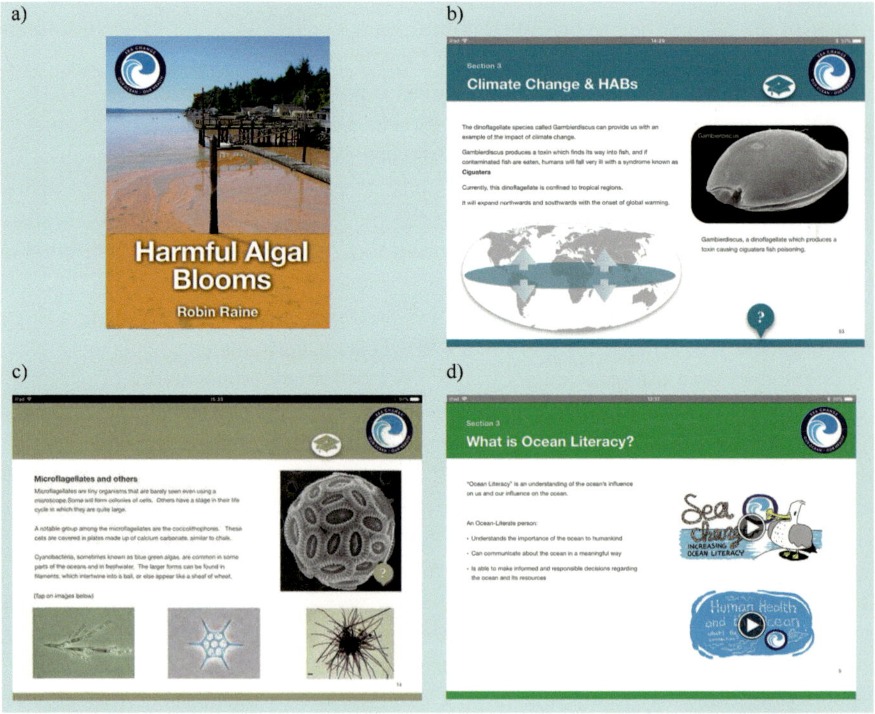

Fig. 11 e-book cover (**a**) and page selection to illustrate images and interactivity (**b–d**)

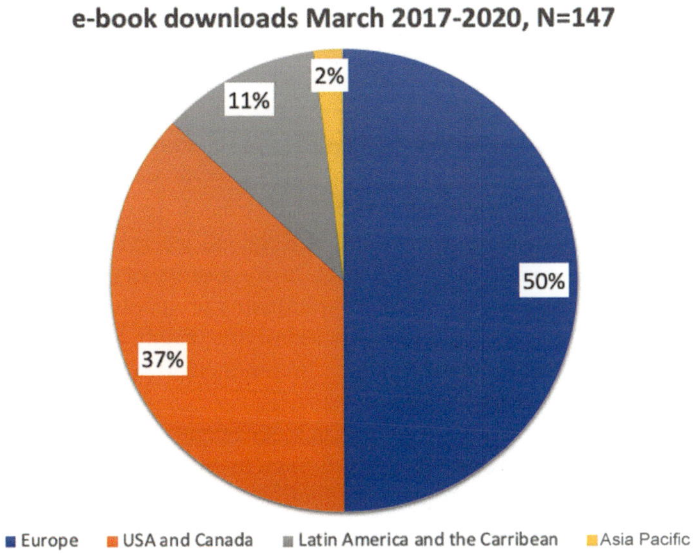

Fig. 12 e-book downloads March 2017–2020

Galway, Ireland website (https://www.nuigalway.ie/colleges-and-schools/arts-social-sciences-and-celtic-studies/news/). It has also been promoted through Twitter feed/direct mail with links to science foundations, and a national listserv, Sharing Science (sharingscience@googlegroups.com) for science teachers which has over 1000 members. The reach through the iBookstore provides for the accessible free download of the e-book in 51 countries worldwide. The graph below indicates HABs e-book downloads across Europe and Worldwide over the past three years (Fig. 12).

To date, data from this research has been translated into one peer-reviewed article (McHugh et al. 2020) and a book chapter (Domegan et al. 2019). In terms of contributing further to legacy and dissemination, NUI Galway worked with Galway Atlantaquaria on their 'Our Ocean Our Health' exhibit, in which the *Harmful Algal Blooms* e-book is showcased. This exhibit was launched on 11 November 2017, and it is intended as one of their permanent exhibitions. Galway Atlantaquaria welcomes 80–85,000 visitors each year, and it is hoped that a portion of these visitors may interact with the e-book (positioned in a secure housing unit), play the video clips, read content, engage with interactive quizzes, with the hope of positively influencing their ocean literacy.

Finally, the e-book was designed so that teachers may use it as a science teaching resource despite the absence of marine science content in formal science curricula. Both our pilot and follow-up research suggest that teachers appreciate how the e-book can be incorporated into teaching in a way that offers opportunities to introduce marine issues and increase ocean literacy (McHugh and McCauley 2020). We believe that because the e-book was designed to work *with* the science curriculum, and not as an additional burden to it, it will be taken up by teachers and students and could contribute to a sea change of ocean understanding.

Reflection on Social Inclusion Process and Initiative Outcome

In terms of diverse audience participation, the e-book received feedback from multiple and diverse stakeholders, including teachers, teacher educators, students and scientists across participating countries in the European Sea Change project throughout the design stage. In line with participation protocols, multiple conversations were held with all stakeholders to elicit design strength, impediment, suggestions for inclusion/alteration, in the explicit awareness that we were working 'with' partners towards a unified design output.

With regard to an unintentional exclusion mechanism—the available breadth of the pilot schools in the partner nations was limited by time and resources. A broader pilot may have further refined the final product.

In terms of the consideration of local social changes—diverse EU stakeholders played an active role in feedback from the beginning, from consideration of the relevant curricula in each country to the readability of the language readability and future translation possibilities. It is anticipated that with greater take up at the school level, and once the e-book is translated into modern European languages, it offers significant potential to allow science teachers to teach the current science curriculum by using marine examples. This innovative approach also may provide opportunities to broaden students' personal interest in learning about science due to the use of unexpected and unconventional marine examples.

Conclusion

Given the rising increase of e-resources in informal and formal learning, it is imperative that we consider the design of these resources, a design process that gives a proactive voice to relevant stakeholders from the beginning rather than post-commentary. An inclusive experience allows for meaningful exchanges between all parties, as co-learners and change agents (Bardus et al. 2018). Each exchange party/stakeholder has something to give and gain from the union. A collective process emerges that connects and empowers participants through individual agency and co-creating value (Bardus et al. 2018). This collective action was encouraged through both the gaming pedagogy and e-book design initiative, and the inclusion of diverse partners within and between stakeholder countries supported the consideration of social changes and social access issues that may arise. The project description is intended as a useful resource for future developers of resources and the importance of explicit contribution and participation from all relevant parties.

This process illustrates the potential to effect behavioural change and to grow ocean literacy with a modest budget. There are more than 1600 CoderDojo organisations in 75 countries that could quite easily take up a similar ocean literacy game design challenge to exponentially expand the students' understanding of marine issues through gamification. Also, as the iBook software is free, similar e-Books could also be designed quite readily targeting a variety of marine topics. Furthermore, either resource can feed into formal education, where there is an absence of marine content across many European countries. The availability of teaching resources that are designed to work with standard science teaching to stealthily introduce marine issues would likely be welcomed especially given the attractiveness of the design of the technology in the hands of students, and the possibility of providing innovative ways for teachers to teach science. While there have been diverse initiatives to strengthen science communication and literacy, we believe that our research into innovative approaches to ocean literacy, resulting in gamification, and the co-creation of an e-book learning resource provide an example of how to begin to introduce new discussions and understandings of marine issues that broaden ocean literacy.

Acknowledgements The research leading to these results has received funding from the European Union's Horizon 2020 Framework Programme for Research and Innovation (H2020-BG-2014-1) under grant agreement No. 652644. This publication reflects the views of the authors, and the European Union cannot be held responsible for any use which might be made of the information contained therein.

References

Alghamdi AH, Li L (2013) Adapting design-based research as a research methodology in educational settings. Int J Educ Res 1(10):1–12

ASTI (2013) Why education matters: the importance of education to Ireland's economy and society. CRM Publications, Dublin. Available via https://www.esri.ie/pubs/JACB201327.pdf. Accessed 3 June 2015

Bardus M, Domegan C, Suggs S, Egberg B (2018) Engaging students and communities through service learning and community-academia partnerships: lessons from the field of social marketing education. In: Pinheiro MM, Estima A, Marques S (eds) Evaluating the gaps and intersections between marketing education and the marketing profession, IGI Global

Bingimlas KA (2009) Barriers to the successful integration of ICT in teaching and learning environments: a review of the literature. Eurasia J Math Sci Technol Educ 5(3):235–245

Bocconi S, Kampylis P, Punie Y (2013) Framing ICT-enabled innovation for learning: the case of one-to-one learning initiatives in Europe. Eur J Educ 48(1):113–130

Boston University School of Public Health (2016) Behavioural change models: the transtheoretical model (stages of change). Available via http://sphweb.bumc.bu.edu/otlt/MPH-Modules/SB/BehavioralChangeTheories/BehavioralChangeTheories6.html. Accessed 12 June 2018

Britton E, Kindermann G, Domegan C, Carlin C (2020) Blue care: a systematic review of blue space interventions for health and wellbeing. Health Promot Int 35(1):50–69

Burden K, Hopkins P, Male T, Martin S, Trala C (2012) iPad Scotland evaluation. University of Hull. Available via http://www2.hull.ac.uk/ifl/ipadresearchinschools.aspx. Accessed 18 Mar 2018

Cava F, Schoedinger S, Strang C, Tuddenham P (2005) Science content and standards for ocean literacy: a report on ocean literacy. Available via http://www.cosee.net/files/coseeca/OLit04-05FinalReport.pdf. Accessed 3 June 2015

Charlier PS, Charlier RH (1971) A case for oceanography at the Inland School. Sci Educ 55(1):15–20

Clark W, Luckin R (2013) iPads in the classroom. What The Research Says. Available via https://cpb-ap-se2.wpmucdn.com/global2.vic.edu.au/dist/5/48534/files/2015/08/ipads-in-the-classroom-report-lkl-v24yz4.pdf. Accessed 30 May 2018

Clifton A, Mann C (2011) Can YouTube enhance student nurse learning? Nurse Educ Today 31(4):311–313

Connaway LS, White D, Lanclos D (2011) Visitors and residents: what motivates engagement with the digital information environment? Proc Am Soc Inform Sci Technol 48(1):1–7

Davison K, Domegan C, McCauley V, McClune W (2008) A review of science outreach strategies north and south: with some recommendations for improvement. Centre for Cross Border Studies, Belfast. https://aran.library.nuigalway.ie/handle/10379/5147. Accessed 30 May 2018

Domegan C, McCauley V, Davison K (2010) Realising the management challenges for science communication outreach: a social marketing perspective. Ir J Manag 19(2):89–108

Domegan C, McHugh P, McCauley V, Davison K (2019) Co-creating a sea change campaign for ocean literacy in Europe: a digital interactive tool for environmental behavioural change. Social marketing in action: cases from around the world. Springer, New York, pp 393–409

European Marine Board, EMB (2017) Marine citizen science: towards an engaged and ocean literate society (Policy Brief). Report 5: October. Available via http://marineboard.eu/sites/marineboard.eu/files/public/publication/EMB_Policy_Brief_5_Marine_Citizen_Science_0.pdf. Accessed 10 Mar 2018

Fauville G, Dupont S, von Thun S, Lundin J (2015) Can Facebook be used to increase scientific literacy? a case study of the Monterey Bay Aquarium Research Institute Facebook page and ocean literacy. Comput Educ 82:60–73

Fauville G, McHugh P, Domegan C, Mäkitalo Å, Friis Møller L, Papathanassiou M, Alvarez Chicote C, Lincoln S, Batista V, Copejans E, Crouch F, Gotensparre S (2018) Using collective intelligence to identify barriers to teaching 12-19 year olds about the ocean in Europe. Mar Policy 91:85–96

Fernández-López Á, Rodríguez-Fórtiz MJ, Rodríguez-Almendros ML, Martínez-Segura MJ (2013) Mobile learning technology based on iOS devices to support students with special education needs. Comput Educ 61:77–90

Fortner RW, Lyon AE (1985) Effects of a Cousteau television special on viewer knowledge and attitudes. J Environ Educ 16:12–20

Garcia-Soto C, van der Meeren GI, Busch JA, Delany J, Domegan C, Dubsky K, Fauville G, Gorsky G, von Juterzenka K, Malfatti F, Mannaerts G, McHugh P, Monestiez P, Seys J, Węsławski JM, Zielinski O (2017) Advancing citizen science for coastal and ocean research. In: French V, Kellett P, Delany J, McDonough N (eds) European Marine Board Position Paper, 23. European Marine Board, Ostend. Available via http://www.vliz.be/en/imis?module=ref&refid=285795. Accessed 8 May 2020

Guest H, Lotze HK, Wallace D (2015) Youth and the sea: ocean literacy in Nova Scotia, Canada. Mar Policy 58:98–107

Heemskerk I, Volman M, Admiraal W, ten Dam G (2012) Inclusiveness of ICT in secondary education: students' appreciation of ICT tools. Int J Inclusive Educ 16(2):155–170

Hynes S, Norton D, Corless R (2014) Investigating societal attitudes towards the marine environment of Ireland. Mar Policy 45:57–65

Kim JH (2014) Apple's iBooks author: potential, pedagogical meanings, and implementation challenges. In: Transforming K-12 classrooms with digital technology. IGI Global, pp 119–139

Liu T-C, Peng H, Wu W-H, Lin M-S (2009) The effects of mobile natural-science learning based on the 5E learning cycle: a case study. Educ Technol Soc 12(4):344–358

Markos A, Boubonari T, Mogias A, Kevrekidis T (2017) Measuring ocean literacy in pre-service teachers: psychometric properties of the Greek version of the Survey of Ocean literacy and

Experience (SOLE). Environ Educ Res 23(2):231–251. https://doi.org/10.1080/13504622.2015.1126807
Martins Gomes D, McCauley V (2012) Science outreach and science education in the primary level: conceptual and pedagogical challenges faced. Literacy Inf Comput Educ J 1(2):930–938
Martins Gomes D, McCauley V (2013a) Science outreach and science education at primary level in Ireland: a mixed methods study. In: Abstracts of the 2013 New Perspective for Science Education Conference, Florence, Italy, 14–15 March 2013
Martins Gomes D, McCauley V (2013b) Science outreach and science education: the analysis of dilemmas faced to promote the creation of the third space. J Theory Pract Educ 1:372–383
Martins Gomes D, McCauley V (2016) Dialectical dividends: fostering hybridity of new pedagogical practices and partnerships in science education and outreach. Int J Sci Educ 38(13–14):2259–2283. https://doi.org/10.1080/09500693.2016.1234729
Massarani L, Merzagora M (2014) Socially inclusive science communication. JCOM: J Sci Commun 13(2):1–2. Available via https://jcom.sissa.it/sites/default/files/documents/JCOM_1302_2014_C01.pdf. Accessed 8 May 2020
McCauley V, Davison K (2015) Micro and macro pedagogy for science education. Chem Action 106(Winter):9–12
McCauley V, Davison K, Byrne C (2015) Collaborative lesson hook design in science teacher education: advancing professional practice. Ir Educ Stud 34(4):307–323. https://www.tandfonline.com/doi/full/10.1080/03323315.2015.1114457
McCauley V, Martins Gomes D, Davison K (2018) Constructivism in the third space: challenging pedagogical perceptions of science outreach and science education. Int J Sci Educ, Part B 8(2):115–134. https://doi.org/10.1080/21548455.2017.1409444
McCauley V, McHugh P, Davison K, Domegan C (2019) Collective intelligence for advancing ocean literacy. Environ Educ Res 25:280–291. https://doi.org/10.1080/13504622.2018.1553234
McFadden DL (1973) Teaching in the tidepools: west coast children study the marine sciences at first hand. Oceans 6(5):44–49
McGarr O (2009) The development of ICT across the curriculum in Irish schools: a historical perspective. Br J Edu Technol 40(6):1094–1108
McHugh M, McCauley, V. (2016) Designing physics video hooks for science students. Phys Educ 51(1). http://iopscience.iop.org/article/10.1088/0031-9120/51/1/015015/meta
McHugh M, McCauley V (2017) Hooked on science. Sci Sch Eur J Sci Teachers 39:55–58 Available via https://www.eso.org/public/archives/schools/pdf/sis_0039.pdf. Accessed 10 May 2020
McHugh M, McCauley V, Davison K, Raine R, Grehan A (2020) Anchoring ocean literacy: participatory iBook design within secondary science classrooms. Technol Pedagogy Educ. https://doi.org/10.1080/1475939X.2020.1715241
McHugh P, Domegan C, Devaney M, Hastings G (2015) A set of sea change guiding principles and protocols. EU Sea Change Project
McHugh P, Domegan C, Duane S (2018) Protocols for stakeholder participation in social marketing systems. Soc Mark Q 24(3):164–193. https://doi.org/10.1177/1524500418761626
Mumtaz S (2000) Factors affecting teachers' use of information and communications technology: a review of the literature. J Inf Technol Teacher Educ 9(3):319–342
National Marine Educators Association (2020) Ocean literacy: the essential principles and fundamental concepts of ocean sciences for learners of all ages (Version 3). Available via https://static1.squarespace.com/static/5b4cecfde2ccd188cfed8026/t/5eb99cc530a3d76767dc7aea/1589222614314/OceanLiteracyGuide_V3_2020.pdf
Ocean literacy Network (2015) Ocean literacy essential principles and fundamental concepts. Available via http://oceanliteracy.wp2.coexploration.org/long_history/. Accessed 26 June 2020
Pegrum M, Oakley G, Faulkner R (2013) Schools going mobile: a study of the adoption of mobile handheld technologies in Western Australian independent schools. Australas J Educ Technol 29(1). https://doi.org/10.14742/ajet.64
Picker L (1980) What is marine education? Sci Child 18:10–11. Available via https://eric.ed.gov/?id=EJ232945. Accessed 10 May 2020

Pörtner HO, Karl DM, Boyd PW, Cheung WWL, Lluch-Cota SE, Nojiri Y, Schmidt DN, Zavialov PO (2014) Ocean systems. In: Field CB, Barros VR, Dokken DJ, Mach KJ, Mastrandrea MD, Bilir TE, Chatterjee M, Ebi KL, Estrada YO, Genova RC, Girma B, Kissel ES, Levy AN, MacCracken KS, Mastrandrea PR, White LL (eds) Climate change, impacts, adaptation, and vulnerability. Part A: global and sectoral aspects. Contribution of Working Group II to the Fifth Assessment Report of the Intergovernmental Panel on Climate Change. Cambridge University Press, Cambridge, UK and New York, NY, USA, pp 411–484

Prensky M (2001) Digital natives, digital immigrants part 1. Horiz 9(5):1–6

Reeves TC (2006) Design research from a technology perspective. Educ Des Res 1(3):52–66

Santoro F, Selvaggia S, Scowcroft G, Fauville G, Tuddenham P (2017) Ocean literacy for all: a toolkit, vol 80. UNESCO Publishing

Schoedinger S, Uyen Tran L, Whitley L (2010) From the principles to the scope and sequence: a brief history of the ocean literacy campaign. National Marine Educators Association. Special Report No. 3, pp 1–7. Available via http://coexploration.org/oceanliteracy/NMEA_Report_3/NMEA_2010-2-History.pdf. Accessed 10 May 2020

Sea Change Consortium (2015) Sea change: funding application. Project No. 652644. H2020-BG-2014-1: Ocean literacy—engaging with society—social innovation. European Commission, pp 1–49

Steffes EM, Duverger P (2012) Edutainment with videos and its positive effect on long term memory. J Adv Mark Educ 1(20):1

Szeto E, Cheng AY-N (2014) Exploring the usage of ICT and YouTube for teaching: a study of pre-service teachers in Hong Kong. Asia Pac Educ Researcher 23(1):53–59

Tran LU, Payne DL, Whitley L (2010) Research on learning and teaching ocean and aquatic sciences. NMEA Special Report 3:22–26. Available via https://cdn.ymaws.com/members.marine-ed.org/resource/collection/9B85E578-8E65-4F88-935E-586B984CD3F0/NMEA_2010-6-Learning.pdf

Weckler A (2013, May 31) Parents being forced to fork out 500 euro for School iPads, Irish Independent. Available via https://www.independent.ie/life/family/learning/parents-being-forced-to-fork-out-500-for-school-ipads-29546495.html. Accessed 27 Nov 2017

White DS, Le Cornu A (2011) Visitors and residents: A new typology for online engagement. First Monday 16(9)

Sail Training Has Set Sail on a Course Towards Ocean Literacy

Laura Ellen Lyth

Abstract Sail training is a non-formal, alternative education space, providing adventurous experiential learning on-board ocean-going sailing vessels for young people up to 25 years old. It facilitates the growth of key competencies through the provision of personal development and active citizenship, whilst in an arduous environment. This is achieved by encompassing youth mobility, heterogeneity, broadening horizons, and assisting the learning of new skills. Whilst raising the awareness of new possibilities through working as crew members to complete a seagoing voyage, it embraces adaptability and change; advocating social cohesion and cultural awareness with an objective to build positive relationships in the local community and beyond. This chapter introduces the maritime alternative education space of sail training with an overview of its evolution, adding context to the definition above. This is followed by how current operations can provide activities related to the Ocean Literacy Framework (OLF). Next is an analysis of an ocean literacy audit of the Tall Ships America (TSA) fleet. Lastly, an introduction to the UN Decade for Ocean Science for Sustainable Development and the Sustainable Development Goals is provided to identify how the European sail training fleet currently focuses on sustainability and their SDG 14: life below water-associated activities.

Keywords Alternative education · Ocean literacy · Sail training

Introduction: Blue Outdoor Education—Sail Training's Relationship with Ocean Literacy

The ocean, It is the unconscious lifeblood on earth. It covers over 70% of our planet and approximately 95% of the earth's water is in our oceans. It is what makes our planet unique compared to other planets, and what makes our planet habitable for all living species. As the dominant living species, not only do humans rely upon the ocean to create a suitable environment to survive, we also rely upon it for beauty,

L. E. Lyth (✉)
Océanie Ltd, Scarborough, UK
e-mail: l.lyth@oceanie.co.uk

K. C. Koutsopoulos and J. H. Stel (eds), *Ocean Literacy: Understanding the Ocean*, Key Challenges in Geography, https://doi.org/10.1007/978-3-030-70155-0_8

clothing, commerce, food, fuel, minerals, and pharmaceutical products. The irony of single-use plastics is that this convenient, synthetic, petroleum by-product which modern society relies upon originates from the ocean, where it is causing the most pollution and detrimental impact on the planet. We also use the ocean for defence, heritage, migration, trade, and recreation. Yet, despite our complete dependence on the ocean, it remains the least explored and understood habitat in our grasp.

However, what some may not realise is that the ocean is an education platform too. Historically, sail training stems as far back as 1856 with the formation of the Navy Lads Brigade (now the Sea Cadets Corps) shortly after the Crimean War. A critical objective of the ethos of sail training has been to support disadvantaged young people to redirect their lives into lifelong employment through a life of adventure and/or at sea. Today, sail training is in approximately 40 countries, with over 200 operators and approximately 570 vessels ranging from yachts to fully rigged tall ships.

It is conceivable that modern sail training's foundations were set in the 1930s due to the technological advances in the maritime domain. The transition from sail to steam powered vessels created a surplus fleet of sailing vessels. One of the most notable modern sail training pioneers, the Australian Alan Villers, purchased the old school ship *George Stage* from Denmark in 1934 (Mystic Seaport 2020), re-naming her *Joseph Conrad.* Lyth (2014) explained that the concept behind Villers' model was to have zero cargo on-board, with only youth crew who had paid to be on-board. There was also a selection of non-paying youths, allowing as many as the budget could facilitate. It was considered that non-paying youths were 'at risk' and in need of what was then called 'character building'. The harsh discipline imposed by the sea along with working manually operated ships whilst sailing round the world was believed to further personal development and create a beneficial experience, whilst redirected those disadvantaged by circumstance. Villers, along with his northern hemisphere contemporaries—Irving and Exy Johnson, pioneered the modern sail training concept.

Modern sail training's core ideology emerged during the post-war era (1945–1979). Inspired by its visionary Alan Villers, the UK in particular saw the formation of the:

- Sail Training Association (now Tall Ships Youth Trust and Sail Training International) [1956].
- London Sailing Project (now Rona Sailing Project) [1960].
- Ocean Youth Club (now Ocean Youth Trust South, Ocean Youth Trust Scotland and Ocean Youth Trust North) [1960].
- Jubilee Sailing Trust [1978].
- Association of Sail Training Organisations (ASTO, co-founded by Lord Dulverton) [1979].

Along with the Sea Cadets, the cardinal organisations of UK sail training were established. In addition to the UK developments, Tall Ships America was founded in 1973. This set the foundations of the Transatlantic (and now global) sail training sphere that exists today.

Whilst remaining loyal to the principal objective, sail training's focus has evolved towards the personal development of young people from diverse backgrounds, promoting "*... the development and education of young people of all nationalities, cultures, religions and social backgrounds through the sail training experience*" (Sail Training International 2020). It focuses on the experience and global citizenship. McCulloch (2004) explains this is achieved through "*taking young people to sea as a means to social and educational objectives* ... [whereby] ... *the core purpose is not to teach seamanship but to use seafaring as a context for education. Sail training might be said to provide a context for learning, but to form only a modest and perhaps not ultimately very significant part of the content of that learning*".

It was not until the late 1990s that scholars began to recognise the positive impact of sail training as an outdoor education space. One of the first academic papers was the "*Case Study of 'Blue watch' on STS Leeuwin*" (Gordon et al. 1996). Since then, there has been a gradual increase in sail training-related research, primarily focusing upon how it delivers experiential education in the outdoor learning arena. This particularly came to fruition during the early to mid-2000s. The University of Edinburgh conducted one of the esteemed collections of studies, including the 'Sail Training Programme Evaluation Self-Assessment Toolkit'. However, the theme of sail training as a research subject is still embryonic, with the number of identified research projects up to early 2020 being:

- 1993–2003: 13 papers.
- 2003–2013: 26 papers.
- 2013–2020: 40 papers.

Particularly since 2008, there has been an increased desire to identify the relationship between sail training, formal education, employability, and its overall impact to encourage further stakeholder engagement. This saw an increased global responsibility to identify its ethical and sustainability practices. This included how sail training implements ocean literacy. By characterising where and how sail training has the potential to create ocean literate citizens who are well equipped to address the many and complex issues associated with the sustainable use of ocean resources, there is potential to enhance the value proposition of sail training. In addition, through advancing ocean literacy in support of the UN Decade of Ocean Science for Sustainable Development (the UN Ocean Decade), collaborative working between sail training and ocean literacy stakeholders could emulate the goals of each other.

By optimising this maritime alternative education space through ocean literacy activities, there could be scope to accelerate the operational capabilities of sail training as an alternative education provider. In addendum, it could also provide opportunities to overcome knowledge limitations and public participation barriers currently associated with ocean literacy. Identifying sail training as an exemplar public participation maritime resource could place it at the forefront of the geographies of education and marine geography.

Sail Training as an Alternative Education Space: Learning Through Experience at Sea

As mentioned in section "Introduction: Blue Outdoor Education—Sail Training's Relationship with Ocean Literacy", the sail training learning environment focuses on learning through experience and personal development, enhancing emotional intelligence and associated behaviours. This creates an environment that can be incredibly beneficial to young people who have become disengaged with mainstream education, employment, or training. Sail training could be considered as an alternative to mainstream learning arenas due to this different educational approach. It possibly has an innovative curriculum, which is flexible in its delivery technique. This is partly due to its operational environment and advocating fluidity in its programme implementation. Where possible, it focuses on modifying the individual trainee's requirements, whilst remaining focused on the collective goal of completing the voyage (Aron 2003; Carnie 2003; Foley and Pang 2006; Magadley and Amara 2018; Olive 2003; Raywid 1994; Smith and Thomson 2014; Tierney 2016; Vadeboncoeur 2009).

Allison and Seaman (2017) state that "*experiential education is often characterised as an alternative to traditional or didactic education … reflecting on the small group experiences, observations of group and individual behaviours and articulation of individual learning among the group*". This suggests that it often involves emotionally intense, immersive small group experiences often in wilderness or remote settings. It also relates to the geographies of education ideology in that informal learning environments such as homes, neighbourhoods, community organisations, and workspaces consider the importance of spatiality in the production, consumption, and implications of formal education systems from pre-school to tertiary education (Bauer 2015; Holloway and Heike 2012).

An alternative education space is an environment "*that deliberately teaches to a different curriculum and tries not to look and feel like a school*" (Kraftl 2014). They are all based on a different set of beliefs about how, what, and the best methods young people learn (Kraftl 2014; Sliwka 2008). Examples include home education, special educational needs/disability schools, public schools, democratic schools, religious schools, pupil re-integration units, and charter schools, etc. Watkin Beck (2016) suggests some may also have an idealistic start point of freedom. Regardless of the target market of a sail training operator, the key goal of the sail training experience is often to re-engage/reinforce commitment to returning to mainstream education, employment, and/or integration back into society.

It could be suggested that due to its experiential learning environment, sail training adopts a holistic approach to learning, engaging the physical, emotional, mental, and spiritual levels of a person. However, Petty (2009) suggests that "*experience in itself does not guarantee learning. In order to learn from experience, we must reflect on our experiences; try to relate them to theory; and then plan how we might do better next time. After carrying out this plan we need to reflect again, and so the process continues*". This process is illustrated in Kolb's Experiential Learning Cycle (see Fig. 1).

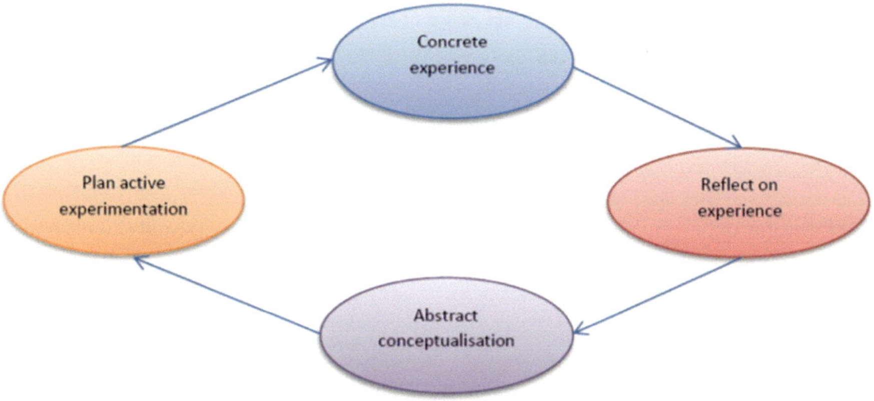

Fig. 1 Kolb's experiential learning cycle (Petty 2009)

Regarding adolescent development and sail training, McCarthy (2011) and McCulloch (2004) suggest that "*it is the structure and context of the activities rather than the activity itself that determines whether the outcomes are positive or negative*". Plus, "*sail training might provide a context for learning, but to form only a modest and perhaps not ultimately very significant part of the content of that learning*" (Cote et al. 2008; Holt and Sehn 2008). This concept supports the Red Square Schools ideology, which the sail training operator of *Ocean Spirit of Moray,* Gordonstoun School, Scotland, is a member of (Gordonstoun School 2020). Round Square Schools (2020) adhere to six ideals, which resonate with the guiding principles witnessed within UK sail training's culture through the promotion of active citizenship in a multicultural, transnational environment. Aguiar (2015) extends this through identifying seven challenge values, claiming, "*A sailing ship provides a totally unique context for adventure in today's world, touching almost every aspect of the human experience in a way that just cannot be found anywhere else*".

This approach has the potential to create a democratic and open communications environment. Likewise, through the inclusion of ocean literacy and direct exposure to the marine environment, this can further promote environmentalism. The nature of sail training is to deliver an adventurous experience whilst facilitating to the needs of the young people during their time on-board, with a spirit of challenge. This aims to advocate confidence, courage, discipline, leadership, teamwork, service, and sustainability, which can be embraced by young people. Sail training could be the birthplace of the authentic marine citizen.

Creating Ocean Literate Citizens Through Experiencing the Marine Environment

As suggested by McCulloch, the key ideology behind sail training is that it is "*less training for the sea than through the sea*" (McCarthy 2011; McCulloch 2002, 2004). This suggests that sail training is not just the process of learning to sail; it is a much bigger experience as through adventure at sea it offers personal development, particularly for young people (Wright 2012).

As identified in the ASTO Theory of Change, it facilitates and implements the development of social capital, "*such as overcoming challenges to achieve goals, working as a team ... learning new skills*" (Noble et al. 2017). However, what makes sail training unique is the physical environment that it operates within, creating a distinctive experience and specific challenges/factors that may provide a significantly greater impact.

Due to sail training being immersed into the marine environment, effectively 'bringing it to life'; the sail training experience also has the potential to be a fundamental facilitator of ocean literacy. Through identifying and enhancing this added value to the sail training ethos, it has scope to put it at the forefront of alternative, experiential, outdoor, and ocean literacy education. It also embraces the responsibility educators have to equip "*the next generation with the necessary skills and knowledge to combat life challenges*", as identified in the UN Sustainable Development Goals—see section "An Overview of the UN Sustainable Development Goals" (Bohatka et al. 2017).

As previously identified, "*to foster and mobilise citizenry involvement in marine environmental issues, citizens are encouraged to understand the ocean's influence on them and their influence on the ocean...* [the ocean] *plays several crucial roles that support the livelihood of humans, and regulate the Earth's climate*" (Dupont and Fauville 2017; Fauville et al. 2018). To develop a passion for the sea, marine awareness needs to be taught from a young age. There is anecdotal evidence from sail training operators that young people participating in sail training voyages do develop an enthusiasm for the sea. This is also supported by Santin and Santoro (2017) suggesting that educating in its broader understanding serves as a high potential channel to reach young citizens, ensuring the preservation of the marine environment. Every human being needs to be aware of their requirement to participate in the conservation and protection of marine life (Widiarti 2010).

Linking Sail Training to Ocean Literacy Framework

It could be suggested that sail training has the potential to facilitate the seven principles of the 'Ocean Literacy Framework'. Three of these principles can be achieved by most sail training operators; these are principles one, three, and six. The other four principles—two, four, five, and seven—can be achieved by operators who chose to adopt an 'Education under Sail®' approach similar to many operators in the USA and Canada; whereby citizen/marine science activities are implemented as part of and in addition to the overarching sail training programme. Each of the seven principles shall now be briefly discussed; highlighting how sail training operators can and may choose to implement each of the principles.

Principle One: The Earth Has One Big Ocean with Many Features

Trainees have the potential to experience different locations and perspectives, witnessing geographical landscapes and transiting past various geological features, possibly for the first time. This in turn brings to life the connection the ocean has to major waterways and watersheds, particularly when visiting ports that are located on rivers, river estuaries, and natural harbours. When included in activities such as passage planning, trainees begin to understand the relationship the ocean has to the earth's circulation system regarding wind, tides, the Coriolis Effect, evaporation, and precipitation. This significantly relates to Principle three (NMEA 2020; Santoro et al. 2017).

Principle Two: The Ocean and Life in the Ocean Shape the Features on Earth

By adopting marine science activities as part of an Education under Sail® programme, trainees learn how the ocean and life in the ocean shape the features on earth through activities such as water quality testing (salinity, dissolved oxygen, turbidity, temperature, clarity nutrients [nitrate and phosphate], and pH). Observations focusing on how the shape of the land determines surface water flow can also be conducted. Trainees could also conduct shoreline processes and surveys investigating coastal features of visited ports, e.g. the evidence of sediment drift and its causes, cliff material and its relationship to coastal erosion, sand/beach structure, detecting wave diffraction and refraction, and identify how it impacts the coastal area (in addition to what impact these events may have on the vessel such as navigation, pilotage, and stability). This can also lead to a discussion about plate tectonics, and the slow effects of isostatic rebound. Trainees can also discuss how sea-level changes impact coastal communities (and in turn mariners such as loss of ports and the impact on tidal ranges) (Image 1).

Image 1 Conductivity, temperature, and depth water sampling on-board *Pelican of London* © Shield Media Services

Principle Three: The Ocean Is a Major Influence on Weather and Climate

Because sailing relies heavily on the natural environment for propulsion, an understanding of oceanic and atmospheric processes controlling weather patterns is required. Young people get an opportunity to experience first hand how these processes influence their ability to navigate their 'sea home' to achieve a successful voyage. Lessons provided in a school environment related to energy, water, and carbon cycles as well as solar radiation can be witnessed and experienced in a real life context. This can assist a young person to understand the relationship of land-based weather patterns. These weather observations can be reported to government agencies such as NOAA or the Metrological 'Met' Office. It can contribute to grasp how major climatic episodes such as hurricanes and cyclones can affect areas at significant distances away from their original source. Lastly, an appreciation of how phenomena such as the El Niño Southern Oscillation climate pattern can create major global weather changes across the world by increasing the sea surface temperature in the Pacific can be obtained. All of these effects can have dramatic biological, chemical, economic, physical, and social consequences (Image 2).

Image 2 "European Tsunamis" discussion on-board *Pelican of London* © Shield Media Services

Principle Four: The Ocean Made the Earth Habitable

Related to some of the activities suggested in principle two, marine biology and ecology activities can be undertaken such as investigating the effects of marine plants (phototrophs) such as phytoplankton, seaweed, and seagrass on human health (oxygen production). How the water quality also impacts phototrophs and marine animals including herbivores (e.g. zooplankton, sea urchins, and manatees), small carnivores (e.g. squid, sardines, snapper), and the top marine predators such as fish, mammals, and birds can also be explored. As the ocean is also a regulator of the climate and a source of water, this also links to sail training activities associated with principles one, three, and six. Marine chemistry can be explored through biosynthesis and chemosynthesis, ocean acidification, and impacts of ocean chemistry imbalances (Image 3).

Principle Five: The Ocean Supports a Great Diversity of Life and Ecosystems

Focusing upon ocean stewardship, activities that can be explored include bathymetry and seafloor mapping, investigating sounds using lead lining and sonar, exploring

Image 3 Trainees conducting sampling surveys on-board *Challenge Wales* © Challenge Wales

food webs and food chains (which also relates to principle two and four activities) including plankton's role in the food chain. Other examples are bottom sample surveying the benthic zone with a petite sonar dredge and searching for and identifying organisms that indicate water quality. The identification of Invasive Aquatic Species (IAS) and how they impact their adopted area(s) and what mariners do to prevent transferring IAS from their native habitats can be investigated. Marine wildlife observations can lead to discussions regarding their dependence on plants and animals living in water (see principle four). This leads to opportunities to introduce the concept of marine diversity and the evolutionary processes, expanding to surveying population size of identified species within a vessel's operational region, including river estuaries, bays, coastal waters, or the high seas. There is scope to introduce the effects of bioaccumulation and biomagnification that can result from the overuse of pesticides. Trainees can also explore the physical properties of seawater and distribution, the impact of ocean chemistry imbalances can be researched, introducing the chemistry of biosynthesis, and chemosynthesis and ocean acidification.

When in port and coastal waters, there could be the opportunity to examine intertidal habitats, learning how the intertidal zone is impacted by human activities, and how we are affected by it. Trainees are able to study the effects of the Holocene glacial retreat on coastal areas, linking this to the polar ice caps and ice sheets melting and how it will impact the ecosystem. As well as how our actions on land impact

Image 4 Trainees collecting data samples on-board *Pelican of London* © Shield Media Services

the ocean environment (e.g. deforestation and industrialisation). Through various sampling exercises, trainees can also gain an understanding of what micro-plastics are and how they are affecting marine ecosystem. Trainees can conduct marine debris surveys and learn the decomposition rates of these materials (Image 4).

Principle Six: The Ocean and Humans Are Inextricably Interconnected

Young people have an opportunity to discover how the marine environment can be a source of inspiration and discovery, as well as a location for recreation and rejuvenation. Through visiting foreign locations, they discover how the marine environment can be a significant influence on a nation's/region's heritage and culture. When navigating around hazards such as offshore oil and gas installations and wind farms, young people may witness how humans are using marine resources to provide energy back on land. Passages may also provide the opportunity for young people to encounter marine life such as dolphins, fish, porpoise, and whales (some for the first time). This can develop an understanding of how the ocean is a food source and may also influence purchasing decisions such as buying sustainable and dolphin friendly caught fish and seafood. As human activity can be a source of ocean pollution, young people could visualise how this impacts the marine ecosystems and gain

Image 5 Trainees collecting Secchi Disk data on-board *Challenge Wales* © Challenge Wales

an understanding of the shipboard pollution controls. Young people may also visualise land-based pollution, such as single use plastics either at sea or in coastal areas as they come into port. This may emphasise to young people why it is everyone's responsibility to care for the ocean and their behaviour on land has a direct effect upon it (Image 5).

Principle Seven: The Ocean Is Largely Unexplored

The most decisive way for sail training operators to make a direct contribution to achieving this principle is through partnership and collaboration. Sail trainers are mariners and mariners live/work at sea. Sail training operators could contribute to the issues associated with the ocean being largely unexplored if the empirical data created through the activities suggested above (and more) were fed to larger third-party organisations such as academia, government agencies, and private entities. This sharing of raw data could make real contributions to resolve the current knowledge gaps in ocean exploration, potentially inspiring future generations to become the next ocean investigators, innovators, modellers, scientists, and surveyors too. Although there is evidence to suggest that relationships are well established in the USA and

Image 6 *Pelican of London* © Shield Media Services

Canada, it appears that this is an area that needs development within Europe and beyond. However, operators such as *Statsraad Lehmkuhl* are taking the lead with their One Ocean initiative (see section "Linking Sail Training to SDG 14").

The more operators understand the direct relationship and value of embracing ocean literacy into their programmes, developing collaborations and partnerships with third-party stakeholders; the amount of empirical data produced by the sail training domain ought to increase. This could then provide further recognition through contributing to the Blue Economy and ultimately blue growth (Image 6).

An Analysis of Ocean Literacy Activities in the Tall Ships America Fleet

The raw data from an independent internal audit of the TSA fleet, by Tebeau (2020) in the first quarter of 2020 was provided for further analysis. This audit evaluated the number of vessels facilitating programmes that link to the Ocean Literacy Principles and the USA's Next Generation Science Standard. Phase one of this project provided data to identify trends and gaps regarding how Ocean Literacy Principles are currently facilitated across the TSA fleet.

Table 1 Number of vessels within the TSA fleet audited providing ocean literacy education

	Number in the TSA fleet		% of the TSA fleet		% of survey trainees		Vessel ratio
No initial assessment or audit[a]	31		21.2		N/A		1: 4.7
Initial assessment made[b]	115		78.8		100.0		1: 1.3
OL not identified[c]	85		58.2		73.9		1: 1.7
OL identified[d]	30		20.5		26.1		1: 4.9
OL identified, not audited[e]		10		6.8		8.7	1: 14.6
OL identified, audited[f]		20		13.7		17.4	1: 7.3
Total number of vessels in TSA fleet[g]	146		100		100		

g = a + c + d; d = e + f; c = b − d

Number of Vessels in the Tall Ships America Fleet Audited

The data presented in Table 1 and Fig. 2 identifies that:

- Almost 80% of the fleet had an ocean literacy initial assessment.
- Approximately one-third of the fleet have been identified to provide some sort of OL education as part of their sail training programme
- Over half of the fleet have been identified as not providing ocean literacy education.
- 1: 4.9 vessels have been identified to provide ocean literacy education, compared to 1: 1.7 vessels been identified as not providing ocean literacy education.
- One in five vessels has yet to undertake an initial assessment, almost the same ratio as ocean literacy education vessels (rounded up to the next whole vessel).

Number of Vessels Audited that Provide Ocean Literacy Principles Education

The data in Table 2 identifies the number of vessels audited providing education activities facilitating the Ocean Literacy Principles and which elements of each principle are provided. The trends recognised from this data are that:

- Principle five is the most popular to teach, followed by principle six then principle one.

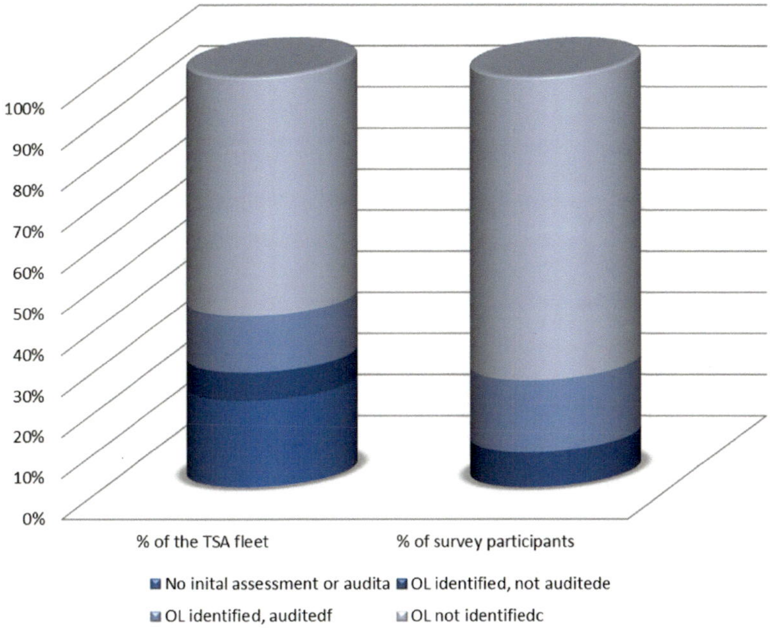

Fig. 2 Number of vessels within the TSA fleet audited providing ocean literacy education

Table 2 Number of audited vessels that provide activities for each of the Ocean Literacy Principles

Ocean Literacy Principle	Total number of vessels	Specific element of each principle achieved									Ave. no. of vessels
		a	b	c	d	e	f	g	h	i	
1	12	5	7	8	6	10	8	1	5	0	6
2	8	5	4	6	4	4					5
3	7	7	1	2	2	1	1	1			2
4	6	5	1	2							3
5	20	19	18	6	18	19	19	8	9	12	14
6	18	5	2	2	11	8	8	5	8		6
7	3	0	0	0	2	2	1				1

- Within principle five, elements 5a, 5e, and 5f are achieved the most, followed by 5d and 5b.
- Only one vessel of the survey sample does not facilitate principle five.
- Principle seven is the least achieved principle, with three of the elements not catered for. Cross referencing to Fig. 3, this is also only provided to those above compulsory education age.
- Over 50 of the audit fleet facilitate at least one element of principle one, five, and six.

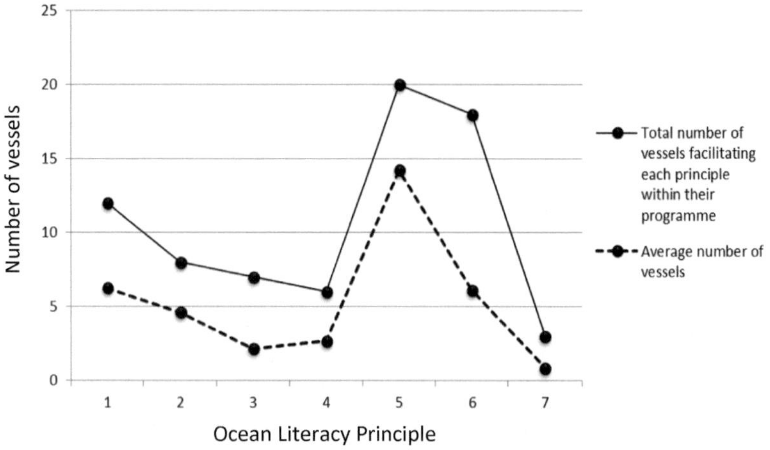

Fig. 3 Vessels providing Ocean Literacy Principles education

- Of the vessels that facilitate each principle, at least 50% facilitate one element of each principle.
- Of the vessels that facilitate each of the principles, only principle five is facilitated by at least 50% of the audited fleet.
- At least 50% of the total number of vessels facilitating each principle teaches at least half of the elements of principle one, two, and five.

There were two vessels that did not identify which elements of each of the principles they provided. Therefore, these vessels were included in the number of vessels providing principle education, but not in the breakdown of each element of the principles. Consequently, 10% of leeway ought to be considered when analysing the results of this data set.

What the data also suggests is that although ocean literacy providers within the audited fleet may contribute significantly to principle five and reasonably to principle one, two, and six, there is a significant gap in facilitating the other principles. Further analysis would be needed to identify what could be the reason behind this shortfall.

Number of Vessels Facilitating Ocean Literacy Principles Education Per School Grade/Age Group

As a comparison, this set of data focuses on the number of vessels that provide OLF activities for each school grade/age group. This provides an indication of what age groups are currently been provided for and where development could be needed to improve ocean literacy education across the age groups educated by sail training operators. The audit results suggest:

Table 3 Age of school grades in the USA school system

Grade	K*	1	2	3	4	5	6	
Age (years)	5–6	6–7	7–8	8–9	9–10	10–11	11–12	
Grade	7	8	9	10	11	12	u/g**	A***
Age (years)	12–13	13–14	14–15	15–16	16–17	17–18	18–21	21+

*Kindergarten; **u/g—undergraduate; A*** adult

- Grades 5–12 participate in activities related to all the principles.
- The target age group appears to the grades 6–8, followed by 4 and 9–12. However, there is approximately a 40% reduction in the number of vessels that cater for grade 4 and 9–12.
- Principles five and six are the most popular for grades 3–5 and 9–12.
- Only two vessels provide OLF-related education for undergraduate students and one vessel caters for adult learning.
- The most popular principle to teach is principle five, followed by principle six and one.
- Principle three is not introduced to grades K–2.
- Principle seven is not introduced to grades K–4.
- The most popular grade to teach ocean literacy activities is grade 6.
- Principle seven is the least popular across this data sample.

What cannot be determined from this information is if the sail training programmes contribute to a long-term education programme through partnerships with mainstream education institutions. Nor does it identify if trainees return to participate in programmes as they get older to further their knowledge through the sail training experience. A deep dive analysis of individual operators would be required to investigate this further (Tables 3 and 4).

Although it appears that sail training is educating young people across the entire OLF, the data from this initial audit may challenge this. It suggests that the sail training domain could be under achieving in creating ocean literate citizens. This is not to say that an individual operator could not be performing to a high standard. Instead, it is as a collective, the sail training domain may not be performing to its optimum capability. Further detailed investigation would need to be undertaken, focusing upon individual operators to obtain a more in-depth understanding.

However, in relation to the target age group, sail training operators appear to make good progress. They deliver above average in respect of educating grades 4–12 (9–18 years), supporting the need to develop awareness from a young age (see section "Creating Ocean Literate Citizens Through Experiencing the Marine Environment") (Figs. 4 and 5).

Table 4 Number of audited vessels that provide Ocean Literacy Principles education to each school grade/age group

Ocean Literacy Principle	School grade/age group								
	K	1	2	3	4	5	6	7	8
1	2	2	2	6	6	6	7	7	7
2	2	2	4	3	3	3	4	4	4
3	0	0	0	2	2	2	7	7	7
4	2	2	2	3	3	3	6	6	6
5	4	6	6	8	10	10	14	14	14
6	3	5	5	6	9	9	13	12	12
7	0	0	0	0	0	2	2	2	2
Mean average	2	2	3	4	5	5	8	7	7

Ocean Literacy Principle	School grade/age group						Mean average	
	9	10	11	12	u/g	A		
1	5	5	5	5	2	0		5
2	2	2	2	2	2	0		3
3	3	3	3	3	0	0		3
4	2	2	2	2	0	0		3
5	10	10	10	10	2	1		9
6	9	9	8	8	2	0		8
7	1	1	1	1	2	0		1
Mean average	5	5	4	4	1	0		
Mean average per school grade/age group)							4.8	
Mean average per Ocean Literacy Principle								4

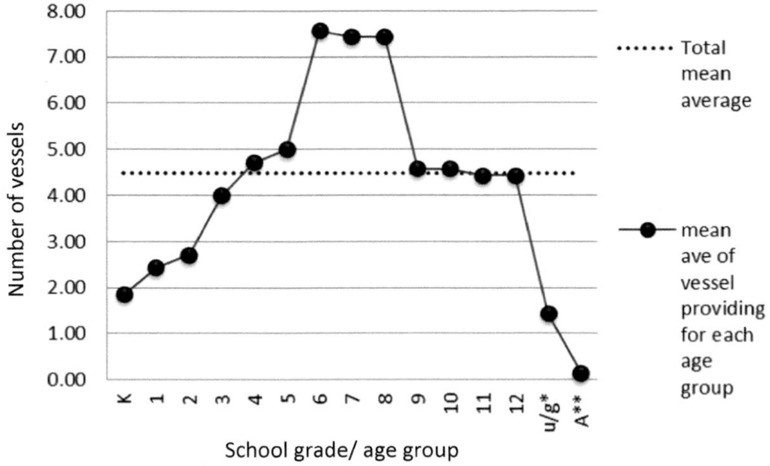

Fig. 4 Mean average of vessels providing Ocean Literacy Principles education for each school grade/age group

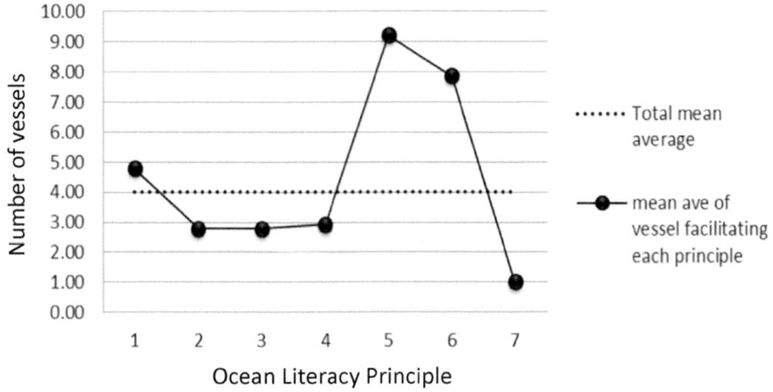

Fig. 5 Mean average vessels facilitating Ocean Literacy Principles education

Shaping Maritime Alternative Education to Be Ocean Literate Through Supporting Sustainability

Internationally, the sail training domain has a long-term commitment to sustainability and protecting the marine environment. In collaboration with Sail Training International (STI), the Foundation for Environmental Education created the voluntary scheme 'STI Blue Flag Scheme' in 2010. Sail Training International (2018) states that there are five overarching principles for operators to adopt, relating to responsible waste and water management, pollution prevention, recycling, and responsible purchasing in particular regarding protected species or underwater archaeological artefacts. However, despite the resonance of ocean literacy and connection to SDG 14, at the time of publication, the direct link between the STI Blue Flag scheme, SDG 14, and the OLF has yet to be identified.

An Overview of the UN Sustainable Development Goals

In 2015, the United Nations SDGs were adopted by 193-member state countries. Building on from the eight Millennium Development Goals, the 17 SDGs form part of the wider 2030 Agenda for Sustainable Development. These goals focus to address issues associated with the economy, the environment, and social impact. Within these 17 goals, there are 169 targets, called deliverables to assist the UN and member states to track goal progress throughout the agenda's lifespan. Thomson (2015) suggests that the SDGs *are "a multi-year process involving civil society, governments, the private sector and academia*". The SDGs are an integrated 'Call to Action', recognising that the deeds focused in one area will affect the outcome of other areas too. UNDP (2020) highlights because development is interconnected; balance is needed within social, economic, and environmental sustainability. The SDGs are expected to be a framework for the member states agendas and policies throughout the timescale of the 2030 Agenda for Sustainable Development. To achieve the SDGs, member states require entities at national, regional, and local levels such as academia, business, civil society, NGOs, and the science community to contribute; with that an awareness of the national delivery strategy is crucial across all levels of society.

Understanding the Purpose of the UN Decade of Ocean Science for Sustainable Development

In preparation for the 5–9 June 2017 high-level UN Ocean Conference which supported the implementation of SDG 14, the UN voluntary commitments register was opened on 17 February 2017. This conference paid special attention to the health of the ocean, advancing the implementation of SDG 14. UNESCO (2019a) identifies that on 5 December 2017 the United Nations proclaimed a Decade of Ocean Science for Sustainable Development (2021–2030). It provides a common framework ensuring that ocean science can fully support countries' actions to sustainably manage the ocean. The UN General Assembly mandated that the IOC-UNESCO will coordinate the Decade's preparatory process (2018–2020). The primary objective of IOC-UNESCO as directed by the UN General Assembly (UNGA) was "*to prepare and coordinate the development of an Implementation Plan for the Decade during the Preparatory Phase (2018-2020)*" (UNESCO 2019b). The Roadmap for the UN Ocean Decade is the initial guide for the implementation plan. Version two was published on 10 June 2018 and the formation of the implementation plan is supported by four interlinked mechanisms (Executive Planning Group, Stakeholder Forum, Regional Workshops, and Global Planning meetings).

Linking Sail Training to SDG 14

It could be suggested that presently, the European sail training fleet appears to be less familiar with OLF and is more focused upon their contributions to the UN Ocean Decade through the SDGs. Sail training operators that advocate SDG 14 by adopting ocean science activities as part of their education programme enable trainees to develop a better understanding of the oceans systems and processes. By encouraging exploration, experimentation, and discovery, these operators create opportunities to facilitate ocean literacy through the SDGs. By adopting such activities, not only can operators contribute to SDG 14, they can advance ocean literacy by supporting the UN Ocean Decade as ocean literacy is a goal supporting sustainable outcomes.

Although it could be assumed that sail training can contribute to several SDGs, particularly SDG 4: quality education, sail training activities equally can intrinsically link to SDG 14, reiterating the intergraded Call to Action of the SDGs (see section "An Overview of the UN Sustainable Development Goals"). Table 5 highlights examples of European sail training operators whose current activities could or do contribute to SDG 14 deliverables.

To advance ocean literacy within Europe, initiatives have developed by a small number of UK sail training operators. Two examples are the 'Ocean Literacy for Sail Trainers' programme developed by the Island Trust to educate sail training professionals and volunteers about the OLF and Challenge Wales has launched an accredited learning programme through Agored Cymru, raising awareness of environmental issues through sail training. Also, *Statsraad Lehmkuhl*, one of the largest tall ships in the world, sets sail on their 18-month 'One Ocean' voyage in August 2021 as part of the UN Ocean Decade. In preparation for this, the vessel has installed scientific research equipment. Although the vessel makes way under sail for 70% of the time, an energy-efficient battery has been fitted, further reducing their CO_2 emissions to become the most environmentally friendly vessel in their class (Image 7).

Table 5 European sail training organisations that contributing to SDG 14

SDG deliverable and description (United Nations 2020)		Examples of contributing sail training organisations
14.1.	By 2025, prevent and significantly reduce marine pollution of all kinds, in particular from land-based activities, including marine debris and nutrient pollution	Challenge Wales, UK Cirdan Trust, UK Sea Cadets, UK
14.2	By 2020, sustainably manage and protect marine and coastal ecosystems to avoid significant adverse impacts, including by strengthening their resilience, and take action for their restoration in order to achieve healthy and productive oceans	The Island Trust, UK
14.3	Minimise and address the impacts of ocean acidification, including through enhanced scientific cooperation at all levels	Bark Europa, Netherlands Gulden Leuw, Netherlands Stad Amsterdam, Netherlands Statsraad Lehmkuhl, Norway
14.A	Increase scientific knowledge, develop research capacity and transfer marine technology, taking into account the Intergovernmental Oceanographic Commission Criteria and Guidelines on the Transfer of Marine Technology, in order to improve ocean health and to enhance the contribution of marine biodiversity to the development of developing countries, in particular small island developing States and least developed countries	
14.C	Enhance the conservation and sustainable use of oceans and their resources by implementing international law as reflected in UNCLOS, which provides the legal framework for the conservation and sustainable use of oceans and their resources, as recalled in paragraph 158 of The Future We Want	It is a legal requirement for all vessels, to comply with MARPOL (IMO 2020). In addition, those registered on the Sail Training International/Foundation of Environmental Education's Blue Flag Scheme could be a significant contributor

Image 7 *Challenge Wales* and *Adventure Wales* © Challenge Wales

Conclusion

It could be suggested that sail training and ocean literacy have an interlinking relationship that is so interwoven; it could be intuitive for sail training operators and professional crew to develop ocean literate citizens. However, because it is instinctive, professional crew may not be completely aware of how they deliver ocean literacy within their current syllabus. Whilst some operators have begun to identify and facilitate links to the OLF and SDG 14, as a whole sail training may not be operating at its full capability in support of creating ocean literate citizens. The limited amount of previous research, the audit discussed in section "An Analysis of Ocean Literacy Activities in the Tall Ships America Fleet", the introduction to SDG 14 activities in section "Shaping Maritime Alternative Education to Be Ocean Literate Through Supporting Sustainability", initiatives being developed by individual operators and this chapter overall may begin to overcome some of these knowledge gaps.

However, as individuals, we cannot protect the ocean if we don't start from within. To comprehend where we need to change in order to empathise with the ocean's suffering, we need to identify how we connect to the ocean, physically, emotionally, and psychologically long term. To become the change agents who reverse the detrimental impact imposed thus far, we need to harmonise and react collectively. When implementing activities such as those suggested in section "Creating Ocean Literate Citizens Through Experiencing the Marine Environment", sail training could show young people otherwise unaware of or familiar with the ocean how to make that deeper connection.

However, this unification comes with many difficulties. As a small section of the maritime domain and an unseen education provider, it can sometimes be overlooked by society. Particularly within the UK, many sail training organisations are charities or have limited funds. This could consequently undermine its ability to deliver ocean literacy effectively and add value to its position as a leading alternative education space. Should policy makers, funding providers, and other third-party stakeholders decide to collaborate and invest in sail training programmes; these barriers for development could be overcome.

The need for empirical data and research especially related to ocean literacy is still limited. As mentioned in section "Introduction: Blue Outdoor Education—Sail Training's Relationship with Ocean Literacy", sail training research overall is still embryonic. This is especially the case in respect of ocean literacy activities. The hope is that this chapter could serve as a 'Call to Action' to those who are inspired to conduct their own research activities related to sail training, ocean literacy, and alternative education to do so. Internally, it would be prudent for the sail training domain to conduct a deep dive elevation of the fleet, including gap analysis, SWOT analysis, and review of the detailed current capabilities and shortfalls regarding the provision of ocean literacy education (see section "An Analysis of Ocean Literacy Activities in the Tall Ships America Fleet"). A critical evaluation of the STI Blue Flag Scheme, linking its goals, requirements, and outcomes to the OLF and SDG 14 could also be beneficial (see section "Shaping Maritime Alternative Education to Be Ocean Literate Through Supporting Sustainability").

Moving forward, developing partnerships, particularly in relation to the UN Ocean Decade to identify where operators already provide ocean literacy education and how it can be advanced within their own programme(s) would be advantageous. By doing so, sail training operators could increase their scope to deliver OLF principle seven activities and contribute to SDG 17: partnerships.

Where operators can collaborate with formal education providers, management and policy makers, more use of the cross curricula links that recognise and build upon the importance of outdoor learning within the marine environment can be achieved. By actively injecting ocean literacy into formal education supported by sail training programmes, there could be scope to create the most authentic form of blue school.

Through collaborative and partnership working, sail training and ocean literacy could be more effective together in creating ocean literate citizens. It could also positively shape the next generation of youth and modern society, creating and maintaining a more sustainable future for all. After all, without the ocean, there would be no sail training, therefore we have a duty of care over it, are accountable for our actions that affect it, and have a responsibility to teach others to respect it.

References

Aguiar LA (ed) (2015) Sail tall ships! A directory of adventure and education under sail, 21st edn. American Sail Training Association/ Pucino Print Consultants, North Kingston, RI, USA

Allison P, Seaman J (2017) Experiential education. In: Peters MA (ed) Encyclopaedia of educational philosophy and theory. Springer Nature Singapore Pte Ltd, Singapore, p 1

Aron LY (2003) Towards a typology of alternative education programs: a compilation of elements from the literature. The Urban Institute, Washington, DC

Bauer I (2015) Approaching geographies of education differANTly. Child Geographies 13(5):620–627. https://doi.org/10.1080/14733285.2015.1044197

Bohatka L, Bardone E, Brikker M (2017) Ark of inquiry recommendations and guidelines. Ark of inquiry consortium

Carnie F (2003) Alternatives in education—a guide. Routledge Falmer, London

Cote J, Strachan L, Fraser-Thomas J (2008) Participation, personal development, and performance through sport. In: Holt N (ed) Positive youth development through sport. Routledge, London, pp 38–39

Dupont S, Fauville G (2017) Ocean literacy as a key toward sustainable development and ocean governance. In: Nunes P, Svensson L, Markandya A (eds) Handbook on the economics and management of sustainable oceans. Edward Elgar Publishers, Cheltenham, pp 519–537

Fauville G, McHugh P, Domeganc C, Mäkitalo Å, Møllerd LF, Papathanassiou M et al (2018) Using collective intelligence to identify barriers to teaching 12–19 year olds about the ocean in Europe. Mar Policy 91:85–96

Foley RM, Pang LS (2006) Alternative education programs: program and student characteristics. High Sch J 89(3):10–21

Gordon S, Harcourt-Smith K, Hay K, Priest S (1996) Case study of "Blue watch" on STS Leewin. J Adventure Educ Outdoor Leadersh 13(1):4–7

Gordonstoun School (2020) Ocean Spirit of Moray. http://www.gordonstoun.org.uk/unique-curriculum/outdoor-education/ocean-spirit-of-moray. Accessed 29 Apr 2020

Holloway SL, Heike J (2012) Geographies of education and learning. Trans Inst Br Geogr 37(4):482–488

Holt N, Sehn Z (2008) Processes associated with positive youth development and participation in competitive youth sport. In: Holt N (ed) Positive youth development. Routledge, London, p 30

IMO (2020) International convention for the prevention of pollution from ships (MARPOL). http://www.imo.org/en/About/Conventions/ListOfConventions/Pages/Int. Accessed 29 Apr 2020

Kraftl P (2014) What are alternative education spaces—and why do they matter? Geography 99:128–138

Lyth LE (2014) Sail training what is it all about: history, ethos and diversity of the organisations and crew. Liaison Officer's supporting guide. Available via Océanie Ltd. Training support guide. https://www.oceanie.co.uk/presentations. Accessed 26 May 2020

Magadley W, Amara M (2018) Alternative education in the Arab-Palestinian Society in Israel: a panacea. Al-Qasemi Journal of Islamic Studies 3(1):95–118

McCarthy L (2011) Learning processes taking place during residential adventure activity programmes. Dissertation, University of London

McCulloch K (2002) Four days before the mast: a study of sail training in the UK. Dissertation, University of Edinburgh

McCulloch K (2004) Ideologies of adventure: authority and decision making in sail training. J Adventure Educ Outdoor Learn 4(2):185–197

Mystic Seaport (2020) JOSEPH CONRAD: squared rigged ship. http://www.mysticseaport.org/locations/vessels/conrad. Accessed 22 May 2020

NMEA (2020) Ocean literacy framework. https://static1.squarespace.com/static/5b4cecfde2ccd188cfed8026/t/5eb99cc530a3d76767dc7aea/1589222614314/OceanLiteracyGuide_V3_2020.pdf. Accessed 25 June 2020

Noble J, Kenley A, Patel S (2017) Sail training theory of change, evidence review and measurement framework. https://uksailtraining.org.uk/images/stories/Sailtrainingworksevidence/public-ASTO-Theory-of-Change-measurement-framework.pdf. Accessed 29 June 2020

Olive E (2003) Real alternatives. Reclaiming Child Youth 12(2):66–67

Petty G (2009) Teaching today, 4th edn. Nelson Thornes, Cheltenham

Raywid MA (1994) Alternative schools: the state of the art. Educ Leadersh 52(1):26–31

Round Square (2020) Ideals. https://www.roundsquare.org/about-us/what-we-do/ideals/. Accessed 29 Apr 2020

Sail Training International (2018) Help us protect our marine environment. https://sailtraininginternational.org/app/uploads/2018/01/Blue-Flag-Scheme-application-form.pdf. Accessed 28 Apr 2020

Sail Training International (2020) Introduction. https://sailtraininginternational.org/sailtraining/introduction/. Accessed 28 Apr 2020

Santin S, Santoro F (2017) The Ocean as a teaching tool: the first MOOC on Ocean literacy. In: Geophysical Research Abstracts, 19. EGU General Assembly, Vienna, Austria

Santoro F et al (eds) (2017) Ocean literacy for all: a toolkit. Paris IOC Manuals and Guides, 80, IOC/ UNESCO and UNESCO Venice Office

Sliwka A (2008) The contribution of alternative education. In: OCED, Innovating to learn, learning to innovate. OECD/CERI, Paris

Smith A, Thomson MM (2014) Alternative education programmes: synthesis and psychological perspectives. Educ Psychol Pract 30(2):111–119. https://doi.org/10.1080/02667363.2014.891101

Tebeau P (2020) Ocean Literacy and education under Sail, Email to L. Lyth. Accessed 9 Mar 2020–25 June 2020

Thomson S (2015, September 16) What are the sustainable development goals? In: Ocean dialogues, world economic forum. https://www.weforum.org/agenda/2015/09/what-are-the-sustainable-development-goals/. Accessed 29 June 2020

Tierney G (2016) Agency outside the margins: Identity development of youth attending alternative high schools. Dissertation, University of Washington

UNDP (2020) What are the sustainable development goals? In: Ocean dialogues, World Economic Forum. https://www.undp.org/content/undp/en/home/sustainable-development-goals.html. Accessed 29 June 2020

UNESCO (2019a) What is the United Nations Decade of Ocean Science for Sustainable Development? https://www.oceandecade.org/about?tab=our-planning#top_of_page. Accessed 29 June 2020

UNESCO (2019b) United Nations Decade of Ocean Science for Sustainable Development (2021–2030) Preparatory Process. https://en.unesco.org/ocean-decade/preparatory-process. Accessed 29 June 2020

United Nations (2020) Sustainable Development Goal 14: conserve and sustainably use the oceans, seas and marine resources for sustainable development. https://sustainabledevelopment.un.org/sdg14. Accessed 29 Apr 2020

Vadeboncoeur J (2009) Spaces of difference: the contradictions of alternative educational programs. Educ Stud 45(3):280–299

Watkin Beck C (2016) Alternative education—on a new-totalitarian road? Home education in Norway, Project. https://www.researchgate.net/publication/312592953_Alternative_education_-_on_a_new-totalitarian_road. Accessed 29 June 2020

Widiarti R (2010) Marine environmental education for children—Indonesia. Sharing innovative experiences: examples of successful experiences in coastal community development, UNDP/Special Unit for South-South Cooperation/IOI Volume, 16:85–97

Wright R (2012) Occupational therapy and sail training. Occup Ther News 20(3):29

Applications

The Importance of Ocean Literacy in the Mediterranean Region—Steps Towards Blue Sustainability

Melita Mokos, Maria Cheimonopoulou, Panayota Koulouri, Monica Previati, Giulia Realdon, Francesca Santoro, Athanasios Mogias, Theodora Boubonari, Alessio Satta, and Christos Ioakeimidis

Abstract Ocean Literacy (OL) is considered to be important for raising awareness of the people concerning conservation, restoration and sustainable use of the ocean and its resources. Addressing environmental issues related to the Mediterranean Sea and increasing OL can be the first step to achieve Sustainable Development Goal 14 (focusing on the ocean) within the UN Agenda 2030 in the Mediterranean region. The adaptation of the Ocean Literacy Framework to the specificities of the Mediterranean Sea can introduce knowledge about different natural, geographical and social components of its marine life and society. This can help different stakeholders (e.g.

The views and opinions of Maria Cheimonopolou in this chapter are her own and do not necessarily reflect those of her institution.

M. Mokos (✉)
Department of Ecology, Agronomy and Aquaculture, University of Zadar, Zadar, Croatia
e-mail: melita.mokos@emsea.eu

M. Cheimonopoulou
Hydrobiological Station of Pella, Ministry of Rural Development and Food, Edessa, Greece

P. Koulouri
Institute of Marine Biology, Biotechnology and Aquaculture, Hellenic Centre for Marine Research, Crete, Greece

M. Previati
Underwater Bio-Cartography (U.BI.CA S.R.L.), Genoa, Italy

G. Realdon
UNICAMearth Group, University of Camerino, Camerino, Italy

F. Santoro
Intergovernmental Oceanographic Commission of UNESCO, Venice, Italy

A. Mogias · T. Boubonari
Democritus University of Thrace, Alexandroupolis, Greece

A. Satta
Mediterranean Sea and Coast Foundation, Cagliari, Italy

C. Ioakeimidis
UN Environment Programme/ Mediterranean Action Plan (UNEP/MAP) Coordinating Unit, Barcelona Convention Secretariat, Athens, Greece

K. C. Koutsopoulos and J. H. Stel (eds), *Ocean Literacy: Understanding the Ocean*, Key Challenges in Geography, https://doi.org/10.1007/978-3-030-70155-0_9

teachers, educators, scientists, policy- and decision-makers, private sector) to better understand the influence that people have on the Mediterranean Sea, and the influence that the Mediterranean Sea has on them. This chapter gives an insight into the main pressures on the natural environment of the Mediterranean Sea, the legal framework for its protection and sustainability, the importance of integrated coastal zone management and marine protected areas in relation to ocean literacy and the role of education in the creation of an ocean-literate society right across the Mediterranean region. The geographical approach contributes significantly to the exploration and understanding of the relationship between the environment and human communities. Formal and non-formal education in different scientific fields, e.g. geography, biology, etc., as well as increased awareness about the interrelation between people and the Mediterranean Sea, could lead to increased protection and conservation of marine wildlife, sustainable management of Mediterranean marine resources and therefore sustainable blue development of the region.

Keywords Ocean Literacy · SDG 14 · Mediterranean Sea · Mediterranean Sea Literacy · Environmental Education · UN Decade of Ocean Science for Sustainable Development

Introduction

The "Sea in the middle of the Earth", the Mediterranean Sea, is the largest and deepest enclosed sea on earth, the cradle of western civilization and one of the most important global biodiversity hotspots, with iconic species worthy of conservation. It is a natural laboratory for geologists, naturalists, biologists and other scientists, and an inspiration for photographers, writers and people who love nature; an extraordinary and fragile treasure chest of biodiversity which needs to be protected. The Mediterranean Sea is also a crucial route for the global economy and trade, geopolitically important, home for approximately 500 million people and a holiday destination attracting more than 300 million tourists per year.

The unique geographic and oceanographic features of the Mediterranean Sea combined with anthropogenic pressures such as coastal urbanization, tourism, overfishing, marine aquaculture, pollution and climate change (Fernandes et al. 2017; Grigorakis and Rigos 2011; Piroddi et al. 2017) have inevitably lead to crucial alterations in the Mediterranean Sea environment. In particular, these pressures affect species, biological communities, ecosystem functioning and its capacity to provide essential goods and services to the society (Guidetti et al. 2014). A geographical approach to these issues can help us to visualize their spatial distribution on different scales, from the global to the local, as well as their potential impact on society and the ability to provide solutions. Geographic tools, such as world and geological maps, bathymetric maps, global ocean circulation models, Geographic Information Systems (GIS), Global Positioning System (GPS), European Atlas of the Sea, Google Earth and Google Maps, can make valuable contributions to our knowledge concerning the relationships between marine natural elements and societal phenomena and processes in the Mediterranean region, thus bridging the environmental and social sciences.

In 2017 the United Nations convened a high-level Our Ocean Conference to support the implementation of Sustainable Development Goal 14 (SDG 14): Conserve and Sustainably Use the Oceans, Seas and Marine Resources, of the 2030 Agenda for Sustainable Development. One outcome of this conference was an intergovernmentally agreed declaration, a "Call for action", who's Article 13.a reads as follows: "*Support plans to foster ocean-related education, for example as part of education curricula, to promote ocean literacy and a culture of conservation, restoration and sustainable use of our ocean*", hence emphasizing the importance of ocean literacy. This demonstrates the strong commitment of the UN to conserve and manage ocean and marine resources for sustainable development both now and in the future. Moreover, the UN has declared a Decade of Ocean Science for Sustainable Development 2021–2030 to support and achieve SDG 14, which simultaneously supports other SDGs (Ryabinin et al. 2019). The Decade aims to achieve major scientific and technological progress by generating seven societal outcomes, one of which is "an inspiring and engaging ocean where society understands and values the ocean in relation to human wellbeing and sustainable development", which includes considerable advancement and increase of ocean literacy in society, from education and school curricula, to decision-makers and the public at large (UN 2020). Moreover, the Intergovernmental Oceanographic Commission of UNESCO (IOC-UNESCO) is currently developing the Ocean Literacy Strategy—ocean literacy for the UN Decade of Ocean Science for Sustainable Development—in order to advance ocean literacy during the UN Decade.

In order to achieve SDG 14 in the Mediterranean region, citizens need to know and be aware of both the sea-related benefits and the threats that might cause the loss of those benefits. Education, which is essential if SDGs are to be achieved, has its own dedicated Goal 4, which aims to "*ensure inclusive and equitable quality education and promote lifelong learning opportunities for all*". Therefore, empowering citizens to make environmentally responsible decisions through ocean literacy education and Mediterranean Sea Literacy (MSL) can make a substantial contribution towards the achievement of SDG 14 in the Mediterranean.

The Mediterranean Sea Literacy Guide was developed by the regional Mediterranean group of the European Marine Science Educators Association (EMSEA-Med), a group consisting of scientists and educators whose aim is to relate the concept of ocean literacy to the Mediterranean region. It is based on ocean literacy principles and concepts adapted to the specificities of the Mediterranean Sea (the latest version of the Mediterranean Sea Literacy guide is available at: http://www.emsea.eu/default.php). MSL guide introduces knowledge about different natural, geographical and social components of marine life and society related to the Mediterranean Sea. Its goal is to help different parts of society including teachers, educators, scientists, policy/decision-makers and the private sector, to better understand the vital importance of the two-way interaction between the Mediterranean Sea and its regional human societies. Raising awareness and creating an ocean-literate society

Fig. 1 NGO voluntary teaching about marine animals' identification for creating an ocean-literate society (project from Liguria, Italy, since 2012) (Photo by: Informare)

can contribute to achieving conservation, restoration and a sustainable blue economy in the "Mare Nostrum".

This chapter will give an insight into the main pressures on the natural environment of the Mediterranean Sea, the legal framework for its protection and sustainability, the importance of integrated coastal zone management and marine protected areas in relation to ocean literacy and the role of education in the creation of an ocean-literate society across the Mediterranean (Fig. 1).

Anthropogenic Pressures Affecting the Mediterranean Sea and Its Resources

Coastal Urbanization

In ancient times, many urban areas around the Mediterranean Sea were located inland from the coast for defensive reasons (Greek urbanization model; UNEP 2001). However, in recent times this well-established pattern changed rapidly, as secondary urban areas were built along the shoreline (UNEP 2001). These new settlements, along with others established initially on the coast (UNEP 2001), have been growing

Fig. 2 An example of the coastal urbanization in Genoa, Italy (Photo by: M. Stefanolo)

ever since, in many cases in a rapid and uncontrolled manner, thus fundamentally transforming Mediterranean coastal environments (Fig. 2).

Nowadays, urbanization is an important driver of change in land use in the Mediterranean basin (Garcia-Nieto et al. 2018). Approximately one-third of the Mediterranean population is situated in the coastal area. The population of the coastal countries is predicted to grow from 466 million in 2010 to 529 million by 2025 (UNEP/MAP 2016).

Demographic growth, rural depopulation and tourism development are among the growth factors of the coastal urbanization phenomenon (Enne et al. 2009; UNEP 2001). High urban occupation leads directly and/or indirectly to soil loss, coastline erosion, reduction of water resources, pollution of groundwater, surface and seawater biodiversity loss, ecosystem fragmentation, soil and groundwater salinization, irreversible loss of fragile, coastal ecosystems (e.g. wetland areas, dune systems), desertification, high flood risk, etc. (Enne et al. 2009; Malak et al. 2011; UNEP/MAP/PAP 2001). In addition, the densely populated, low-elevation coastal Mediterranean zone, along with its fragile ecosystems, is expected to be highly impacted by sea-level rise (Wolff et al. 2018), thus affecting the development of coastal planning, e.g. tourism, human migrations (Galassi and Spada 2014).

Tourism in the Mediterranean

The Mediterranean basin is one of the most popular tourist destinations in the world with over 30% of total global tourism occurring in this region. While tourism in the Mediterranean contributes largely to the region's economic production, it also plays a critical role in the deterioration of the marine environment (Randone et al. 2017).

Coastal and maritime tourism therefore exerts immense pressure on the Mediterranean region. The development of tourism in coastal and marine regions changes the original features of the visited destinations which had attracted tourists in the first place. Decades of mass tourism have led to the decline of previously pristine areas, thus threatening the health of the iconic Mediterranean coasts. Unfortunately, according to Zahedi (2008), nations prioritize immediate economic benefit before long-term environmental sustainability and protection.

Marine and coastal ecosystems are threatened by mass tourism development, which is one of the main drivers of ecosystem degradation in the region, intensifying the littoralization process, and consequently, resulting in the loss of natural resources and the accumulation of waste (Fosse and Le Tellier 2017; Randone et al. 2017; Zahedi 2008).

Cruise ship tourism in the Mediterranean makes up 18.7% of all world cruise destinations (Dowling and Weeden 2017; Ocean Atlas 2017). The United Nations Environment Programme (UNEP) has identified tourist ships as one of the main pollution sources in the marine environment (Carić and Mackelworth 2014). Even though modern ships have reduced their environmental impacts, they are still a significant source of air, noise and marine pollution. Ships' biofouling and ballast water release systems constitute major pathways for the introduction of invasive species into the marine environment of the Mediterranean Sea. In addition, recreational boating has a significant ecological impact on the environment through habitat degradation via anchoring (Fig. 3), construction of marinas and ports, production of waste water and litter, noise, etc. For example, anchoring in the beds of the endemic species

Fig. 3 Mediterranean endemic species *Posidonia oceanica* and *Pinna nobilis* are some of the most threatened species caused by human impacts such as anchoring (Photo by: H. Čižmek)

of *Posidonia oceanica* (Linnaeus) Delile (1813) can have a severe negative impact on this particular habitat and on related organisms (Montefalcone et al. 2006, 2008). Anchors can also damage individual species as evidenced by the mass mortality event of the noble pen shell, *Pinna nobilis* (Linnaeus, 1758) (Fig. 3) around the western Mediterranean, or the Mediterranean pillow coral, *Cladocora caespitosa* (Linnaeus, 1767) (Vázquez-Luis et al. 2017), and impact community assemblages such as the coralligenous bio-concretions (Gerovasileiou et al. 2009; Milazzo et al. 2002). In addition, mechanical destruction of seagrass beds can cause the erosion of organic carbon stocks conserved in the sediment, which may lead to increased atmospheric CO_2 (Serrano et al. 2016).

Overfishing and Marine Aquaculture

Fish stocks in the Mediterranean Sea have been declining for decades. Targeted or multi-species fisheries are the most common threat to marine fishes, directly affecting 33% of native species in the Mediterranean Sea, and another 18% indirectly which are caught as bycatch species (Malak et al. 2011). In particular, during the last 60 years there was a reduction in the abundance of fish species (~34%) and top predators (~41%) (Piroddi et al. 2017). Most stocks continue to be fished beyond biologically sustainable limits. The European hake, *Merluccius merluccius* (Linnaeus, 1758), is by far the most overexploited species in the Mediterranean, followed by red mullet, *Mullus barbatus* (Linnaeus, 1758), and sardine, *Sardina pilchardus* (Walbaum 1792) (FAO 2018). The iconic and commercially important Atlantic bluefin tuna, *Thunnus thynnus* (Linnaeus, 1758) is listed as endangered in the IUCN Red List of Threatened Species, due to its population decline resulting from decades of overfishing and mismanagement in the Mediterranean (Malak et al. 2011). Nowadays, the Mediterranean and the East Atlantic bluefin tuna stocks show signs of population growth. Although there is uncertainty regarding the level of this recovery, it demonstrates that effective management of international fisheries regarding highly valuable species, overexploited for decades, is still possible (Fromentin and Rouyer 2018). In addition, native Mediterranean cartilaginous fish (e.g. sharks, rays) face a 53% risk of extinction, as they constitute a retained valuable fisheries bycatch (IUCN 2016).

Overfishing causes a reduction in density, biomass and reproductive potential of fish stocks, as well as dramatic changes in the structure and functioning of food webs and in the physical properties of the seafloor (Guidetti et al. 2014). Bottom trawling, in particular, directly causes a reduction in the complexity and availability of benthic habitats (Malak et al. 2011), a decrease in benthic biomass and biodiversity and affects the functioning and productivity of benthic ecosystems (Eigaard et al. 2017). Non-resilient deep-water Mediterranean benthic ecosystems are especially vulnerable (Paradis et al. 2017), as traditional fishing grounds have been shifting to deeper habitats over the last 50 years.

Food resource depletion resulting from overfishing also impacts marine mammals, turtles and birds (Soriano-Redondo et al. 2016; UNEP/MAP 2012). Characteristic

Fig. 4 *Corallium rubrum* is a characteristic Mediterranean habitat-forming species impacted by overfishing (Photo by: Informare)

Mediterranean habitat-forming species, such as the red coral, *Corallium rubrum* (Linnaeus, 1758) (Fig. 4) and *P. oceanica* meadows, as well as inshore rocky habitats, are also impacted (Cattaneo-Vietti et al. 2016). Overfishing of natural fish stocks is a contributing factor which has led to a rapidly growing mariculture sector in the Mediterranean region. Over the recent decades, Gilthead sea bream, *Sparus aurata* (Linnaeus, 1758) sea bass, *Dicentrarchus labrax* (Linnaeus, 1758), have become the most commercially important finfish followed by molluscs as aquaculture species (Grigorakis and Rigos 2011).

Aquaculture production around the Mediterranean and the Black Sea, coming mostly from marine and brackish waters, reached more than 2.3 million tonnes in 2013, having increased by 164% since 1993 (Massa et al. 2017). Consequently, intensive farming of marine animals impacts the Mediterranean marine environment in many ways, such as genetic interactions between native and escaped cultured fish, introduction of alien species, transfer of diseases, release of organic wastes, habitat alteration, etc. (Grigorakis and Rigos 2011).

Pollution

Eighty percent of pollutants in the Mediterranean Sea (UNEP/MAP-MEDPOL/WHO 2008; UNEP/MAP 2012) come from land-based sources. Marine and coastal pollution can be linked to the presence of nutrients, organic matter, microorganisms, heavy metals, persistent organic pollutants (POPs), oil pollution, litter as well as types of energy as underwater sound (European Commission 2017; UNEP/MAP 2012).

Nutrients, Organic Matter and Microorganisms

Eutrophication is the result of nutrient inputs (e.g. dissolved nitrogen and phosphorus) into Mediterranean waters, which predominantly originate from municipal sewage and agricultural fertilizer run-off. It causes the decline of macrophytes leading to their replacement by short-lived algal species, as well as causing radical changes in phytoplankton communities which may result in harmful algal blooms (HABs) (UNEP/MAP 2012). Some micro-algae responsible for HABs produce toxins that may bioaccumulate in organisms, with adverse effects on shellfish, fish, marine birds and mammals including humans (Ferrante et al. 2013; UNEP/MAP 2012). Temporary and prolonged bans on the harvesting and sale of mussels resulting from HABs, have frequently affected molluscan aquaculture in the Mediterranean Sea (UNEP/MAP 2012). Within the Mediterranean Sea, eutrophication is a localized phenomenon, occurring mainly in semi-enclosed coastal areas as in the North Adriatic Sea (nutrient inputs from river Po) (Boesch 2019) and not in oligotrophic open waters (UNEP/MAP 2017).

Organic matter originating from eutrophication processes, mostly from urban and industrial waste waters, has a synergetic effect in depleting oxygen by their decomposition and causing a reduction in light penetration in marine and coastal Mediterranean waters. Benthic communities, including seagrass meadows (Fig. 5), are the first to suffer from significant loss of biodiversity in Mediterranean areas such as those adjacent to sewage outfalls or within urbanized bays (UNEP/MAP 2012). These changes cause a deterioration in water quality and consequently have an impact on tourism.

Sewage effluents containing human and animal excreta (wildlife and domestic animals) are cited as possible pathogenic contaminants of Mediterranean recreational waters (Fewtrell and Kay 2015; UNEP/MAP-MED POL/WHO 2008), commonly causing human enteric illness (Kamizoulis and Saliba 2004; UNEP/MAP-MED POL/WHO 2008).

Phytoplankton species under stress conditions, as well as algal blooms, may be responsible for the formation of marine mucilage affecting Mediterranean coastal areas, especially in the northern Adriatic Sea (Carroni et al. 2015; Danovaro et al.

Fig. 5 Seagrass meadow of *Posidonia oceanica* in the shallow water of coastal area of Liguria, Italy (Photo by: Informare)

2009). The consequences for the marine environment are adverse, resulting in lower ecosystem resilience and damage to tourism and fisheries.

Heavy Metals, Persistent Organic Pollutants (POPs) and Oil Pollution

Atmospheric deposition, run-off from metal-contaminated sites and urban and industrial waste waters represent the major sources of toxic metals (e.g. mercury, lead, cadmium) in the Mediterranean Sea. Heavy metal concentrations (e.g. mercury) in Mediterranean fish have been found to be twice as high as those found in the same species living in the Atlantic Ocean (UNEP/MAP 2012). Risks to Mediterranean ecosystems are also present from the effects of bioaccumulation, not only of toxic metals but also of persistent organic pollutants (POPs) in shellfish and/or top predators such as the bluefin tuna (Chiesa et al. 2016). Disruption of the endocrine and reproductive systems of marine organisms (e.g. Mediterranean swordfish, *Xiphias gladius* (Linnaeus, 1758) is among the recorded effects of POPs, leading to the increase of ecological stress in marine and coastal organisms in general (UNEP/MAP 2012).

Oil pollution in the Mediterranean Sea is linked to major shipping routes in open waters and oil-related facilities (e.g. refineries, terminals and ports) in nearshore waters. The latter generally exhibit higher concentrations of polycyclic aromatic hydrocarbons (PAHs), the most toxic compounds of crude oil, in marine organisms and sediments surrounding these facilities (UNEP/MAP 2012). PAHs are known to have multiple effects at the genetic, cellular, biochemical and physiological levels of various species (UNEP/MAP 2012). The extraction of large oil and gas reserves, recently discovered in the eastern Mediterranean, increases pollution risks with unknown effects on the unique deep-sea eastern Mediterranean ecosystems (Liu et al. 2017).

Marine Litter

In the Mediterranean Sea, while marine litter has been an issue of concern since the 1970s, today it poses a critical, complex and multidimensional problem for the region (UNEP/MAP 2015). Marine litter is found washed ashore along the coastline, floating in the water column and also lying on the seafloor. Plastic materials dominate on the beaches (Fig. 6), accounting for over 80% of the marine litter found there (ICC 2016; UNEP/MAP 2017). Microplastics (<5 mm) have been found in the Mediterranean in concentrations which are among the highest in the world (Suaria et al. 2016).

Ingestion of, or entanglement in, marine litter is among the most significant impacts on marine life. In particular, ingested microplastics by zooplankton enter marine food webs. Zooplanktivorous predators (mesopelagic fish, baleen whales, some sharks) are exposed to microplastics by direct ingestion of contaminated zooplankton and/or accidental ingestion during feeding activity (Fossi et al. 2016). Furthermore, microplastics are carriers of toxic chemicals that bioaccumulate, having

Fig. 6 Frequent image on the Mediterranean beaches: plastic bottle that is destined to fragment into smaller pieces known as microplastics (Photo by: Informare)

Fig. 7 Mediterranean red gorgonians *Paramuricea clavata* (Risso, 1826) (Photo by: Informare)

a major toxicological impact on marine organisms (Fossi et al. 2016; Romeo et al. 2015) and possibly on humans (Seltenrich 2015). Ingested marine debris was found in up to 80% of pelagic turtles *Caretta caretta* (Linnaeus, 1758) in the central Mediterranean in a recent study (Casale et al. 2016). Accidental entanglement has been identified as one of the most important threats to the survival of the world's most endangered seal species, the Mediterranean monk seal, *Monachus monachus* (Hermann, 1779) (Karamanlidis et al. 2008). More than half of the recorded debris in the deep-sea floor in the northwestern Mediterranean come mostly from fishing gear, directly impacting benthic organisms, primarily gorgonian corals (Fig. 7), followed by black corals and sponges (Angiolillo et al. 2015). Consequently, Mediterranean marine litter has widespread impacts on marine biodiversity and Mediterranean Sea ecosystem services.

Underwater Sound

Sound travels five times faster in water than in air and consequently covers longer distances (Rako-Gospic and Picciulin 2019). Therefore, it is utilized by marine organisms, i.e. for communication, prey location, mating and navigation (Hildebrand 2009; Rako-Gospic and Picciulin 2019). However, the level of underwater noise caused by anthropogenic activities has alarmingly increased worldwide in the last decades causing different impacts on marine life (Hildebrand 2009; Peng et al. 2015) and therefore has been recognized as a source of pollution by the European Commission (Marine Strategy Framework Directive, 2008/56/EC). It is mainly caused by marine traffic, coastal and offshore works (e.g. harbours, wind farms, oil and gas wells drilling), seismic surveys (e.g. hydrocarbon extraction), naval exercises (e.g. sonars and detonations), fishing, oceanographic experiments and geophysical mapping (Hildebrand 2009; Maglio et al. 2016; Peng et al. 2015).

Potential effects of anthropogenic sound sources target a variety of marine organisms and range from body malformations during larval development to reduction of growth and reproductive rates, auditory damage and hearing loss, displacement from feeding or breeding areas and stranding of certain species (Anguilar de Soto et al. 2013; Carroll et al. 2017; Peng et al. 2015).

The first rough overview of spatial occurrence of noise-producing anthropogenic activities in the Mediterranean Sea was performed for cetacean conservation (e.g. geophysical surveys, coastal and offshore industrial projects, military operations, marine traffic,) and several noise hotspots were identified(Maglio et al. 2016). The Ligurian Sea, the Straits of Sicily and the Northern part of the Hellenic Trench are the areas where cetacean habitats overlap with noise hotspots (Maglio et al. 2016). Military operations (e.g. naval exercises that use mid-frequency active sonars) are responsible for stranding events of Cuvier's beaked whales, *Ziphius cavirostris* (Cuvier, 1823) along the coasts of the Mediterranean Sea for the last two decades (Maglio et al. 2016). Cephalopods of the Mediterranean Sea, such as the European squid, *Loligo vulgaris* (Lamarck, 1798), common cuttlefish, *Sepia officinalis* (Linnaeus, 1758), common octopus, *Octopus vulgaris* (Cuvier, 1797) and southern shortfin squid, *Illex coindeti* (Vérany, 1837), experience acoustic trauma when exposed to high-intensity and low-frequency sound which can be caused by research surveys (André et al. 2011). In the Adriatic Sea, it has also been observed thatacoustic communication of the Mediterranean finfish is possibly affected by boat noise (Codarin et al. 2009). Further studies in the field of marine noise are needed to define impacts of underwater sound on marine biota in the Mediterranean Sea.

Climate Change and Ocean Acidification

Climate change is affecting the Mediterranean region, and it is expected to become drier and warmer over the course of the twenty-first century. Climatic models predict a pronounced decrease in precipitation and an increase in frequency of extremely

Fig. 8 Mediterranean endemic calcifying coral *Cladocora caespitosa* (Linnaeus, 1767) (Photo by: Informare)

high-temperature events in the Mediterranean region, as well as rapid mean warming (Giorgi and Lionello 2008), which will also affect the marine environment. Mediterranean warming has impacts on growth, survival, fertility, early life history, reproduction, migration and phenology of pelagic and benthic organisms, ranging from phytoplankton to marine vegetation, invertebrates and vertebrates (Lacoue-Labarthe et al. 2016; Marbà et al. 2015). In addition, endemic marine species, which are characterized by a limited capacity to adapt to ocean warming, are severely affected (Marbà et al. 2015) (Fig. 8). Disease outbreaks related to climatic events are becoming more frequent in the Mediterranean Sea: faunas are shifting (geographical distribution) and invasive species are spreading and becoming established within an already highly impacted marine biota (Lejeusne et al. 2010).

Climate change is also raising the sea level in the Mediterranean, which will impact densely populated coasts by the flooding of low-lying coastal areas, erosion of beaches and salt intrusion into freshwater aquifers (Aral and Chang 2017; Lichter et al. 2010; Mabrouk et al. 2013). Additionally, the increase of evaporation along with damming of rivers has led to an increase in salinity since the 1960s (Borghini et al. 2014).

Anthropogenic-induced emissions of carbon dioxide are primarily responsible for the acidification of Mediterranean Sea waters. A remarkably decreasing annual trend in the Mediterranean pH has been documented and can be interpreted as an indicator of acidification within the Basin (Flecha et al. 2015). There is evidence that Mediterranean acidification negatively affects the survival, growth and early life history of shellfish (Lacoue-Labarthe et al. 2016). Though fish, as motile species, are more resilient to acidification, they are nonetheless also affected by it. Fisheries and aquaculture, which are of important economic value in the Mediterranean, will, most likely, also be impacted.

Microalgae, seaweed communities and seagrass meadows will be impacted by acidification, causing the potential loss of important habitats for a wide range of organisms and possibly favouring non-indigenous algal species (Lacoue-Labarthe et al. 2016).

Vermetids and calcifying corals are particularly vulnerable to acidification, which most likely will lead to the loss of biodiversity and shore erosion in the case of vermetid reefs (Lacoue-Labarthe et al. 2016). At the community scale, a shift to more carbon dioxide-tolerant species could lead to a proliferation of jellyfish and anemones, which are resilient to, or actively benefit from warming and/or acidification (Lacoue-Labarthe et al. 2016).

Another consideration is that changes in seawater chemistry can modify the bioavailability of contaminants, favouring bioaccumulation in some sea organisms and consequently in human consumers (Lacoue-Labarthe et al. 2016).

The resulting lack of healthy marine ecosystems can cause severe socio-economic impacts. Decline of fisheries and aquaculture production as well as quality decline of coastal tourism will lead directly and/or indirectly to loss of employment, reduction of food security, human health and wealth of Mediterranean countries (Cramer et al. 2018; Weatherdon et al. 2016). Regional differences also tend to increase, particularly between the North and the South of the Mediterranean region (Linares et al. 2020; Werz and Hoffman 2017). Alterations of Mediterranean marine and coastal environments due to human pressures, but especially due to climate change along with population growth and sociopolitical instabilities, can trigger further migration of human populations and jeopardize human security in the Mediterranean region (Cramer et al. 2018; Werz and Hoffman 2017).

Biodiversity Loss: A Major Consequence of Pressures

The Mediterranean Sea is considered to be a biodiversity hotspot with approximately 17,000 species of marine organisms (Bianchi and Morri 2000; Coll et al. 2010, 2012; Cuttelod et al. 2008). Both its geological history and environmental conditions have led to this great diversity of marine life (Bianchi and Morri 2000; Mannino et al. 2017). However, the Mediterranean Sea is "under siege" when it comes to marine biodiversity, as stated in Coll et al. (2012). The current state of biodiversity reflects the cumulative effects of the anthropogenic pressures affecting the Mediterranean coastal and marine environment, as despite this richness of life the existence of many species is threatened, with some of them subject to multiple pressures (UNEP/MAP 2012).

The Mediterranean Sea is one of the regional seas most impacted by different anthropogenic pressures, which were previously mentioned, in addition to shipping. Biofouling and ballast water transportation along shipping routes have triggered the introduction of non-indigenous species in the Mediterranean Sea. Further non-indigenous species have been introduced via the Suez Canal and through aquaculture. The establishment of non-indigenous species, which may be further encouraged by climate change, is considered one of the main causes of biodiversity loss in the

Fig. 9 A loggerhead turtle, *Caretta caretta*, accidentally caught by a beach seine net in Elounda Bay, Crete, Greece (ICZM project of the IMBBC) (Photo by: C. Dounas)

Mediterranean (Coll et al. 2010; Galil 2007; Otero et al. 2013) as they have the potential to affect many aspects of marine and other aquatic ecosystems.

Many species are currently considered to be threatened in the Mediterranean Sea (UNEP-MAP RAC/SPA 2010), such as the Mediterranean monk seal *Monachus monachus* (Hermann, 1779) and the European eel *Anguilla anguilla* (Linnaeus, 1758). Top predators in the Mediterranean, such as sharks, are facing a particularly high risk of extinction caused by accidental killing, intensive fishing activities and pollution, all of which represent severe threats towards these species (Cuttelod et al. 2008). Incidental entanglement in fishing gears also impacts marine mammals, turtles and birds (Fig. 9), which are long-lived species with low reproductive rates and delayed sexual maturity (Soriano-Redondo et al. 2016; UNEP/MAP 2012).

Conservation actions are needed to protect and preserve this rich biodiversity. Individual species protection measures, networks of Marine Protected Areas (MPAs), conservation of the wider environment using an ecosystem-based approach, additional monitoring and research, communication and education all play complementary roles in preserving Mediterranean biodiversity (Cuttelod et al. 2008) (Fig. 10).

The importance of biodiversity conservation has been reinforced recently during the COVID-19 pandemic, as some diagnostic tests used for detecting the SARS-Cov-2 virus are based on an enzyme isolated from bacteria living in Mediterranean hydrothermal vents (Belkin et al. 1986).

Fig. 10 An example of a Mediterranean coralligenous assemblage, one of the most biodiverse habitats (Photo by: Informare)

Pressures and Ocean Literacy

Pressures, predominantly caused by humans, threaten Mediterranean marine ecosystems as well as their resources and services and, consequently, societies of the region. They are derived not only from coastal activities but also from those inlands. Humans—whether they live, work, holiday by the sea or away from it—influence, most of the time in a negative way, the Mediterranean Sea. Ocean literacy as a common practice can empower people to make informed decisions and take responsible actions, thus contributing to a healthy Mediterranean Sea and support its many different societies and cultures (Stoll-Kleemann 2019). To achieve sustainability of the Mediterranean Sea, different stakeholders and sectors as well as citizens need to understand the relationship between their activities and the resulting pressures they put upon marine ecosystems, their resources and services.

Threats that are pressing marine ecosystems are not always understood by the general public (Kopke et al. 2019; Lotze et al. 2018), which can be a barrier for achieving a pro-environmental societal change (Kopke et al. 2019). Scientific outreach and dissemination can be used to overcome this barrier and favour informed and responsible decisions, concerning marine ecosystems, of all involved stakeholders (researchers, policy makers, local communities, etc.) (Mea et al. 2016). However, knowledge itself doesn't always lead directly to ocean citizenship but knowledge together with public awareness are necessary prerequisites for behavioural and societal changes (Fletcher and Potts 2007). Furthermore, by combining knowledge provision as well as individuals' emotional involvement in choices regarding ocean-related topics, individuals are more likely to engage in pro-environmental behaviours and attitudes that would lead to advanced ocean literacy (Jefferson et al. 2015; Kollmuss and Agyeman 2002; Stoll-Kleemann 2019). A recent study carried out by Ashley et al. (2019) showed that OL initiatives and activities lead to an increase in knowledge, awareness and attitudes of the people which consequently support actions that reduce negative impacts on the marine environment.

A plethora of educational and citizen science projects have been performed in Mediterranean countries, though the impact of such OL activities is not always thoroughly evaluated (for a detailed list of the projects see "Ocean-Related Educational Projects and Networks in the Mediterranean" subsection of this chapter). These projects promote activities relevant to OL, which provide knowledge concerning human pressures on the marine environment and aim to change behaviours and attitudes of students, teachers, stakeholders and citizens. Several of them (e.g. CIGESMED, COMBER, Green Bubbles, Harmony, MARLISCO, MELTEMI, PERSEUS, Reef Check Italia, Sea for Society, Spot the jellyfish) identify human pressures (e.g. marine litter, tourism) as well as the resulting biodiversity loss and climate change through participatory approaches and involvement of different target groups of participants (e.g. volunteer divers, students, citizens). This way, the projects aim to enhance understanding of the links between natural and anthropogenic pressures and ecosystem functioning of the Mediterranean Sea. The role of Marine Protected Areas (see also section "Marine Protected Areas" of this chapter for the role of MPAs in OL) is also fundamental for promoting OL activities (e.g. MPA-adapt, MPA-engage).

Projects that have been carried out within the OL framework and worth mentioning are: (a) ResponSEAble, aimed to map European marine research and knowledge, to understand the complex human–ocean relationships and to provide a wide range of media and outreach activities; (b) Sea Change, aimed to establish a fundamental "Sea Change" in the way European citizens view their relationship with the sea, by empowering them, as Ocean-Literate citizens, to take direct and sustainable action towards a healthy ocean, healthy communities and ultimately a healthy planet; and (c) MARINE_ECOMED, an Erasmus + project aiming at creating an international strategic partnership in order to promote sustainable marine management and communication strategies in the Mediterranean Region by developing educational materials (for details see "Ocean-Related Educational Projects and Networks in the Mediterranean" subsection).

Additionally, the "Mediterranean Education for Sustainable Development" (MEdIES), the major educational and training initiative of the Mediterranean Information Office for Environment, Culture and Sustainable Development, was launched in Johannesburg during the World Summit on Sustainable Development in 2002. The aim of MEdIES is to facilitate the educational community to contribute in a systematic and concrete way to the implementation of Agenda 21 and Agenda 2030, through the successful application of innovative educational programmes in all Mediterranean countries (for details see "Ocean-Related Educational Projects and Networks in the Mediterranean" subsection).

Legal Framework for the Protection and Sustainability of the Mediterranean Sea

The UN Environment's Regional Seas Programme (1974) has created a unique approach to the protection of the coastal and marine environment. The aim of the programme is to promote cooperation among neighbouring countries to foster the "shared seas" approach, and to develop concerted actions for their protection. In 1975, the Mediterranean Action Plan (MAP) was adopted and this, the first-ever Regional Seas Programme under the UN Environment umbrella, was followed by the Convention for the Protection of the Mediterranean Sea against Pollution (Barcelona Convention) (1976). Its main objectives are related to the assessment and control of marine pollution, sustainable management of natural marine and coastal resources, protection of the natural and cultural heritage, solidarity among Mediterranean coastal states and improvement of the quality of life. The Barcelona Convention has seven protocols: dumping at sea; prevention and emergency response to pollution by oil and other harmful substances; land-based pollution; specially protected areas and biodiversity; protection from pollution deriving from offshore activities; transboundary movement of hazardous wastes and integrated coastal zone management.

Under these conventions, the Mediterranean countries must address multiple challenges to improve the sustainability of the marine environment, taking into consideration its physical, geographical and oceanographic characteristics, as well as its geopolitical context. Overall governance of the region should take into account its extremely heavy maritime traffic, the legal and illegal transportation of humans, the over-exploitation of living and non-living marine resources, without forgetting the need to protect its immense cultural heritage and its importance as a hotspot for marine biodiversity.

With regard to shipping, all Mediterranean countries are members of the International Maritime Organization (IMO), which is the United Nations specialized agency with responsibility for the safety and security of shipping and the prevention of marine and atmospheric pollution by ships (MARPOL Convention).

In the Mediterranean region there are also some interesting sub-regional arrangements such as: (a) the Agreement concerning the Creation of a Marine Mammal Sanctuary in the Mediterranean, created between France, Italy and Monaco which has resulted in the establishment of the Pelagos Sanctuary (Fig. 11); (b) the RAMOGE Agreement, applicable to the area between Marseille (France) and La Spezia (Italy), which provides for scientific, technical, legal and administrative cooperation to be decided jointly on actions to be undertaken for integrated management of the coastline; and (c) the Agreement on the Conservation of Cetaceans in the Black Sea, the Mediterranean Sea and the contiguous Atlantic Area (ACCOBAMS), which aims to protect cetaceans, and which has been ratified by most Mediterranean States.

There are two regional fisheries organizations in the Mediterranean: (a) the General Fisheries Commission for the Mediterranean (GFCM), which promotes the development, conservation, rational management and best utilization of living marine

Fig. 11 Cetaceans in the Pelagos Sanctuary, the Mediterranean Marine Mammal Sanctuary (Photo by: Informare)

resources, as well as the sustainable development of aquaculture in the Mediterranean, the Black Sea and connecting waters; and (b) the International Commission for the Conservation of Atlantic Tunas (ICCAT), which concerns the conservation and management of tuna and tuna-like species.

Several key EU policies are also important for the protection and sustainability of the Mediterranean region such as: (a) the Water Framework Directive (WFD; 2000/60/EC), which aims to achieve good water quality status for all EU water bodies (including marine waters up to one nautical mile from shore); (b) the Marine Strategy Framework Directive (MSFD; 2008/56/EC), which aims to achieve Good Environmental Status (GES) for the EU's marine waters and to protect the resources upon which marine-related economic and social activities depend; (c) the Blue Growth Strategy (COM (2012) 494), which is Europe's long-term strategy to support sustainable growth in the marine and maritime sectors as a whole; (d) the Maritime Spatial Planning Directive (MSP; 2014/89/EU), which is a framework for the integrated governance of maritime activities in order to mitigate degradation, restore and sustain critical monetary and social/cultural ecosystem services; (e) the Common Fisheries Policy (CFP; EC/170/83), a set of rules for managing European fishing fleets and for conserving fish stocks, which aims to ensure that fishing and aquaculture are environmentally, economically and socially sustainable and that they provide a source of healthy food for EU citizens; (f) the Habitats Directive (92/43/EEC), which aims to promote the maintenance of biodiversity, taking into account economic, social, cultural and regional requirements; (g) the Directive (EU) 2019/883 of the European Parliament and of the Council of 17 April 2019 on port reception facilities for the delivery of waste from ships, amending Directive 2010/65/EU and repealing Directive 2000/59/EC; and (h) the Single-Use Plastics Directive (EU) 2019/904 of the European Parliament and of the Council of 5 June 2019 on the reduction of the impact of certain plastic products on the environment.

Top-down approaches, such as legal and policy frameworks that define citizens' rights of access to information, consultation and active participation as well as the institutions charged with the application of these rights, are built (OECD 2001). Therefore, they are important for government–citizens relations concerning the protection and conservation of the marine environment as well as sustainable use of its resources and development. On the other hand, bottom-up approaches (e.g. public awareness-raising, citizens' active engagement, individual/societal behaviours and attitudes) towards this direction are also necessary and include the development of tools and practices inevitably connected with ocean literacy (e.g. guides, leaflets, educational materials, workshops for marine educators/students, information centres, events, civil society organizations). The public opinion and actions in shaping the policy agenda (e.g. Single-Use Plastics Directive) are strongly enhanced when a large part of the community becomes ocean literate. The need of large and informed participation of individuals in environmental issues is also important in order to facilitate the required individual/societal behavioural changes and attitudes. This connection invokes a sense of global ownership and responsibility, therefore leading to more stable and sustainable management approaches to ocean governance (French et al. 2015).

Europe has increasingly moved towards an ecosystem-based approach rather than regulating maritime activities through separate sectoral policies, recognizing the complex relationship between environmental problems and human activities (French et al. 2015). To that end, the EU has adopted more than 200 pieces of legislation that affect marine environmental policy and management and has developed an extensive policy framework in order to manage and address the environmental challenges and human activities influencing Europe's seas (French et al. 2015).

Integrated Coastal Zone Management

Integrated Coastal Zone Management (ICZM) is a process which has been embraced by nations around the world as a central concept for the protection and conservation of coastal and marine environment, sustainable use of marine resources and therefore sustainable development, under national jurisdiction in Chapter 17 of Agenda 21, which was adopted at the United Nations Conference on Environment and Development (1992). The goals of this continuous, proactive and adaptive process of resource management are: (a) to maintain essential ecological processes, life support systems and biological diversity in coastal and marine areas; (b) to identify interactions among coastal and ocean uses and the related ecosystems; and (c) to reduce the vulnerability of coastal areas and their inhabitants to natural hazards (Rochette and Billé 2010).

A fundamental issue in this process is the comprehensive understanding of the relationships between coastal resources and their uses, which need to be understood and expressed not only in scientific terms but also in socio-economic ones. Therefore, the ICZM process is designed to overcome the fragmentation in the sectoral management approach, with an integrated methodology in which several dimensions may be involved (e.g. between sectors and levels of governance, across the land–water interface, between nations).

Apart from policies, regulations and management strategies (e.g. Maritime Spatial Planning Directive supporting the ICZM process), public participation is a very important element of the ICZM process, which can be achieved by enhancing knowledge and awareness about the coastal/marine environment and its relevant issues, thus leading to increased public involvement and support for sustainable coastal and marine management strategies.

The ICZM process comes very close to the definition of ocean literacy, which is an understanding of the influence of the ocean on citizens (e.g. goods and services), and the citizens' influence on the ocean (e.g. impacts of human activities on the marine environment). An ocean-literate person is one who is able to make informed and responsible decisions regarding the ocean and its resources.

The elements which characterize the ICZM process (e.g. sustainable development, integration, top-down and bottom-up approaches) are held in common with ocean literacy initiatives, i.e. to combine scientific knowledge with an emotional attachment to nature, and integrate this into the indispensable role of governance for the purpose of promoting ocean sustainability (Santoro et al. 2017).

Even if the ICZM process never explicitly mentions "ocean literacy" it can be clearly considered as a policy that embraces its approach and many of its principles, such as the recognition of the interconnectedness of the ocean and humans, the promotion of individual and collective approaches for the sustainability of the marine environment, and the use of knowledge to promote the full participation of all actors concerned with the use of ocean (French et al. 2015) (Figs. 12 and 13).

In particular, the Protocol on ICZM in the Mediterranean, which complements the existing set of Protocols of the Barcelona Convention for the Protection of the Marine Environment and the Coastal Region of the Mediterranean (UNEP/MAP/PAP 2008), includes objectives and principles (Articles 5 and 6) that are in full accordance with the approach of ocean literacy. Furthermore, the Protocol clearly calls for the active involvement of the public in the establishment, implementation and updating of coastal management strategies and plans (Article 14). Finally, Articles 15 and 25 of the Protocol refer to awareness-raising, training, education and research, revealing the crucial role that the Ocean Literacy Framework can and must play for a new vision of sustainable global development (UNESCO 2015).

Fig. 12 Educational initiative concerning marine biodiversity of Crete, Greece, organized by the IMBBC and attended by children of the second grade of an elementary school (Photo by: C. Dounas)

Fig. 13 A snorkeler involved in a citizen science project of the IMBBC (https://comber.hcmr.gr/) about fish identification in the Cretan Sea, Greece (Photo by: C. Dounas)

Marine Protected Areas

Marine Protected Areas (MPAs) are recognized as the most effective management and conservation tool for providing protection of biodiversity, for increasing the resilience of ecosystems to anthropogenic changes and for ensuring sustainable use of natural heritage. They also provide a range of benefits for fisheries, local economies and the marine environment as their goals include: (a) the conservation of biodiversity and ecosystems; (b) the mitigation and potential reversal of the decline in productivity by protecting breeding, nursery and feeding habits; (c) the establishment of fishing uses, regulations, catch quotas and no-take zones; (d) the raising of the profile of an area for marine tourism and recreation and further broadening economic options; (e) enhancing opportunities for environmental education, training, heritage and culture; and (f) providing pristine environments for research and understanding the natural ecosystem (Abdulla et al. 2008; Australian Government 2003; Gabrié et al. 2012). MPAs must be effectively implemented, enforced and managed. However, the establishment of a network of MPAs is a key mechanism as, through interconnections and interdependencies, they contribute to each other's integrity by decreasing overall vulnerability (Gabrié et al. 2012; PISCO & UNS 2016).

Mediterranean countries have legal obligations to protect the marine environment and to designate MPAs according to various agreements, policies and laws (EEA 2015; Gabrié et al. 2012; PISCO & UNS 2016). Key international instruments include the Convention on Biological Diversity's Aichi Target 11 and the United Nations Sustainable Development Goal 14, both of which call for protecting at least 10% of the oceans and seas in MPAs by 2020. The Specially Protected Area and Biological Diversity Protocol (1995), calls upon countries to establish MPAs. The Regional Activity Centre for Specially Protected Areas (RAC/SPA) was tasked with the responsibility of assessing the natural heritage situation and assisting Mediterranean countries in implementing this protocol.

In the Mediterranean Sea, similar to many coastal ecosystems across the world, marine protected areas (MPAs) have become the primary tool for in situ habitat and biodiversity conservation with more than 1,000 MPAs covering approximately 6.5% of its surface (MedPAN & UNEP-MAP-SPA/RAC 2016; PISCO & UNS 2016) (Fig. 14). However, the Convention on Biological Diversity's Target 11 of an effective conservation of at least 10% of each ecological region in the world has not yet been reached.

In the Mediterranean Sea, the term "Marine Protected Area (MPA)" is used to describe any marine and/or coastal area that is under protection by legal means for the conservation of natural habitats, species or specific natural features. It includes a wide range of areas, established under various designations, at various levels (subnational, national, regional or even international), and providing various degrees of protection, reflecting cultural and political differences among the countries (MedPAN & UNEP-MAP-SPA/RAC 2016). The majority have been classified as multiple-use marine areas, which seek a balance between biodiversity conservation and human use (Abdulla et al. 2008). Historically, designation was primarily driven by the presence of charismatic species (Fig. 15) and unique features, more than on a holistic ecological approach (Abdulla et al. 2008).

Besides the implementation of MPAs in the Mediterranean, Coll et al. (2012) identified that areas with high marine biodiversity in the Mediterranean were mainly located along the central and north shores, with lower values in the south-eastern regions, and areas of potential high cumulative natural and anthropogenic threats were widespread in both the western and eastern basins, with fewer areas located in the south-eastern region. Because the areas where the interaction of high biodiversity and high threats is concentrated overlap only 2% of the existing MPAs, they are the best candidates for the further implementation of management and protection, and thus provide an opportunity for ocean literacy to boost the idea that the MPA network has to be extended.

The community awareness-raising actions concerning marine protected areas have increased all over the world in order to address the increasingly complex threats to marine ecosystems (EEA 2015). The aim of ocean literacy, in this case, is for the broader public to understand the importance of maintaining a healthy marine environment and how they can help by being informed about the role of MPAs for conservation and protection of this environment (Menabit et al. 2017). Ocean literacy in general and MPAs in a specific and local way could also assist services provided by marine and coastal ecosystems, namely: (a) provisioning services (e.g. fisheries, building materials), (b) supporting services (e.g. life-cycle maintenance for fauna, nutrient cycling), (c) regulating services (e.g. carbon sequestration and

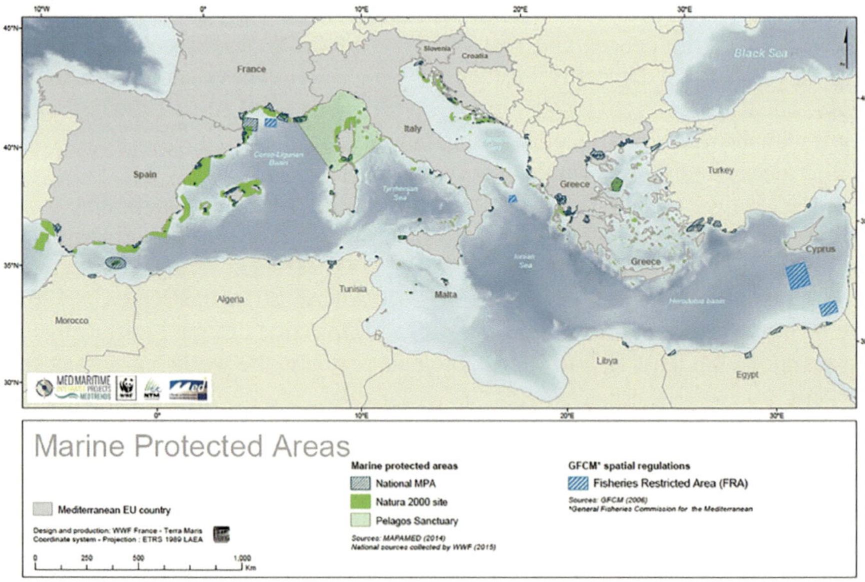

Fig. 14 MPAs and priority areas of conservation in the Mediterranean Sea (*Source* Piante and Ody (2015). Blue Growth in the Mediterranean Sea: the challenge of Good Environmental Status. MEDTRENDS Project. WWF-France. 192 pp.)

Fig. 15 Seahorse *Hippocampus sp.* (Linnaeus, 1758) is classified by International Union for the Conservation of Nature (IUCN) as a near-threatened species in the Mediterranean Sea (Photo by: Informare)

storage, erosion prevention, waste water treatment, moderation of extreme events), and (d) cultural services (e.g. tourism, recreational, aesthetic and spiritual benefits) (Millennium Ecosystem Assessment 2005).

Ocean/marine citizenship offers a potentially reliable tool for monitoring the environmental status of MPAs once-reliable indicators and ad hoc protocols have been designed and citizens have been adequately trained (CIESM 2015). The use of participatory techniques for monitoring changes, as a part of an MPA management plan, provides useful evaluation data and opportunities for administrators and stakeholders to interact and build trust in the management process. The participatory process is also fundamental to: (a) raising awareness; (b) developing the ability to respond to environmental issues; (c) strengthening confidence with institutions; and (d) developing a sustainable local economy.

MPAs have been recognized as suitable places for implementing ocean literacy and several of them are declared for educational purposes (Kasai 2006). The establishment and management of MPAs have provided important educational benefits, because they offer opportunities for people to experience and study relatively pristine/undisturbed marine environments, especially in fully protected MPAs (Kasai 2006). Participants in these learning and leisure activities are more likely to become informed citizens and contribute to future decisions about the marine environment and its resources.

Furthermore, MPAs provide an attraction for tourist visitors seeking local knowledge of the area, while at the same time they provide information, training and support for local people involved in the tourist industry (Australian Government 2003). Finally, MPAs have an important role in educating local communities and visitors about culture, history and heritage associated with the areas they protect, alongside biodiversity conservation and sustainable use (Australian Government

2003). Ocean Literacy Framework is available to support local marine conservation and MPAs (Kasai 2006).

Environmental Education

The role of the Mediterranean Sea is very important in the lives of at least those half a billion people who live around it, because it influences their climate, food, economy and culture; without it, they could not survive. Despite its crucial role, this sea remains largely unexplored while being simultaneously overexploited. Anthropogenic pressures, combined with the unique features of the Mediterranean Sea, are leading directly and/or indirectly to significant alterations of the delicate Mediterranean ecosystems.

For this reason, now more than ever, it is important to improve our knowledge about the sea, and not only to establish sea protection and management plans. People are more inclined to respect what they know well: ocean literacy and citizen science projects are indispensable to educate people, including citizens and tourists, and to involve them in the process of environmental protection (Fig. 16).

The novelty of ocean literacy lies precisely in its aim to connect knowledge, and protection, looking at the sea not only in the scientific and environmental context, but also in the economic, social and cultural contexts.

Fig. 16 Beach cleaning events organized to educate citizens and tourists in Liguria, Italy (Photo by: M'Importa)

How Ocean Literate Are the People of Mediterranean Countries?

It is not easy to answer this question due to the relative recency of ocean literacy as an emergent issue in Europe and in the Mediterranean countries in general. In the United States, the need for and the definition of ocean literacy have been identified since 2002 (Cava et al. 2005), although it required a long and collective effort to result in the completion of the Ocean Literacy Framework in 2009. It was only in 2012 that a European network set its sights on ocean literacy: the European Marine Science Educators Association (EMSEA) was established (Copejans et al. 2012; Dupont and Fauville 2017). In the following years the European Marine Board published its Position Paper 20, claiming that "Europe's maritime ambitions require an ocean- literate population" (European Marine Board 2013) and the Rome Declaration (EurOCEAN 2014) put forward the goal of "Promoting a wider awareness and understanding of the importance of the seas and ocean in the everyday lives of European citizens". Even more recently (2015), within the European Marine Science Educators Association (EMSEA), a regional work-group was founded specifically for the promotion of the ocean literacy in the Mediterranean region. This group, EMSEA Med, has developed the Mediterranean Sea Literacy guide, based on the experience and documents of the Ocean Literacy Framework.

Considering the heterogeneity among European, Asian and African countries, all over the Mediterranean region in various sectors (state of economic development, school systems and curricula, political/governance system, etc.), it is little wonder that research on ocean literacy is still at its infancy. Very few studies have addressed the topic of ocean literacy, often it is only as a part of environmental education, a field with a longer tradition in education as well as in research.

Among the few investigations, Erdoğan et al. (2009) studied the components of environmental literacy in elementary science education in Turkey, finding the mentions of water, water cycle and ocean in Science and Technology textbooks for grades 4–8. In the same year, Yavetz, Goldman and Pe'er (2009) investigated environmental literacy in a sample of Israeli pre-service teachers at the beginning and at the end of their three-year university course. Among the relevant topics, knowledge concerning water distribution on earth, its usage and pollution were addressed. According to the authors, the student teachers, survey showed evidence of poor basic environmental knowledge at the beginning of their course; however, there was significant improvement in their knowledge levels by the end of their course. Curiously, environmental attitudes and behaviours were positive at the beginning of the course, but they did not significantly improve by the end of it. On average, student teachers' environmental literacy was considered to be at an unsatisfactory level for entering the teaching profession (Yavetz et al. 2009).

In 2011, CLAMER, the first EU large-scale investigation addressing ocean issues among the general public, involved three Mediterranean countries: Spain, France and Italy (Buckley et al. 2017). Eighteen percent of respondents chose climate change as the main problem facing the world, with the Spanish sample giving it a ranking even higher and believing it to be caused by human activity. Nevertheless, pollution was rated the most important environmental matter in relation to the coastline or the sea, followed by coastal erosion, sea-level rise, melting ice caps, sea temperature rise and flooding. Spanish and French participants also mentioned tsunamis. Invasive alien species raised concern mostly among Italian and Spanish respondents, while ocean acidification was overlooked by the entire sample. The knowledge level of ocean issues linked to climate change appeared to be satisfactory across those countries surveyed.

Between 2013 and 2015 the Laboratory of Environmental Research and Education of Democritus University of Thrace published the results of two research studies on ocean literacy-related issues carried out on pre-service teachers enrolled in this University. Boubonari et al. (2013) studied ocean pollution, finding moderate knowledge and positive attitudes, with differences according to the distance the respondents lived from the sea. Mogias et al. (2015) investigated Ocean Literacy Principles by means of questionnaires addressing content knowledge and stewardship attitudes. Pre-service teachers evidenced "moderate" knowledge on ocean issues, respectively, good knowledge for water cycle, ocean surface and ocean influence on climate, and poorer knowledge on more specific topics (e.g. global volume of freshwater on earth, deep ocean ecosystems and the role of the ocean in the carbon cycle).

In 2015, Ben zvi-Assaraf and Orion (2005) performed an extensive study on the perception of the water cycle within earth systems among 1000 7th–9th grade Israeli students. Based on a constructivist approach, they found sparse misconceptions (alternative frameworks) on groundwater dynamic nature, cyclic thinking and global magnitude scale. The authors complained about the effect of traditional theoretical teaching about water, and highlighted the need for "explanatory stories" and fieldwork experiences.

More recently Squarcina and Pecorelli (2017) studied Ocean Citizenship in Italy through an analysis of primary school textbooks and interviews with primary school teachers. According to these authors, Italians have a limited awareness of the sea, whether considered as a recreational activity or as an exploitable economic resource. Moreover, the authors criticize the minor role given to Ocean Citizenship in the national school curriculum and in the environmental education guidelines, where focus was primarily on the local environment and ocean-related topics were generally overshadowed by freshwater issues.

Finally, several studies emerged in 2019, evaluating ocean literacy of school students. One of these refers to a cross-cultural study among primary school students in three Mediterranean countries: Greece, Croatia and Italy (Mogias et al. 2019). The results indicated a rather moderate level of knowledge in the total sample, while slight differences were recorded among the three countries, revealing common knowledge and misconceptions. The moderate level of knowledge regarding ocean-related topics was also confirmed by a pilot study on Mediterranean middle school students among the same countries (Cheimonopoulou et al. 2019a). In Italy, an interventional research on primary and middle school students' knowledge, and attitudes towards ocean issues, confirmed the findings of Mogias et al. (2019), i.e. moderate students' knowledge and rather positive attitudes towards marine environment. The study included different teaching activities in primary and middle schools, after which, both knowledge and attitudes revealed a significant increase in primary school but no significant change in middle school, showing the importance of well-designed and carefully implemented didactical intervention (Realdon et al. 2019). Positive results on the increase of students' knowledge and behaviours were evident also in a Greek study concerning a European Maritime Day event with a middle school low sample number (Cheimonopoulou et al. 2019b).

Ocean Literacy in the National School Curricula of Mediterranean Countries

The large variety in languages, educational systems and ways of interacting with the sea across the Mediterranean countries leads to a multifaceted picture of ocean literacy in the different school curricula.

Information relevant to the national school curricula provided here is derived from: (a) drawing data from previous studies, (b) TIMSS 2015 Encyclopedia: Education Policy and Curriculum in Mathematics and Science, issued by IEA—International Association for the Evaluation of Educational Achievement (Mullis et al. 2016), and (c) the opinions of researchers and educators from several Mediterranean countries. The information provided here is not exhaustive due to the difficulty in retrieving accessible and verified information. Before giving further details, it should be noted that environment, ecosystems and biodiversity topics are present in most of the addressed curricula.

Spain: In primary schools, there is no specific mention of the sea, which is presumably included in the topic "basic elements of the physical environment". In middle schools, the hydrosphere and tides are mentioned, with details on "properties of water, the water cycle, the importance of water for life and the impact of human activity on water resources".

France: In primary schools, though there is no specific mention of the sea, there is mention of water in general: "a resource; its states and changes of state; its path in nature; and maintaining its quality for different uses".

Italy: In primary schools, there is no mention of the sea in the science curriculum. Oceans, however, are addressed in the geography curriculum. In middle schools, there is no mention of the sea, which is presumably included in the topic "Earth's structure". In high schools, "hydrosphere" is a prescribed topic, with more details for technical and vocational high schools. Investigations into the actual teaching about the sea in Italian high schools showed evidence of generally poor teachers' interest for the topic, even if compulsory and taught in nearly all schools (Realdon et al. 2016; Sturani 2016). On the other hand, sea-related topics are always addressed, even if briefly, in primary and middle school textbooks and more extensively in high school textbooks.

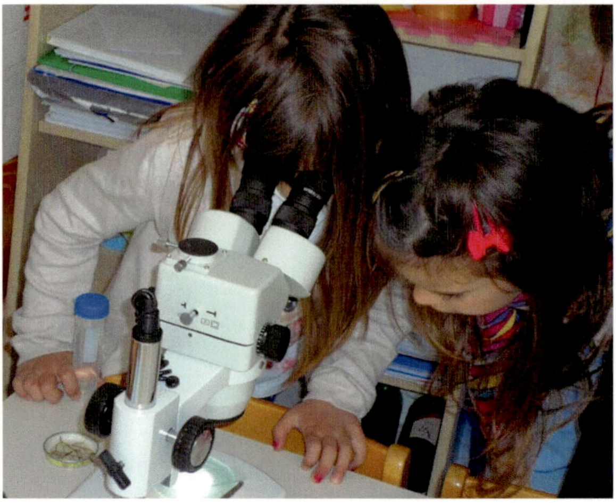

Fig. 17 Education about marine planktonic organisms with children aged 5–6 (Photo by: M. Mokos)

Slovenia: Specific mention of the sea is missing in both primary and middle school curricula, but "substances in nature (e.g., water, soil and air)" are prescribed topics. The Geography curriculum includes "general geography", probably addressing sea-related issues. In 8th grade, teachers can choose further science topics.

Croatia: There are various references concerning water and sea in the primary school curricula. Under Nature and Society subject, the following topics are listed: "water cycle, Adriatic Sea, water as a life condition". Under Geography, "ocean names, sea features (salinity, temperature, tides, waves)" are prescribed. In middle schools, in the Biology curriculum there are topics on: "life conditions in the sea, organisms in the coastal and open sea and on the sea bottom, molluscs, algae, echinoderms, fish"; under Geography: "climate" and "Adriatic Sea"; under Chemistry: "water characteristics"; under Physics: "waves (in general)".

Malta: Given the location, sea issues are present from the early school years (Fig. 17). In primary schools Maltese students develop "awareness of the physical and human elements in Malta, the Mediterranean and the world" and "focus on the contrasts between the Maltese environment and the environment in other Mediterranean countries".

Greece: In the Greek curriculum, at both primary and secondary school level, there is no systematic focus on ocean literacy issues. Nevertheless, there is some basic, although extremely fragmented, information about the sea in the most recent textbooks. Among the topics addressed in primary school textbooks, marine ecosystems and human influence on the sea can be found. Similar topics are also present in more detail in middle school textbooks, with a few references concerning winds, sea temperature and currents under the Geography curriculum (Fig. 18).

Turkey: 7th grade students are taught about ecosystems in general. Marine ecosystems are specifically mentioned—but only in a brief outline. 10th grade students are taught one unit on aquatic biomes. This unit covers freshwater and marine regions.

Fig. 18 Ocean Literacy education projects are also being carried out inland (Veria, Greece) (Photo by: M. Cheimonopoulou)

According to Erdoğan et al. (2009), water, water cycle and ocean topics are addressed in Science and Technology textbooks for grades 4th–8th.

Cyprus: According to Orion and Fortner (2003), since the 1990s there had already been some efforts at science integration and the incorporation of environmental topic areas in the schools of Cyprus. "Several schools across the country participate in the environmental monitoring and data sharing of the GLOBE programme" and "Eco-Schools programme, encouraged and assisted by the Cyprus Marine Environment Protection Association, to adopt curriculum innovations that introduce global environmental issues and local action to elementary students".

Israel: One of the most important environmental issues in the Israeli national school curriculum is the hydrological system. In primary school, general scientific and environmental topics regarding water cycle, energy and geographical zones of the planet include ocean-related content. In lower secondary school, water is addressed in the Geography curriculum as "water cycle within the earth systems". Schools have the possibility to choose certain elective, subsidized (often requiring additional payment from parents) activities and programmes in which they have ocean-related activities coming to the school (EcoOcean) or they can travel to research stations and facilities to learn about marine science (IOLR), or participate in coastal field trips (National Parks Authority, EcoOcean).

Egypt: In primary school, there is only a concise mention of the ocean. In middle school, ocean names and sizes are addressed in Geography. In high school, ocean-related topics are fragmented under different subjects. For example, in Geology and Geography, Egyptian pupils study oceans and rivers and earth topography. Aquatic animals are studied as part of the Biology curriculum; other examples include tides and waves, which could be part of the Physics curriculum, and water composition within the Chemistry course.

Algeria: Ocean and marine topics are briefly mentioned, being used to make students understand certain physical phenomena. The ocean is not yet studied in order to understand its importance, all that it brings to us in our daily life and our impact on it. The new programmes for primary and middle school which started in 2018, however, do emphasize the importance of the protection of the environment in the various levels and courses.

Morocco: No explicit reference to the ocean was found. Only in 4th grade a mention of "Water and the environment, water use and conservation, pollution, and organisms in nature" was found. In the 8th grade, one of the quoted topics is "Earth system", which should include ocean-related issues.

This incomplete report of ocean literacy-related topics in the school systems of the Mediterranean countries highlights some common features in different educational contexts:

- There exists extreme fragmentation of ocean-related topics under different subjects (Earth science, Geography, Chemistry, Physics);
- There is a lack of a global view on "the ocean's influence on people's lives and their influence on the ocean";
- There is widespread acknowledgement of the existence of environmental issues and problems, common along the Mediterranean coasts.

These elements represent challenges and opportunities at the same time, as they evidence the need to disseminate ocean literacy information and teaching resources among the teachers' community.

In addition to the formal teaching of ocean-related topics in schools, there are many non-formal educational opportunities provided by international and regional NGOs active in Mediterranean countries, which provide fieldwork teaching activities for schools and organize summer camps, beach cleaning, contexts, marine events and festivals for students. Mediterranean aquaria and zoos are also providers of educational activities for students and teachers training in their premises, as well as universities and research institutions, within their outreach initiatives. Recently, EU4Ocean, the European Ocean Coalition that connects different organizations, projects and people contributing to ocean literacy and sustainability, has taken the opportunity of a global coordinated initiative by launching the network of European Blue Schools, aimed at integration of ocean literacy in EU schools.

Ocean-Related Educational Projects and Networks in the Mediterranean

In parallel with the introduction of environmental education in school curricula, and following the recommendations of UNESCO-UNEP Belgrade Charter (UNESCO-UNEP 1976) and the European Union (European Community 1993; European Union 1988), large numbers of educational projects addressing marine environment have been developed in Mediterranean countries (Fig. 19).

Fig. 19 Marine educational project focused on biodiversity in the intertidal rocky zone of the Adriatic Sea, Croatia (Photo by: M. Mokos)

Environmental education activities usually start within the local environment, which, in Mediterranean countries, often includes coastal and marine areas. Therefore, we can affirm that ocean-related educational projects have been common along the coast of the Mediterranean, especially the European ones, since the 1980s (MIO-ECSDE 2003, 2004). From the territorial point of view, most of these projects are on nationwide or local levels, with fewer international projects involving different countries (MIO-ECSDE 2004).

Many of these initiatives address the sea as part of the local environment, while some of them focus exclusively on sea-related issues. Their educational targets can range from students to the general public. The involved initiatives can be of different natures: formal school education, including the development and distribution of teaching resources, education research and informal active citizenship campaigns and citizen science projects (European Commission—Directorate General for Maritime Affairs and Fisheries 2018). Concerning the temporal dimension, ocean-related educational projects range from an entire year to single events and celebrations, such as World Oceans Day and European Maritime Day. Financial support for these initiatives can come from governments and local authorities; from intergovernmental institutions and organizations such as the EU and UNESCO; from NGOs, ranging from international (e.g. WWF, IUCN, MIO-ECSDE, FEE) to national and smaller.

In the following paragraphs a few examples of ocean-related educational, ocean literacy and citizen science projects, networks and events exclusively about the Mediterranean Sea are briefly presented. The list is incomplete as many projects are initiated spontaneously or may not be financed and consequently are not included in international databases. Although incomplete, this list shows how ocean literacy and ocean education has been growing in recent years and is becoming a key issue for many international projects.

List of Ocean Literacy-related Projects:

Green Bubbles (2015–2018): a Horizon 2020 EU-funded project dedicated to recreational SCUBA diving, an activity engaging millions of people worldwide. The aim of Green Bubbles, which involved students and teachers through classroom and fieldwork activities, was to maximize the benefits associated with diving while minimizing its negative impacts, thus achieving the environmental, economic and social sustainability of underwater tourism (www.greenbubbles.eu).

MARINE_ECOMED (2018–2021): an Erasmus + project aiming to create an international strategic partnership in order to promote sustainable marine management and communication strategies in the Mediterranean Region. The project is developing educational materials, such as Open Education Resources in Planning and management of marine and coastal areas and ocean literacy, a MOOC on Planning and management of marine and coastal areas and a Handbook on Marine education and communication in the Mediterranean, which will be freely available in the MARINE_ECOMED website (https://www.marine-ecomed.net/).

MARLISCO (2012–2015): The main objectives of the MARine Litter in Europe Seas: Social Awareness and CO-Responsibility project were to increase the awareness of the consequences of societal behaviour in relation to waste production and management on marine socio-ecological systems, to promote co-responsibility among the different actors and to define a more sustainable collective vision of marine environment (www.marlisco.eu).

MEdIES (2002–today): Mediterranean Education for Sustainable Development is the major educational and training initiative of the Mediterranean Information Office for Environment, Culture and Sustainable Development (http://mio-ecsde.org), launched in Johannesburg during the World Summit on Sustainable Development (WSSD 2002). The aim of MEdIES is to facilitate the educational community to contribute in a systematic and concrete way to the implementation of Agenda

21 and Agenda 2030, through the successful application of innovative educational programmes in all Mediterranean countries (www.medies.net).

PERSEUS (2012–2015): Policy-oriented marine Environmental Research for the Southern European Seas was a project aiming to identify the interacting patterns of natural and human-derived pressures on the Mediterranean and Black Seas, assessing their impact on marine ecosystems, and to design an effective and innovative research governance framework based on sound scientific knowledge by using the objectives and principles of the Marine Strategy Framework Directive. The project also addressed students through Perseus@School Network (http://www.perseus-net.eu/site/content.php).

ResponSEAble (2015–2017): a Horizon 2020 project aimed to map European marine research and knowledge, to understand the complex human–ocean relationships and the economic benefits deriving from our seas and the ecosystems they support. The project also provided a wide range of media and outreach activities, including videos, film-making competitions, an educational computer game and other learning materials, a social media campaign and an interactive website (www.responseable.eu).

Sea Change (2015–2018): a Horizon 2020 project aimed to establish a fundamental "Sea Change" in the way European citizens view their relationship with the sea, by empowering them, as Ocean-Literate citizens, to take direct and sustainable action towards a healthy ocean, healthy communities and ultimately a healthy planet (www.seachangeproject.eu).

Sea for Society (SFS) (2012–2015): a European-funded project that engaged stakeholders, citizens and youth in an open and participatory dialogue to share knowledge, forge partnerships and empower actors on societal issues related to the Ocean. In doing so, the project aimed to develop and enrich the concept of the "Blue Society", while preparing mechanisms for future cooperation (www.seaforsociety.eu).

List of citizen science projects

Citizen science projects involve a large number of citizens, often with no formal qualifications, who act as observers concerning a specific scientific issue (e.g. invasive species, jellyfish), creating "geo-referenced information" (e.g. records presented in Google Maps). They are largely untrained and voluntarily collect data which are quality-controlled by scientists. Citizen science projects represent an innovation which contributes to different disciplines (e.g. geography, marine science) and their relation to the general public.

CIGESMED SeasEra: a project that aims to enhance understanding of the links between natural and anthropogenic pressures and ecosystem functioning, and to define and maintain the "Good Environmental Status" (GES) of the Mediterranean Sea also by engaging volunteer divers (www.cigesmed.eu).

COMBER (2010–2013): Citizens' Network for the Observation of Marine BiodivERsity project, initiated under the European funded FP7 project ViBRANT, aimed at engaging citizens in a coastal marine biodiversity observation network. A website has also been developed and functioned for communication and promotion of the network, offering data-entry tools for collecting information (http://www.comber.hcmr.gr).

Harmony: an Interreg Italia–Malta project that aims at protecting marine biodiversity in the Italo-Maltese area with common strategies and the involvement of local communities (https://www.harmony-italiamalta.eu/).

Marine Litter Watch (MLW): MLW includes a mobile application, a web portal and a database for citizen engagement in fighting marine litter and for collecting and sharing comparable data on beach litter. It also provides a platform for marine litter communities to meet, share their knowledge and co-create approaches to marine litter monitoring (https://www.eea.europa.eu/themes/water/europes-seas-and-coasts/assessments/marine-litterwatch).

MELTEMI (2014–2020): an Interreg Balkan-Mediterranean project that promotes a joint effort against marine litter (Albania, Bulgaria, Greece, Cyprus). Through a series of interconnected actions, it foresees the active engagement of the society and key-stakeholders by informing, educating, training and networking them towards an evidence-based policy assessment framework for marine litter (https://meltemi-balkanmed.eu/).

MPA-Adapt: a project guiding Mediterranean MPAs through the climate change era building resilience and adaptation. This is achieved through capacity-building workshops for an effective management, development of risk assessments and an investigation of the potential actions and priorities needed to ensure the adaptability and the resilience of biodiversity and local communities, including fishermen and other stakeholders (https://mpa-adapt.interreg-med.eu/).

MPA-Engage: a project that supports and promotes the role of Mediterranean MPAs as nature-based solutions to adapt and mitigate climate change, even by the stakeholder engagement through participatory approaches, including Marine Citizen Science (https://mpa-engage.interreg-med.eu/).

Reef Check Italia: an organization that created different protocols to monitor the Mediterranean Sea. One of the most important is the Underwater Coastal Environment Monitoring Protocol, intended to assess the ecological status of the Mediterranean marine coastal habitats thanks to the help of short-trained SCUBA divers and snorkelers (www.reefcheckmed.org).

Spot the jellyfish: a campaign in Malta following a citizen science approach and relying on the collaboration of the general public, mariners, divers and especially the younger generations with their teachers and parents, by recruiting their assistance in recording the presence and location of different jellyfish through the use of a specific reporting leaflet (www.ioikids.net/jellyfish).

List of networks

European Marine Science Educators Association (EMSEA): an international non-profit organization committed to boosting ocean literacy in Europe which provides a platform for ocean education in the different European regional seas. The goals are: to raise educators' awareness of ocean issues and of the need for a sustainable future for our coasts, seas and ocean; to promote education and training in the fields of marine sciences; to advance ocean literacy in Europe and worldwide; to improve the quality of marine science education and to function as a platform for sharing and disseminating information and expertise among its members (http://www.emsea.eu/default.php).

Ocean Literacy Italia (OLI): a network for ocean literacy in Italy created in 2017 with almost 100 representatives from academia, NGOs, the private sector and public entities. The purpose of ocean literacy Italia is to promote the incorporation of marine science into Italian school curricula and to raise awareness of the social, economic and cultural importance of the sea and sea-related services to all stakeholders, including decision-makers, individual citizens and private sector representatives (http://www.oceanliteracyitalia.it).

Conclusion

The Mediterranean Sea has always been a centre of diverse human societies and cultures. The use and exploitation of this Sea had started several millennia ago. Human population growth and anthropogenic pressures are characteristics of nowadays' situation of this region more than ever. Existing pressures are causing severe biodiversity and habitat loss. A region known to be a biodiversity hotspot is also a climate-change hotspot with over 460 million people living in it, thus affecting it with their everyday life activities. To ensure the sustainability of Mediterranean marine ecosystems and their resources and the well-being of people living in the region, individuals need to be aware of the impact that everyday life decisions have towards the marine environment, thus they need to be ocean literate.

If ocean literacy in the Mediterranean region means to understand how the Sea influences people, and how people influence the Sea, to become a Mediterranean-Sea-literate person means to know the fundamental principles and concepts that govern the Mediterranean Sea and, consequently, to make informed decisions in everyday life circumstances. There is an urgent need and challenge at the same time for spreading ocean literacy in the Mediterranean region in order to allow diverse societies and cultures, with different behaviours and attitudes, to become informed and to understand critical issues associated with sea-related topics. Ocean literacy adapted to the specific features of the Mediterranean Sea (Mediterranean Sea Literacy) aims to build sea-literate citizens, which can take initiatives, leading to responsible decisions on marine resources and sustainability. The already existing legal framework aims at the protection of the Mediterranean Sea. Integrated Coastal Zone Management supports sustainable use and development of the Mediterranean region. Biodiversity can be protected through the networks of MPAs, which at the same time have an important role in education, thus increasing OL at the local and regional level. Despite all these, the level of ocean literacy among the Mediterranean citizens is still unknown. Research, done so far, indicates that the knowledge level is low to moderate and needs to be improved. Integrating ocean literacy in school curricula and promotion of formal and non-formal education activities that will enhance pro-environmental behaviours and attitudes can contribute to a future ocean-literate generation ("Generation Ocean"). Finally, a geographical approach to these issues can help us visualize their spatial distribution on different scales, from the global to the regional and local, as well as their potential impact on society and the ability to provide solutions.

Acknowledgements We would like to thank EMSEA, especially Peter Tuddenham and Fiona Crouch for their inspiration and support in the creation of the EMSEA-Med group and of the work being done to advance ocean literacy across the Mediterranean region. We would also like to thank Prof. A. Eleftheriou and Dr. C. Dounas for the comments and critical review of the manuscript, M. Eleftheriou for the English spelling and language corrections as well as Dr. B. Goodman, N. Khelil, Dr. M. Snoussi, Prof. A.E. Kideys, A. Kideys, and M.A. Hassaan for providing information about school curricula.

References

Abdulla A, Gomei M, Maison E, Piante C (2008) Status of Marine Protected Areas in the Mediterranean Sea. IUCN, Malaga and WWF, France

André M, Solé M, Lenoir M, Durfort M, Quero C, Mas A, Lombarte A, van der Schaar M, López-Bejar M, Morell M, Zaugg S, Houégnigan L (2011) Low-frequency sounds induce acoustic trauma in cephalopods. Front Ecol Environ 9:489–493

Anguilar de Soto N, Delorme N, Atkins J, Howard S, Williams J, Johnson M (2013) Anthropogenic noise causes body malformations and delays development in marine larvae. Sci Rep 3:2831

Angiolillo M, Di Lorenzo B, Farcomeni A, Bo M, Bavestrello G, Santangelo G, Cau A, Mastascusa V, Cau A, Sacco F, Canese S (2015) Distribution and assessment of marine debris in the deep Tyrrhenian Sea (NW Mediterranean Sea, Italy). Mar Pollut Bull 92:149–159

Aral MM, Chang B (2017) Spatial variation of sea level rise at Atlantic and Mediterranean coastline of Europe. Water (Switzerland) 9:11p

Ashley M, Pahl S, Glegg G, Fletcher S (2019) A change of mind: applying social and behavioral research methods to the assessment of the effectiveness of ocean literacy initiatives. Front Mar Sci 6:20p

Australian Government (2003) The benefits of Marine Protected Areas. Commonwealth of Australia, Department of the Environment and Heritage

Belkin S, Wirsen CO, Jannasch HW (1986) A new sulfur-reducing, extremely thermophilic eubacterium from a submarine thermal vent. Appl Environ Microbiol 51:1180–1185

Ben-zvi-Assaraf O, Orion N (2005) A study of junior high students' perceptions of the water cycle. J Geosci Educ 53:366–373

Bianchi CN, Morri C (2000) Marine biodiversity of the Mediterranean Sea: situation, problems and prospects for future research. Mar Pollut Bull 40:367–376

Boesch DF (2019) Barriers and bridges in Abating Coastal Eutrophication. Front Mar Sci 6:25p

Borghini M, Bryden H, Schroeder K, Sparnocchia S, Vetrano A (2014) The Mediterranean is becoming saltier. Ocean Sci 10:693–700

Boubonari T, Markos A, Kevrekidis T (2013) Greek pre-service teachers' knowledge, attitudes and environmental behaviour toward marine pollution. J Environ Educ 44:232–251

Buckley PJ, Pinnegar JK, Painting SJ, Terry G, Chilvers J, Lorenzoni I, Gelcich S, Duarte CM (2017) Ten thousand voices on marine climate change in Europe: different perceptions among demographic groups and nationalities. Front Mar Sci 4:17p

Carić H, Mackelworth P (2014) Cruise tourism environmental impacts—the perspective from the Adriatic Sea. Ocean Coast Manag 102:350–363

Carroll AG, Przeslawski R, Duncan A, Gunning M, Bruce B (2017) A critical review of the potential impacts of marine seismic surveys on fish & invertebrates. Mar Pollut Bull 114:9–24

Carroni S, Bresciani A, Delaria MA, Meloni F, Navone A, Panzalis P, Heimann K, Ceccherelli G (2015) Ecology of the benthic mucilage-forming microalga *Chrysophaeum taylorii* in the W Mediterranean Sea: substratum and depth preferences. Estuar Coast Shelf Sci 161:38–45

Casale P, Freggi D, Paduano V, Oliverio M (2016) Biases and best approaches for assessing debris ingestion in sea turtles, with a case study in the Mediterranean. Mar Pollut Bull 110:238–249

Cattaneo-Vietti R, Bo M, Cannas R, Cau A, Follesa C, Meliadò E, Russo GF, Sandulli R, Santangelo G, Bavestrello G (2016) An overexploited Italian treasure: past and present distribution and exploitation of the precious red coral *Corallium rubrum* (L., 1758) (Cnidaria: Anthozoa). Ital J Zool 83:443–455

Cava F, Schoedinger S, Strang C, Tuddenham P (2005) Science content and standards for ocean literacy: a report on ocean literacy. Available at: http://coexploration.org/oceanliteracy/documents/OLit200405_Final_Report.pdf

Cheimonopoulou MTh, Mogias A, Realdon G, Mokos M, Koulouri P, Previati M, Boubonari T (2019a). Mediterranean middle school students' knowledge, attitudes, and behaviours towards ocean-related topics: an EMSEA-Med pilot study. Paper presentation, 7th European Marine Science Educators Association, Sao Miguel, Azores, Portugal, September 16–20

Cheimonopoulou MTh, Realdon G, Mogias A, Koulouri P, Mokos M, Previati M, Boubonari T (2019b) Ocean literacy intervention activities: a case study from a European Maritime Day Event (EMD) in Mainland Greece. Paper presentation, 7th European Marine Science Educators Association, Sao Miguel, Azores, Portugal, September 16–20

Chiesa LM, Labella GF, Panseri S, Pavlovic R, Bonacci S, Arioli F (2016) Distribution of persistent organic pollutants (POPs) in wild Bluefin tuna (*Thunnus thynnus*) from different FAO capture zones. Chemosphere 153:162–169

CIESM (2015) MMMPA/CIESM International Joint Conference on Mediterranean Marine Protected Areas: Integrated Management as a Response to Ecosystem Threats—Conference Report [co-edited by Cerrano C, Henocque Y, Hogg K, Moschella P, Ponti M], 15–17 October 2015, Ancona, Italy

Codarin A, Wysocki LE, Ladich F (2009) Effects of ambient and boat noise on hearing and communication in three fish species living in a marine protected area (Miramare, Italy). Mar Pollut Bull 58:1880–1887

Coll M, Piroddi C, Albouy C, Ben Rais Lasram F, Cheung WWL, Christensen V, Karpouzi VS, Guilhaumon F, Mouillot D, Paleczny M, Lourdes Palomares M, Steenbeek J, Trujillo P, Watson R, Pauly D (2012) The Mediterranean Sea under siege: spatial overlap between marine biodiversity, cumulative threats and marine reserves. Global Ecol Biogeogr 4:465–480

Coll M, Piroddi C, Steenbeek J, Kaschner K, Ben Rais Lasram F, Aguzzi J, Ballesteros E, Bianchi CN, Corbera J, Dailianis T, Danovaro R, Estrada M, Froglia C, Galil BS, Gasol JM, Gertwagen R, Gil J, Guilhaumon F, Kesner-Reyes K, … Voultsiadou E (2010) The biodiversity of the Mediterranean Sea: estimates, patterns and threats. PLoS One 5, 36p

Copejans E, Crouch F, Fauville G (2012) The European Marine Science Educators Association (EMSEA): towards a more ocean literate Europe. J Mar Educ 28:43–46

Cramer W, Guiot J, Fader M, Garrabou J, Gattuso J-P, Iglesias A, Lange MA, Lionello P, Llasat MC, Paz S, Peñuelas J, Snoussi M, Toreti A, Tsimplis MN, Xoplaki E (2018) Climate change and interconnected risks to sustainable development in the Mediterranean. Nat Clim Change 8:972–980

Cuttelod A, García N, Abdul Malak D, Temple H, Katariya V (2008) The Mediterranean: a biodiversity hotspot under threat. In: Vié J-C, Hilton-Taylor C, Stuart SN (eds) Wildlife in changing world. IUCN, pp 89–101

Danovaro R, Umani SF, Pusceddu A (2009) Climate change and the potential spreading of marine mucilage and microbial pathogens in the Mediterranean Sea. PLoS One 4:8p

Dowling R, Weeden C (2017) Cruise ship tourism. CAB International, Wallingford

Dupont S, Fauville G (2017) Ocean literacy as a key toward sustainable development and ocean governance. In: Nunes P, Svensson LE, Markandya A (eds) Handbook on the economics and management of sustainable oceans. Edward Elgar & UNEP, Cheltenham, UK, pp 519–537

EEA (2015) Marine protected areas in Europe's seas: an overview and prospective for the future. European Environment Agency

Eigaard OR, Bastardie F, Hintzen NT, Buhl-Mortensen L, Buhl-Mortensen P, Catarino R, Dinesen GE, Egekvist J, Fock HO, Geitner K, Gerritsen HD, Gonzalez MM, Jonsson P, Kavadas S, Laffargue P, Lundy M, Gonzalez-Mirelis G, Nielsen JR, Papadopoulou N, ... Rijnsdorp AD (2017) The footprint of bottom trawling in European waters. ICES J Mar Sci 74:847–867

Enne G, d'Angelo M, Madrau S, Zucca C (2009) Urbanization and desertification in European Mediterranean coastal areas: a case study in north-western Sardinia (Alghero, Italy). Hum Settlement Dev 2:56–61

Erdoğan M, Kostova Z, Marcinkowski T (2009) Components of environmental literacy in elementary science education curriculum in Bulgaria and Turkey. Eurasia J Math Sci Technol Educ 5:15–26

EurOCEAN (2014) Rome declaration. EurOCEAN 2014 Conference, Rome, October 7–9

European Commission (2017) Commission Decision (EU) 2017/848 of 17 May 2017 laying down criteria and methodological standards on good environmental status of marine waters and specifications and standardised methods for monitoring and assessment, and repealing Decision 2010/477/EU. J Eur Union 2017:32

European Commission (2018) Ocean literacy—best practices and user stories. Directorate-General for Maritime Affairs and Fisheries

European Community (1993) Towards sustainability. Off J Eur Communities, C138/7

European Marine Board (2013) Navigating the future. IV. Position Paper 20 of the European Marine Board. Ostend, Belgium

European Union (1988) Resolution of the Council and the Ministers of Education meeting within the Council on the European dimension in education of 24 May 1988. Off J C 177, 06/07/1988 P. 0005–0007

FAO (2018) The state of Mediterranean and black sea fisheries. General Fisheries Commission for the Mediterranean. Rome. 172 pp. Licence: CC BY-NC-SA 3.0 IGO

Fernandes PG, Ralph GM, Nieto A, García Criado M, Vasilakopoulos P, Maravelias CD, Cook RM, Pollom RA, Kovačić M, Pollard D, Farrell ED, Florin A-B, Polidoro BA, Lawson JM, Lorance P, Uiblein F, Craig M, Allen DJ, Fowler SL, ... Carpenter KE (2017) Coherent assessments of Europe's marine fishes show regional divergence and megafauna loss. Nat Ecol Evol 1 (0170)

Ferrante M, Conti GO, Fiore M, Rapisarda V, Ledda C (2013) Harmful algal blooms in the Mediterranean Sea: effects on human health. Euromediterranean Biomed J 8:25–34

Fewtrell L, Kay D (2015) Recreational water infection: a review of recent findings. Curr Environ Health Rep 2:85–94

Flecha S, Perez FF, Garcia-Lafuente J, Sammartino S, Ríos AF, Huertas IE (2015) Trends of pH decrease in the Mediterranean Sea through high frequency observational data: indication of ocean acidification in the basin. Sci Rep 5:16770

Fletcher S, Potts J (2007) Ocean citizenship: an emergent geographical concept. Coastal Manage 35:511–524

Fosse J, Le Tellier J (2017) Sustainable tourism in the Mediterranean: state of play and strategic directions. Plan Bleu, Valbonne (Plan Bleu Paper, 17)

Fossi MC, Marsili L, Baini M, Giannetti M, Coppola D, Guerranti C, Caliani I, Minutoli R, Lauriano G, Grazia Finoia M, Rubegni F, Panigada S, Berub M, Urban Ramírez J, Panti C (2016) Fin whales and microplastics: the Mediterranean Sea and the Sea of Cortez scenarios. Environ Pollut 209:68–78

French V, Chu N-C, Santoro F, Sousa Pinto I, Borges D, McDonough N (2015) Review of ocean literacy in European maritime policy. Sea Change. One Ocean, One Health

Fromentin J-M, Rouyer T (2018) The Eastern Atlantic and Mediterranean bluefin tuna: an archetype of overfishing and rebuilding? In Garcia SM, Ye Y (eds) Rebuilding of marine fisheries. Part 2: Case studies. FAO Fisheries and Aquaculture Technical Paper No. 630/2. FAO, Rome, pp 1–11

Gabrié C, Lagabrielle E, Bissery C, Crochelet E, Meola B, Webster C, Claudet J, Chassanite A, Marinesque S, Robert P, Goutx M (2012) The status of marine protected areas in the Mediterranean Sea. MedPAN & RAC/SPA. Ed: MedPAN Collection
Galassi G, Spada G (2014) Sea-level rise in the Mediterranean Sea by 2050: roles of terrestrial ice melt, stericeects and glacial isostatic adjustment. Global Planet Change 123:55–66
Galil BS (2007) Seeing red: alien species along the Mediterranean coast of Israel. Aquat Invasions 2:281–312
Garcia-Nieto AP, Geijzendorffer IR, Baro F, Rochef PK, Bondeau A, Cramer W (2018) Impacts of urbanization around Mediterranean cities: changes in ecosystem service supply. Ecol Ind 91:589–606
Gerovasileiou V, Sini MI, Poursanidis D, Koutsoubas D (2009, January 15–16) Contribution to the knowledge of Coralligenous Communities in the NE Aegean Sea. In: 1st Mediterranean Symposium on the Coralligenous Conservation and other calcareous bio-concretions. Tabarka
Giorgi F, Lionello P (2008) Climate change projections for the Mediterranean region. Global Planet Change 63:90–104
Grigorakis K, Rigos G (2011) Aquaculture effects on environmental and public welfare—the case of Mediterranean mariculture. Chemosphere 85:899–919
Guidetti P, Baiata P, Ballesteros E, Di Franco A, Hereu B, Macpherson E, Micheli F, Pais A, Panzalis P, Rosenberg AA, Zabala M, Sala E (2014) Large-scale assessment of Mediterranean marine protected areas effects on fish assemblages. PLoS One 9:14p
Hildebrand JA (2009) Anthropogenic and natural sources of ambient noise in the ocean. Mar Ecol Prog Ser 395:5–20
ICC (2016) Ocean conservancy/International Coastal Cleanup. http://www.oceanconservancy.org/
IUCN (2016) The conservation status of sharks, rays and chimaeras in the Mediterranean Sea. https://www.iucn.org/sites/dev/files/content/.../brochure_medredlist_sharks.pdf
Jefferson R, McKinley E, Capstick S, Fletcher S, Griffin H, Milanese M (2015) Understanding audiences: making public perceptions research matter to marine conservation. Oceanic Coastal Manage 115:61–70
Kamizoulis G, Saliba L (2004) Development of coastal recreational water quality standards in the Mediterranean. Environ Int 30:841–854
Karamanlidis AA, Androulaki E, Adamantopoulou S, Chatzispyrou A, Johnson WM, Kotomatas S, Papadopoulos A, Paravas V, Paximadis G, Pires R, Tounta E, Dendrinos P (2008) Assessing accidental entanglement as a threat to the Mediterranean monk seal Monachus monachus. Endanger Species Res 5:205–213
Kasai R (2006) Marine protected areas and marine education: a case study of Tasmania and Victoria. University of Tasmania
Kollmuss A, Agyeman J (2002) Mind the gap: why do people act environmentally and what are the barriers to pro-environmental behavior? Environ Educ Res 8:239–260
Kopke K, Black J, Dozier A (2019) Stepping out of the Ivory Tower for ocean literacy. Front Mar Sci 19:13p
Lacoue-Labarthe T, Nunes ALDP, Ziveri P, Cinar M, Gazeau F, Hall-Spencer JM, Hilmi N, Moschella P, Safa A, Sauzade D, Turley C (2016) Impacts of ocean acidification in a warming Mediterranean. Reg Stud Mar Sci 5:1–11
Lejeusne C, Chevaldonne P, Pergent-Martini C, Boudouresque CF, Perez T (2010) Climate change effects on a miniature ocean: the highly diverse, highly impacted Mediterranean Sea. Trends Ecol Evol 25:250–260
Lichter M, Zviely D, Klein M, Sivan D (2010) Sea-level changes in the Mediterranean: past, present, and future—a review. In: Seckbach J, Einav R, Israel A (eds) Seaweeds and their role in globally changing environments. Cellular origin, life in extreme habitats and astrobiology, vol 15. Springer, Dordrecht
Linares C, Díaz J, Negev M, Sánchez Martínezc G, Debono R, Paz S (2020) Impacts of climate change on the public health of the Mediterranean Basin population—current situation, projections, preparedness and adaptation. Environ Res 182(109107):14p

Liu J, Techtmann SM, Woo HL, Ning D, Fortney JL, Hazen TC (2017) Rapid response of Eastern Mediterranean deep sea microbial communities to oil. Sci Rep 7:11p

Lotze HK, Guest H, O'Leary J, Tuda A, Wallace D (2018) Public perceptions of marine threats and protection from around the world. Oceanic Coastal Manage 152:14–22

Mabrouk BM, Jonoski A, Solomatine D, Uhlenbrook S (2013) A review of seawater intrusion in the Nile Delta groundwater system-the basis for assessing impacts due to climate changes and water resources development. Hydrol Earth Syst Sci Dis 10:10873–10911

Maglio A, Pavan G, Castellote M, Frey S (2016) Overview of noise hotspots in the ACOOBAMS area. Part I Mediterranean Sea, Final Report

Malak DA, Livingstone RS, Pollard D, Polidoro BA, Cuttelod A, Bariche M, Bilecenoglu M, Carpenter KE, Collette BB, Francour P, Goren M, Kara MH, Massutí E, Papaconstantinou C, Tunesi L (2011) Overview of the conservation status of the marine fishes of the Mediterranean Sea. IUCN Gland, Switzerland and Malaga, Spain

Mannino AM, Balistreri P, Deidun A (2017) The marine biodiversity of the Mediterranean Sea in a changing climate: the impact of biological invasions. In: Fuerst-Bjelis B (ed) Mediterranean identities—environment, society, culture. INTECH, London, pp 101–127

Marbà N, Jorda G, Agusti S, Girard C, Duarte CM (2015) Footprints of climate change on Mediterranean Sea biota. Front Mar Sci 13:11p

Massa F, Onofri L, Fezzardi D (2017) Aquaculture in the Mediterranean and Black Sea: a blue growth perspective. In: Nunes PALD, Svensson LE, Markandya A (eds) Handbook on the economics and management of sustainable oceans. Edward Elgar, Cheltenham, Northampton, pp 93–123

Mea M, Newton A, Uyarra MC, Alonso C, Borja A (2016) From science to policy and society: enhancing the effectiveness of communication. Front Mar Sci 3, 17p

MedPAN & UNEP-MAP-SPA/RAC (2016) The 2016 status of Marine protected Areas in the Mediterranean—main findings. MedPAN, RAC/SPA

Menabit S, MureŞan M, Begun T, Pavel B, Seghedi A (2017) "The School differently" learning about Marine Protected Areas—a proactive educational approach towards implementation of measures of marine habitats conservation and protection. Geo-Eco-Marina 23:215–222

Milazzo M, Chemello R, Badalamenti F, Camarda R, Riggio S (2002) The impact of human recreational activities in marine protected areas: what lessons should be learnt in the Mediterranean Sea? Mar Ecol 23:280–290

Millennium Ecosystem Assessment (2005) Ecosystems and human well-being: synthesis. Island Press, Washington, DC

MIO-ECSDE (2003) Sustainable Mediterranean 30. Mediterranean Information Office for Environment, Culture and Sustainable Development

MIO-ECSDE (2004) Sustainable Mediterranean 34. Mediterranean Information Office for Environment, Culture and Sustainable Development

Mogias A, Boubonari T, Markos A, Kevrekidis T (2015) Greek pre-service teachers' knowledge of ocean sciences issues and attitudes toward ocean stewardship. J Environ Educ 46:251–270

Mogias A, Boubonari T, Realdon G, Previati M, Mokos M, Koulouri P, Cheimonopoulou M (2019) Evaluating ocean literacy of elementary school students: preliminary results of a cross-cultural study in the Mediterranean region. Front Mar Sci 6:14p

Montefalcone M, Chiantore M, Lanzone A, Morri C, Albertelli G, Bianchi CN (2008) BACI design reveals the decline of the seagrass *Posidonia oceanica* induced by anchoring. Mar Pollut Bull 56:1637–1645

Montefalcone M, Lasagna R, Bianchi CN, Morri C, Albertelli G (2006) Anchoring damage on Posidonia oceanica meadow cover: a case study in Prelo cove (Ligurian Sea, NW Mediterranean). Chem Ecol 22:207–217

Mullis IVS, Martin MO, Goh S, Cotter K (eds) (2016) TIMSS 2015 encyclopedia: education policy and curriculum in mathematics and science. Retrieved from Boston College, TIMSS & PIRLS International Study Center

OCEAN ATLAS Facts and Figures on the Threats to Our Marine Ecosystems. (2017). Heinrich Böll Foundation Schleswig-Holstein, the Heinrich Böll Foundation (national foundation), and the University of Kiel's Future Ocean Cluster of Excellence
OECD (2001) Citizens as partners: OECD handbook on information, consultation and public participation in policy-making. OECD Publishing, Paris
Orion N, Fortner RW (2003) Mediterranean models for integrating environmental education and earth sciences through earth systems education. Mediterr J Educ Stud 8:97–111
Otero M, Cebrian E, Francour PP, Galil B, Savini D (2013) Monitoring marine invasive species in Mediterranean Marine Protected Areas (MPAs): a strategy and practical guide for managers. IUCN, Malaga, Spain
Paradis S, Puig P, Masque P, Juan-Díaz X, Martín J, Palanques A (2017) Bottom-trawling along submarine canyons impacts deep sedimentary regimes. Sci Rep 7:12p
Peng C, Zhao X, Liu G (2015) Noise in the sea and its impacts on marine organisms. Int J Environ Res Public Health 12:12304–12323
Piante C, Ody D (2015) Blue growth in the Mediterranean Sea: the challenge of good environmental status. MedTrends Project, WWF-France, p 192pp
Piroddi C, Coll M, Liquete C, Macias D, Greer K, Buszowski J, Steenbeek J, Danovaro R, Christensen V (2017) Historical changes of the Mediterranean Sea ecosystem: modeling the role and impact of primary productivity and fisheries changes over time. Sci Rep 7:18p
PISCO & UNS (2016) The science of marine protected areas, 3rd edn, Mediterranean. Partnership for Interdisciplinary Studies of Coastal Oceans and University of Nice Sophia Antipolis
Rako-Gospić N, Picciulin M (2019) Underwater noise: sources and effects on marine life. In: Shepard C (ed) World Seas: an environmental evaluation Volume III: Ecological issues and environmental impacts, 2nd edn. Academic Press, pp 367–389
Randone M, Di Carlo G, Costantini M, Tzanetti T, Haferkamp D, Portafaix A, Smits M, Antoniades V, Kachaner N, Osborne A, Chaudhry T, McPhillips J, Astier C (2017) Reviving the economy of the Mediterranean Sea: actions for a sustainable future. WWF Mediterranean Marine Initiative, Rome, Italy
Realdon G, Mogias A, Fabris S, Candussio G, Invernizzi C, Paris E (2019) Assessing ocean literacy in a sample of Italian primary and middle school students. Rend Online Soc Geol It 49:107–112
Realdon G, Paris E, Invernizzi MC (2016) Teaching earth sciences in Italian liceo high schools following the 2010 reform: a survey. Rend Online Soc Geol It 40:71–79
Rochette J, Billé R (2010) IDDRI, 2010. Analysis of the Mediterranean ICZM protocol: at the crossroads between the rationality of provisions and the logic of negotiations. IDRI
Romeo T, Pietro B, Pedà C, Consoli P, Andaloro F, Fossi MC (2015) First evidence of presence of plastic debris in stomach of large pelagic fish in the Mediterranean Sea. Mar Pollut Bull 95:358–361
Ryabinin V, Barbière J, Haugan P, Kullenberg G, Smith N, McLean C, Troisi A, Fischer A, Aricò S, Aarup T, Pissierssens P, Visbeck M, Oksfeldt Enevoldsen H, Rigaud J (2019) The UN decade of ocean science for sustainable development. Front Mar Sci 6:10p
Santoro F, Santin S, Scowcroft G, Fauville G, Tuddenham P (2017) Ocean literacy for all—a toolkit. IOC/UNESCO and UNESCO Venice Ofce, Paris (IO Manuals and Guides). Ocean Literacy for All—A toolkit, IOC/UNESCO and UNESCO Venice Office, Paris (IOC Manuals and Guides, 80)
Seltenrich N (2015) New link in the food chain? Marine plastic pollution and seafood safety. Environ Health Perspect 123:35–41
Serrano O, Ruhon R, Lavery PS, Kendrick GA, Hickey S, Masqué P, Arias-Ortiz A, Steven A, Duarte CM (2016) Impact of mooring activities on carbon stocks in seagrass meadows. Sci Rep 6:10p
Soriano-Redondo A, Cortes V, Reuez-Gonzalez JM, Guallar S, Bécares J, Rodríguez B, Arcos HM, González-Solís J (2016) Relative abundance and distribution of fisheries influence risk of seabird bycatch. Sci Rep 6:8p

Squarcina E, Pecorelli V (2017) Ocean citizenship: the time to adopt a useful concept for environmental teaching and citizenship education is now. J Res Didactics in Geogr (J-READING) 2:45–53

Stoll-Kleemann S (2019) Feasible options for behaviour change toward more effective ocean literacy: a systematic review. Front Mar Sci 6:14p

Sturani M (2016) Teaching physical geography in Italian high schools (Licei): current situation and perspective. Rend Online Soc Geol It 40:91–98

Suaria G, Avio CG, Mineo A, Lattin GL, Magaldi MG, Belmonte G, Moore CJ, Regoli F, Aliani S (2016) The Mediterranean plastic soup: synthetic polymers in Mediterranean surface waters. Scientific Reports 6:10p

UN (2020) United Nations Decade of Ocean Science for Sustainable Development 2021–2030 Implementation Plan, Version 2.0 July 2020; UNESCO: Paris, France

UNEP (2001) Urbanisation in the Mediterranean region from 1950 to 1995. Sophia-Antipolis, Priority Actions Programme, Blue Plan Papers

UNEP/MAP (2012) State of the Mediterranean Marine and Coastal Environment. Barcelona Convention, Athens

UNEP/MAP (2015) Marine Litter assessment in the Mediterranean 2015. UN Environment/Mediterranean Action Plan

UNEP/MAP (2016) Mediterranean strategy for sustainable development 2016–2025. Plan Bleu, Regional Activity Centre, Valbonne

UNEP/MAP (2017) 2017 Mediterranean quality status report, land- and sea-based pollution, Ecological Objective 10 (EO 10) Marine Litter

UNEP/MAP-MED POL/WHO (2008) Assessment of the state of microbial pollution in the Mediterranean Sea. Technical Reports Series No. 170, UNEP/MAP, Athens

UNEP/MAP/PAP (2001) Coastal zone management in the Mediterranean. Split, Priority Actions Programme

UNEP/MAP/PAP (2008) Protocol on integrated coastal zone management in the Mediterranean. Split, Priority Actions Programme

UNEP-MAP RAC/SPA (2010) The Mediterranean Sea biodiversity: state of the ecosystems, pressures, impacts and future priorities. RAC/SPA, Tunis

UNESCO (2015) Rethinking education: towards a global common good? Paris

UNESCO/UNEP (1976) The Belgrade charter. Connect: UNESCO/UNEP Environ Educ Newsl 1:1–2

Vázquez-Luis M, Alvarez Perez E, Barrajón A, García-March JR, Grau A, Hendriks IE, Jiménez S, Kersting D, Moreno D, Pérez M, Ruiz JM, Sánchez J, Villalba A, Deudero S (2017) S.O.S. Pinna nobilis: a mass mortality event in Western Mediterranean Sea. Front Mar Sci 4:6p

Weatherdon LM, Magnan AK, Rogers AD, Sumaila UR, Cheung WWL (2016) Observed and projected impacts of climate change on marine fisheries, aquaculture, coastal tourism, and human health: an update. Front Mar Sci 3:21p

Werz M, Hoffman M (2017) Climate change and migration in the Mediterranean: challenges for the future. In: IE Med. Mediterranean yearbook 2017: strategic sectors, economy and territory, panorama, pp 270–273

Wolff C, Vafeidi AT, Muis S, Lincke D, Satta A, Lionello P, Jimenez JA, Conte D, Hinkel J (2018) A Mediterranean coastal database for assessing the impacts of sea level rise and associated hazards. Sci Data 5:11p

Yavetz B, Goldman D, Pe'er S (2009) Environmental literacy of pre-service teachers in Israel: a comparison between students at the onset and end of their studies. Environ Educ Res 15:393–415

Zahedi S (2008) Tourism impact on coastal environment. WIT Trans Built Environ 99:45–57

Fostering Ocean-Literate Generations: The Portuguese Blue School

Raquel L. Costa, Bernardo Mata, Fernanda Silva, Patrícia Conceição, and Laura Guimarães

Abstract The last two decades produced a wealth of work revealing the pressing need to increase ocean literacy in support of ocean sustainability and governance. The aim of this chapter is to present the Portuguese Blue School programme, an initiative born in response to the above need, to prepare ocean-literate citizens and support the development of the blue economy. The Blue School PT programme promotes an integrated political strategy for marine education capable of engaging all sea sectors, with the goal of improving the level of ocean literacy in the short and long term. Headed by the Ministry of Sea, the Blue School is supported by universities, governmental and non-governmental organizations, municipalities, industries and companies. Over the different sections we present an overview of the ocean literacy context underlying the creation of the Blue School, the Blue School concept and functioning and the results of its first years of implementation. This programme, fitting the objectives of the United Nations (UN) Sustainable Development Goals (SDG) 4, 14 and 17, accounts for the national strategies for sea and education and the evolving view of ocean literacy, addressing identified barriers to its implementation in schools.

Keywords Portuguese Blue School · Ocean literacy in formal education · Ocean literacy community

Introduction

The ocean inspires and enriches culture, brings people together and acts as a true commercial highway. It is an oasis of biodiversity and a source of inspiration for artists and writers. It represents an engine for tourism, a gym for athletes, and coastal regions

R. L. Costa (✉) · B. Mata · F. Silva · P. Conceição
Directorate General for Maritime Policy, Ministry of Sea, Lisbon, Portugal
e-mail: raquel.costa@dgpm.mm.gov.pt

L. Guimarães
CIIMAR - Interdisciplinary Centre of Marine and Environmental Research and Department of Biology, Faculty of Sciences,
University of Porto, Porto, Portugal

K. C. Koutsopoulos and J. H. Stel (eds), *Ocean Literacy: Understanding the Ocean*, Key Challenges in Geography, https://doi.org/10.1007/978-3-030-70155-0_10

are places of well-being for a large part of the population (European Marine Board 2013). The ocean offers us much more than seafood (Correia et al. 2020). Marine components are used in unexpected "blue" foods like peanut butter. Medicines used in virus-related diseases or cancer, as well as some pain killers, have marine extracts in their composition. The ocean offers us so much and we give so little. The passivity (Fig. 1) and absence of behaviour change in citizens shows that it is necessary to do more than raise awareness, and encourage all citizens and sectors of society to assume informed and responsible attitudes towards the ocean. It is also necessary to go beyond knowledge, i.e. citizens must know how to communicate, act and decide. Such is the essence of ocean literacy (hereinafter also referred to as OL).

Even if there is a historical and cultural perception of the importance of the ocean to our country, mirrored in art, literature, architecture, geography, gastronomy and sport, the recognition of our own influence on the ocean and of the ocean's influence on us remains limited. Developing and promoting active actions and instilling individual and collective responsibilities towards the ocean is fundamental, regardless of where we live, the country where we were born in, our culture, our priorities or our taste.

In Portugal, 40 thousand companies had a direct impact on the Economy of the Sea, employing 126 thousand people and generating a GVA of 3.3 billion euros in 2018 (Direção-Geral de Política do Mar 2009). The Portuguese Exclusive Economic

Fig. 1 Why should we bother? A Charles Saxon Cartoon (New Yorker, 21st of March 1983) illustrating the need to increase ocean knowledge

Zone is also one of the largest in Europe, bearing wealthy resources and representing a significant added value to the national and European blue economy. Additionally, with the extension of the continental shelf, Portugal has the tenth largest maritime area in the world corresponding to 97% of its territory (Ministério do Mar 2019). Recognizing this, over the last decade Portugal has reinforced its commitment to valuing the ocean by creating a Ministry of the Sea and establishing a National Ocean Strategy, which envisions the social, economic and environmental promotion of the ocean and coastal areas for the benefit of the Portuguese population (Direção-Geral de Política do Mar 2014). A new National Ocean Strategy is currently under preparation for the 2020–2030 decade. Ocean literacy should respond with one voice acting as an aggregator of actions, sectors and actors, but only the mixture of knowledge and emotional connection will allow us to assume that we all have individual and collective responsibilities towards the ocean. More than communicating relevant aspects of the ocean to society, it will be necessary to include society itself in this transformation, particularly children and young people, the future active citizens.

Empowering citizens and future generations to take actions and decisions towards protecting ocean health is vital to promote sustainable development and ocean governance. In Portugal, as in around the globe, different initiatives, organizations, international projects and collaborations have been mobilizing to make the society aware of the state of the ocean (Innis and Simcock 2016). But some questions have been arising. Has this effort effectively contributed to change human perception and behaviour in the last 15 years? Were we able to make people understand how their actions, even if done 500 km away from the ocean, will impact its sustainability? Have we managed to involve people and organizations from different socio-economic sectors in this demand? Which initiatives can better promote both short- and long-lasting change? It is essential to create strategies and identify the tools and approaches that can better transform ocean knowledge into long-lasting preparedness and actions for ocean governance.

Due to the Decade of Ocean Science for Sustainable Development (UNESCO 2020) and the National Ocean Strategy, the next decade can generate a unique momentum to create a behavioural change tsunami in our society. One of the keys to an effective and permanent change towards the ocean is to implement creative and collaborative ocean-inspiring projects in and with school communities. Here we present the Portuguese Blue School, created as an answer to the global needs identified above, and support the development of the national blue economy. Over the next sections, we present the ocean literacy context that inspired the creation of the Blue School, the concept itself, and the results of the first years of implementation, up to the adoption of the concept by schools in other countries.

Sailing Through the History of Ocean Literacy in Portugal

In a coastal country like Portugal, where the sea is steeped in art, literature, architecture, popular wisdom, economy, gastronomy and sport, the ocean is part of each of us and our society. Our vision is imbued in the ocean, although individually and collectively, Portuguese people have little awareness of their impact on the ocean. Portugal has a territorial dimension that is 97% sea, placing us as the tenth country with the largest maritime area in the world and the third in Europe. This new, unknown territory requires a long-term strategy in which OL plays an important role. Future generations have an increased responsibility to sustainably manage this maritime territory.

Hence Portugal was the first country to translate and adapt the Ocean Literacy Principles (National Geographic Society et al. 2007) to the Atlantic reality, an initiative coordinated by Ciência Viva in 2011 (Ciência Viva 2020). Ocean literacy was highlighted in the National Strategy for the Sea 2013–2020 (Direção-Geral de Política do Mar 2014) and in public policies 2020–2023 (Governo de Portugal 2020), and commitments related to ocean literacy under the Sustainable Development Goal 14 (Our Ocean 2018) were made.

In Portugal OL is seen as a broad and comprehensive concept that considers scientific and environmental aspects but is also fingerprinted by our history and culture. Ocean literacy needs to be capable of covering areas such as economy, technology, entrepreneurship, history, geography, gender equality, culture, value or sustainability. For the development of the blue economy and the sustainability of our sea, technological development and entrepreneurship must be considered in ocean literacy, and this implies an urgent paradigm shift and a new interventional and conscious perception of the ocean. This need is in line with the Intergovernmental Oceanographic Commission (IOC)-UNESCO's new vision for ocean literacy (Santoro et al. 2017) with an inclusive and holistic approach more readily adaptable to and adoptable in different geographic, historic and cultural settings. This new approach is of easy use by any actor promoting ocean literacy, inside or outside formal educational contexts (Fig. 2), and closer to the broad and holistic ocean literacy view in Portugal.

Ocean literacy is defined by the IOC-UNESCO as a means to:

- Learn about ocean issues from a multiplicity of knowledge areas;
- Identify and understand personal and global perspectives;
- Apply decision-making processes to complex issues that affect individual, community or global well-being.

Overall, the multiple visions and ways for implementing ocean literacy produced different approaches that have enriched the concept itself and consolidated its importance. However, raising awareness about the relationship between the ocean and humankind is hardly enough. Numerous awareness initiatives are taking place around the world, most of which are temporary because they depend on project financing. From educational resources, websites and games to activities for different society sectors, the offer is varied and interesting but generally without conclusive results

Fig. 2 Ocean literacy integrated multi-perspective approach proposed by the IOC-UNESCO (Santoro et al. 2017)

on how OL is affecting people perception and behaviour. Little transformation in communities or the society is evident. Effective change in attitudes implies crossing complex barriers. Each country or community has different priorities related to the sea to tackle, and consequently, each country, region or place faces different obstacles. If the change in individual behaviours is explained by the link between people and the subject, with the emotional component playing a decisive role, how can we effectively change the behaviours of groups, sectors or communities?

Answering these and other relevant questions relies on defining medium-term strategies to consolidate ocean literacy as was done in Canada (COLC 2020) and will be done in Portugal during the next decade. Each country needs to build an integrated inner strategy, aligned with the global needs, for developing inclusive and holistic OL in order to reach all society sectors. It is crucial to develop an implementation plan with defined objectives and goals. It is necessary to encourage research in this area of knowledge to identify gaps and support the decisions and strategies to be followed and, most importantly, to monitor and assess the ocean literacy implementation programmes developed.

Ocean literacy actions in Portugal have been predominantly devoted to children and youngsters. The educational offer available to schools, identified very early as insufficient, has been multiplying, showing an increasing diversity of contents, stakeholders and sectors. The effort of the stakeholders to promote ocean literacy actions among the school community is meritorious, especially considering the common lack of financial and human resources specifically assigned to these actions. Even so, considering the multiplicity of initiatives in Portugal, the country lacks a political strategy to unite these efforts and, above all, capable of defining objectives, indicators and targets to be achieved. The Blue School PT programme was developed by the Ministry of the Sea (Escola Azul 2020) to overcome some of these needs considering ocean literacy among the younger generation.

Blue School

First Seeds

The concept of Blue Schools has been discussed in European marine education forums and conferences for several years. Most notably, the conferences of the annual European Marine Science Educators Association (EMSEA 2020) have always taken an interest in a programme capable of engaging schools in ocean-related themes, while promoting the sea sector involvement in ocean literacy.

In 2015, the Sea Change project (Sea Change 2020) set the development of a European Blue School concept as a long-term initiative, aiming to empower school communities to become responsible and aware, while acting towards marine sustainability through learning. An initial survey led to the consideration of a tiered approach as suitable to the establishment of Blue School, namely by creating national/regional committees, setting implementation and dissemination plans and establishing a dedicated team to support teachers (Sea Change consortium, *personal communication*).

Nonetheless, as the result of additional consultations and working groups, the main conclusion considered that it was not possible to come up with a single European concept due to the clear diversity of European cultural contexts and to environmental education programmes already well-established in some countries (Sea Change consortium, *personal communication*). This final report also recognizes that the implementation of Blue School needs a long-term political and financial support to ensure a dedicated team and a platform to manage the schools involved. The establishment of Blue Schools is seen as a lengthy and complex process, requiring a solid stakeholder network, connection with pre-existent projects and financial support for the follow-up.

Meanwhile, it was in Portugal that the ideal setting to effectively establish a Blue School programme was created. With the support of the Ministry of the Sea the ocean literacy team led the way, assembling a national working group to discuss and shape a Portuguese Blue School concept and set the foundations for its actual implementation nation-wide. The first meetings, held by the end of 2015, included the most relevant national players in marine education and ocean literacy: representatives from the Ministries of Sea, Education and Science; universities (marine sciences and education research centres); museums; aquariums; non-governmental organizations; the Portuguese National Commission for UNESCO; marine educators; school directors and teachers.

The Portuguese Blue School concept was developed and fine-tuned over the next years, being launched as a pilot-programme in the 2017/18 school year. The Ministry of Sea supports the initiative guaranteeing an ocean literacy team that ensures its implementation and development. Relevant stakeholders were included in the Blue School Scientific and Pedagogical Commission and other marine and maritime players came onboard to participate as partners offering experiences and activities to Blue Schools. The Blue School also welcomed the partnership with municipalities, supporting the dissemination and implementation of the programme at a local level.

The structure of the Blue School programme was primarily designed to address the relevant issues identified by the national working group, which reflected the lack of a true educational policy where the sea is regarded as a structural priority.

The Portuguese sea sector was already working on projects, activities and educational resources about the ocean, but its dissemination and broad acceptance was the real problem. The actions of different sea players were scarcely known, and this lack of outreach hampered their adoption by both schools and marine institutions. Teachers were not aware of the existing opportunities, and the institutions' work on outreach activities ended up producing underwhelming impacts and limited networking towards this pressing common aim.

This lack of communication was noticeable on the scarce articulation between entities, who acted mostly dispersed and isolated. Strengthening cooperation among institutions represented a fundamental challenge of the Blue School programme. Only working together would it be possible to build a structured, continuous and effective strategy for ocean literacy.

In the field, the main challenge relied on most school's excessive focus on students' academic success. There are some schools who resort to partnerships and thrive in promoting alternative teaching methodologies, but they are still a minority. In most cases, schools and teachers opt for more traditional teacher-centred methods, without any sort of deviation from the school curriculum. These schools often show an impeding resistance to develop school projects involving diversified partnerships.

The scarce inclusion of ocean issues in classes was also identified. Even when schools decided to focus on marine topics, teachers lacked specific knowledge and ocean literacy training, which represented a clear obstacle. This was commonly reflected on the difficulties most teachers usually found when tackling ocean issues:

- *Finding resources and activities (in native language) about the ocean to use in the classroom;*
- *Identifying resources adequate to curricula topics;*
- *Selecting curricula topics to address ocean subjects in a meaningful way;*
- *Adequate homework and follow-up on activities developed by partner institutions.*

The Blue School programme was thus created within this framework as a strategy to promote a project and problem-based approach in schools, with a stronger emphasis on interdisciplinarity and on developing students' critical thinking, behaviour and action towards the ocean (UNESCO 2014; OECD 2018; Santoro et al. 2017). It was also faced as an opportunity to promote and articulate the educational offer of different entities, allowing them to participate in a single and coordinated ocean literacy strategy for schools (Governo de Portugal 2020).

Over these first 3 years, the programme has been tested and adjusted in a collaborative way with the entire community of Blue Schools. These changes and adaptations, made alongside those who effectively implement the programme in the field, have produced results that overcome our best expectations. Based on its success, the European Commission's Directorate-General for Maritime Affairs and Fisheries (DG MARE) launched in 2019 a tender for the development of a European Ocean Alliance Platform where one of the tasks pointed towards the creation of a European network of Blue Schools, specifically stating that: "This action will be based on the existing successful stories such as in Portugal or Malta" (European Commission 2019). The European Network of Blue Schools was launched in September 2020 with the purpose to inspire teachers to challenge their students to develop a "Find the Blue" project that links them to the ocean (European Commission 2020).

What Is a Blue School?

The Blue School programme distinguishes the schools working on ocean literacy, certifying them as able to offer an ocean curriculum to their communities. The certification is awarded by the Ministry of the Sea based on the integrated work developed by the Blue School candidate and the follow-up of the Coordination of the programme. The Blue School concept was settled, as previously mentioned, by a working group that assembled major Portuguese ocean literacy stakeholders. Out of the collective experience gathered from each one's work with schools and children in ocean-related topics, certain issues were clearly identified; some of them related to the intricacies of the school system and others to the sea sector's own approach to education and ocean literacy:

- *The Portuguese school system relies on traditional teaching methods and teachers don't have the time or budget to develop different sorts of activities.*
- *Schools and teachers struggle to create projects that go beyond the classroom setting.*
- *Resorting to external partnerships, schools may provide activities about the ocean for their students. But there is no follow-up by the teachers on these topics as they do not feel comfortable to teach about the sea.*
- *Most educational actions about the ocean created by marine institutions lack an adequate outreach and are mostly unknown to schools.*

- *The different entities and professionals of the sea sector do not invest in promoting a clear and open communication approach with schools and the society, neglecting to play an active role in ocean literacy.*

It was out of these considerations and aiming to surpass these patent limitations that the eight Blue School criteria emerged (Table 1).

The Blue School concept and its eight criteria are aligned with the Ocean Literacy Framework (National Geographic Society et al. 2007); with national orientations from the National Ocean Strategy (Direção-Geral de Política do Mar 2014) and the Development Education Guidelines issued by the Ministry of Education (Martins et al. 2017; Ministério da Educação 2017); with international guidelines from the European Commission (European Commission 2018), the Organization for Economic Co-operation and Development (OECD 2018) and the United Nations Educational, Scientific and Cultural Organization (UNESCO) (UNESCO 2014, 2015). It is also in line with Objectives and expected Societal Outcomes of the United Nations Decade of Ocean Science for Sustainable Development, and thus with the Agenda 2030 Sustainable Development Goals 4 (Quality Education), 14 (Life Below Water) and 17 (Partnerships for the Goals). All these international strategies reinforce the importance of multiple literacies, critical thinking, informed decision-making, students' autonomy, communication and interdisciplinarity (OECD 2018; European Commission 2018; UNESCO 2015).

Also considered were the barriers to the implementation of ocean literacy in schools identified by the Sea Change project (Fauville et al. 2018). The Sea Change project (Sea Change 2020) developed a study to draw a pan-European portrait of the barriers to teach about the ocean in formal education. Focusing on the 12- to 19-year-old age group, national consultations with education stakeholders were carried out in eight countries (Belgium, Denmark, Greece, Ireland, Portugal, Spain, Sweden and United Kingdom) to identify the barriers and discuss options to overcome them (Fauville et al. 2018).

A European Influence Map was obtained where barriers are interconnected and influence each other. The main purpose of this map and study was twofold:

- To contribute to improve education stakeholders' awareness and understanding of the main challenges in marine education;
- To provide a roadmap for policy makers to develop applicable solutions and mobilizing actions that could mitigate the barriers to teach about the ocean.

This Multistage European Influence Map groups the barriers into higher-order themes defining their influence and impact on teaching 12- to 19-year-old youngsters about the ocean. From the map analysis, it is possible to conclude that "Awareness and perceived knowledge" exerts the highest level of influential factors. On the other hand, "Connections between humans and the ocean" and "Blue economy" have the lowest level of influence for teaching 12- to 19-year olds about the ocean (Gotensparre et al. 2017). Table 2 represents an adaptation of the Multistage European Influence Map where each barrier theme connects to the corresponding Blue School asset that works as a response to overcome it.

Table 1 The Portuguese Blue School. Criteria for certification of Blue Schools

Blue School Criteria	Blue School Criteria description
A Blue School must explore an ocean topic	Each Blue School should tackle a problem about the ocean that is meaningful to the school's local community, during two school years. This two-year requisite guarantees clear continuity and promotes more complex projects or actions
A Blue School must select Blue School ambassadors	Blue School students should be given more responsibility and autonomy. The school should select at least two students as ambassadors, playing an active role in the definition, development and communication of the school's actions
A Blue School must engage students of different grades	A Blue School should be able to engage students of different grades in ocean-related issues, promoting interaction and collaborative work between different age groups
A Blue School must address the ocean in different subjects or disciplines	Blue School promotes an inclusive perspective of ocean literacy, including all STEM disciplines, sports, arts and humanities. This criterion promotes interconnection between different subjects, in accordance with the national educational standards and international guidelines for quality education
A Blue School must communicate its actions	Portuguese schools are not used to communicate, so this requisite represents an incentive for the rightful dissemination of the Blue School projects or actions
A Blue School must interact with the partners' network	Blue Schools must involve the sea sector. They must resort to at least one member from the Blue School partners' network, which presents them with specific educational offers
A Blue School must involve the local community	School projects or actions should go beyond the classroom and beyond the school, reflecting a clear connection with local communities. The link with the school's social environment is crucial to promote the students' emotional and intellectual investment in the project
A Blue School must encourage students to act, intervene and decide	A Blue School project should instigate not only knowledge but also the actual application of that knowledge. To transform knowledge into action. To inspire attitudinal changes and positively affect our society global perception and attitudes towards the ocean. This is the ultimate requisite

Table 2 Contribution of the Portuguese Blue School to overcome the meta-barrier themes identified in the European consultation carried out within the Sea Change project

Higher-order themes	Barriers	Portuguese Blue School Concept	Rationale
Awareness and perceived knowledge	Barriers that relate to the concept of ocean literacy: either lack of understanding of the concept or some of the elements of ocean literacy	Each Blue School must explore an ocean topic in articulation with the international Ocean Literacy Framework	When applying to become a Blue School, each school must connect its project or challenge with both the UNESCO multi-perspective approach on ocean literacy and the Ocean Literacy Principles The Coordination guides and helps schools during the application process Ocean literacy training workshops are provided to Blue School Teachers
Policies and strategies	Policies and strategies affect how schools are run and the school budgets	The Blue School network	In addition to the Coordination itself, the Blue School network includes more than 89 partner institutions and 12 municipalities. This is a collaboration of close to 100 entities helping and facilitating the implementation of ocean literacy in schools
Engagement	Barriers relating to the importance of engaging students and speaking their language to increase interest and awareness of the ocean	A Blue School must encourage students to act, intervene and decide	One of the ultimate goals of a Blue School should be to instigate not only knowledge but the actual application of that knowledge, to transform knowledge into action

(continued)

Table 2 (continued)

Higher-order themes	Barriers	Portuguese Blue School Concept	Rationale
Formal education sector	All the barriers that originate from the formal education sector	A Blue School must address the ocean through different subjects	The ocean should be worked through multiple subjects in articulation with the formal curriculum Various hands-on activities to be performed in different formal and non-formal contexts are offered by the Blue School partners A problem-based approach underlies the Blue School concept enhancing students' interest in educational activities
The ocean itself	The very nature of the ocean makes it difficult to experience or understand	A Blue School must interact with the partners' network	The school interaction with the sea sector, marine educators or marine scientists included in the partners' network represent a good platform to overcome this barrier and bring the sea experience to the students in different ways
Collaboration	Collaboration between different marine education actors	The Blue School network	The Blue School partners' network gathers institutions working on ocean literacy from all over the country. This network represents a good way of facilitating networking between these entities and particularly on National and Regional Meetings, where teachers, partners and municipalities are gathered to discuss ocean literacy

(continued)

Table 2 (continued)

Higher-order themes	Barriers	Portuguese Blue School Concept	Rationale
Connections between humans and the ocean	It is about how the ocean matters to both individuals and society	A Blue School must involve the local community	The relations established between each Blue School with partners and specifically with the local community should be able to foster a better connection with the ocean and more efficiently multiply the message
Blue economy	The barriers all relate to Blue Society, which is the long-term strategy to support sustainable growth in the marine and maritime sectors as a whole	A Blue School must interact with the partners' network	The Blue School programme is led by the Ministry of Sea, with the support of municipalities, the sea sector and universities, among others. These are entities engaged to encourage widespread acceptance and a better understanding of the meaning of a Blue Society, the importance of the Blue Economy to Portugal, through their work with Blue Schools It also contributes to attract the interest of the students for professional careers linked to the ocean, as socially and economically rewarding

Although there are significant differences in the barriers identified in each country (Gotensparre et al. 2017; Governo de Portugal 2017), this connection between Multi-stage European Influence Map and Blue School displays how this programme, through its concept and structure, can overcome these barriers.

The guidelines issued by the Ministry of Education (Martins et al. 2017; Ministério da Educação 2017) to promote a twenty-first-century education cite specific actions, the following being an inspiration and clear influence underlying the definition of the Blue School criteria:

- Approach the teaching subjects by connecting them with problems and situations from the students' daily lives or from their socio-cultural and geographical environment, using diversified materials and resources;
- Provide experimentation of different tools and techniques, articulating different areas of knowledge, promoting observation activities and critical thinking;
- Develop cooperative learning activities, oriented towards the integration and exchange of knowledge, with the promotion of projects;
- Resort, with a critical perspective, to information and communication technologies and to diversified sources of information;
- Promote, systematically and intentionally, activities that allow students to make choices, confront points-of-view, solve problems and take decisions founded on values and principles;
- Value students' proactive work and self-initiative, promoting active intervention in school and in the local community.

The launch of the Blue School programme would temporally coincide with the pilot stage of the Ministry of Education's Programme for Curricular Autonomy and Flexibility (Governo de Portugal 2017), which brought to schools and teachers a refreshed educational mindset, with an openness for transversal and autonomous work in initiatives and programmes like the Blue School.

In summary, a Blue School actively engages the school community in ocean literacy, integrating the local community into its actions and interacting with the Blue School partners' network. A Blue School encourages its students to become more aware of the ocean by inspiring them to emotionally bond with the sea and to translate their knowledge into effective attitudinal changes. It promotes the development of critical thinking and initiative in students, as well as an integrated and holistic view of the ocean. A Blue School uses different fields of knowledge in transversal, innovative and creative learning. Among the certified Blue Schools, Santa Maria Maior School is just one excellent example of a fully engaged school community. Located in Viana do Castelo, a city with a long history related to shipbuilding, this school addresses ocean topics in all subjects and school activities, so becoming a Blue School was an obvious step. Their aim is to develop a curriculum that summons, in an articulated and transversal way, knowledge and skills, involving students in a participatory learning way.

All students were involved in the project, from 6 to 18 years old, and the school interacted with several Blue School partners in the development of many activities for different age groups. Some partners took their activities to the school while others welcomed students to their facilities. One of the most highly regarded experiences was when students from Santa Maria Maior high school were accepted for an internship at CIIMAR (Interdisciplinary Centre of Marine and Environmental Research) Laboratories. In the school, intra- and inter-class scientific debates guided by the teachers and/or researchers were staged.

Taking advantage of Viana do Castelo location and social connection with the ocean, the students had classes at local beaches. The ocean was addressed in several subjects from science to arts, sports, literature, history and geography (Figs. 3 and

Fig. 3 In Santa Maria Maior, the ocean was addressed in several subjects from science to arts, sports, literature, history and geography

4). Using the ocean as background for all disciplines gave each subject an amplified meaning, also by connecting them to the specifics of the local environment.

The school chose 4 Blue School ambassadors who presented the project in two international scientific conferences (Fig. 5), the Blue School National Meeting and several other national meetings. To involve the local community the students organized a conference and exhibition at school with a group of former codfish fishermen. During the exhibition, which was called: "Heroes that time doesn't erase - One day aboard a codfish lugger ...", the fishermen were able to share their life experiences with students and community. Santa Maria Maior School communicated its project to the community in local newspapers, but the school has also a TV channel available on a website for the broader community, where several news reports about the project were developed and broadcasted (Fig. 6).

As of June 2020, at the end of the programme's third year, the Blue School network comprised more than 200 schools working on ocean literacy, with close to 30,000 students, 89 partner institutions and 12 municipalities involved.

The Blue School partner network includes institutions providing mostly hands-on educational activities or mentoring in ocean-related issues, establishing multidisciplinary partnerships. These educational offers about the ocean are available in the Blue School platform (https://escolaazul.pt/atividades) and cover different areas of knowledge, thus underlining the different perspectives that the ocean can be worked on. From the more than 350 actions available, one can find options to work on OL through literature, sports, arts, history, sciences, technology, culture, etc. The Blue School platform allows to filter activities according to age group, location and category, which naturally represents a tremendous asset for teachers seeking to find the right opportunities to work with their students. Most of these actions are presented

Fig. 4 Students were part of an internship at CIIMAR (Interdisciplinary Centre of Marine and Environmental Research) Laboratories

as a privileged offer to Blue Schools (discounts, exclusive activities, easier access, scientific support, direct contact with people from the sea sector).

For the partner institutions, the Blue School provides them with more visibility to the schools that are implementing structured and continuous projects about the ocean. But above all, being part of Blue School means coming aboard a truly integrated strategy to promote ocean literacy among younger generations.

A Blue School Scientific and Pedagogical Commission is also instituted, comprising 12 institutions. The Commission supports the analysis of the schools' applications and the Blue School Coordination team in the conceptual and structural development and evolution of the Blue School programme.

Finally, with the natural growth of the Blue School, an increasingly important role in the success of the programme is played by municipalities. Currently, 12 Portuguese municipalities have expressed their commitment to OL by partnering Blue School and promoting the programme's implementation at a local level. They specifically support and promote the educational projects in their Blue Schools and identify local institutions to integrate the partner network.

Fig. 5 The students presented their Blue School project in two international conferences, the Blue School National Meeting and some local events

Fig. 6 The school has a TV channel broadcasted for the local community where several activities were shared

How to Become a Blue School

Application to Become a Blue School

Submission of an application is the first step a school must take to become a Blue School. The Blue School programme is voluntary and the application is made on the Blue School platform (www.escolaazul.pt) by a Coordinating Teacher who becomes the school's focal point (Fig. 7). Much of the interaction between the school and the Blue School National Coordination (hereafter referred to as Coordination) is done through this platform which is where the schools can also search for scientific support and ocean educational actions offered by partners' network and present evidence on how their projects are developing.

In the platform the Coordinating Teacher must fill in an application form, describing and submitting the school project for consideration. This form is purposely complex, requiring some reflection by the applicant school. One of the issues identified by the working group tailoring the creation of this programme laid on the difficulty many Portuguese schools feel when structuring a school project. This application form represents an attempt to tackle this situation, as it asks the teacher/school to reflect and explain how their challenge will achieve the eight criteria necessary to get the Blue School certification.

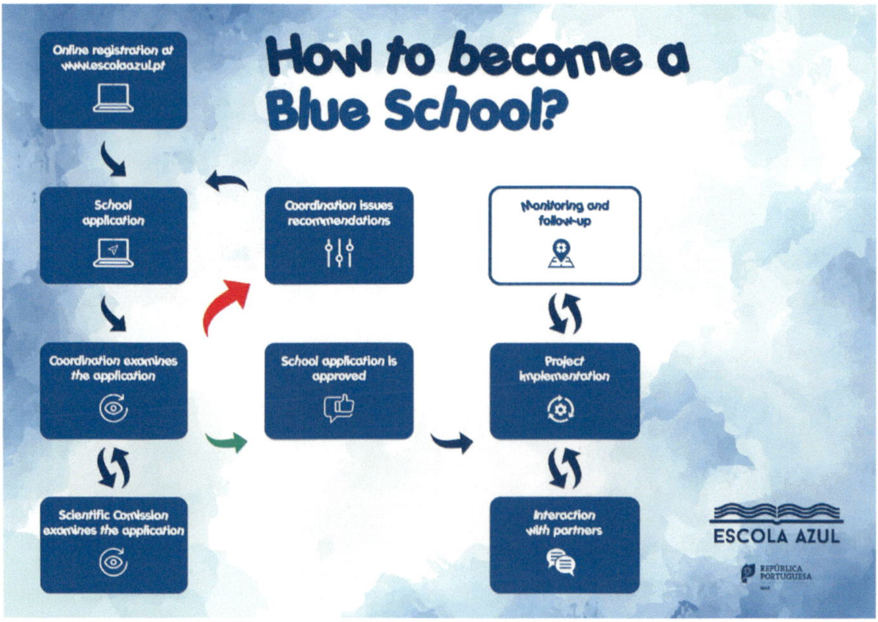

Fig. 7 Workflow for the certification of a Blue School

Fig. 8 Blue School challenges for 2019/2020. Schools must select one of these challenges when applying to become a Blue School

The application form represents the basic structure that a school must follow and works as a guideline to ensure the school develops a consistent proposal for the two years ahead. The school is asked to choose an ocean-related issue they want to work on. For that purpose, the application form has five pre-established challenges or a free ocean challenge (Fig. 8). These pre-established topics were added to the application to help schools build their project in a more focused way, and to present them with meaningful working topics, other than the obvious ones. Schools are always encouraged to choose a topic which is relevant not only for the school community but also for their local community, one of the key factors for the success of their project.

In order to frame their challenge and objectives in the Blue School concept, schools are also asked to identify what—IOC-UNESCO ocean literacy perspectives (Santoro et al. 2017), Ocean Literacy Principles (National Geographic Society et al. 2007) and UN Sustainable Development Goals (UNESCO 2020)—their actions will attempt to explore. By combining the UNESCO ocean literacy perspectives and the Ocean Literacy Principles in the same frame (Fig. 9) teachers are encouraged to see ocean literacy not only as something to be developed in science classes, but as a comprehensive concept capable of encompassing different knowledge areas.

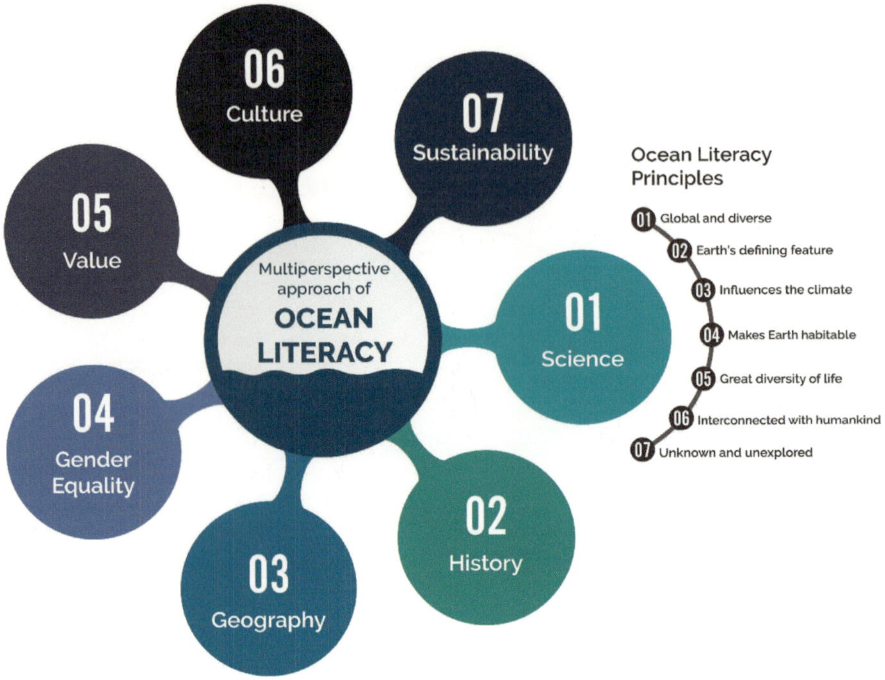

Fig. 9 Ocean literacy interpretation adopted by the Blue School National Coordination. When defining their challenge and objectives schools must identify perspectives and principles their work will explore

Application Analysis

Once submitted, the application is analysed by the Blue School National Coordination. This marks the beginning of an interaction between the Coordination and the school. Unless the project meets all the criteria at first (which seldom happens), the Coordination makes suggestions to help the school structure, adjust or redefine their proposal, find suitable partners to interact with, outline objectives and define strategies to achieve all the criteria. The Blue School Coordination never refuses submissions. After this first interaction it is up to schools to make the necessary changes and continue the dialogue until the application is structured enough to be accepted. Once the applications are soundly structured and able to achieve the Blue School eight criteria, they are also analysed by one of the Scientific and Pedagogical Commission entities which act as a final reviewer. Entities are chosen depending on the main topic schools want to tackle, as well as their geographical location (Fig. 7).

This personalized interaction during the application analysis has produced very positive results, since it provides the school and the Coordinating Teacher with a follow-up and a level of confidence and motivation that is later reflected on the quality of the projects.

Challenge Implementation

Once the school challenge is accepted, schools may start its implementation and must comply with some annual requirements (Fig. 10).

On the Blue School platform, the Coordinating Teacher obtains access to a section where he/she can update any relevant information on the project implementation and share with the Coordination any difficulty it faces. This is also where the school appoints its Blue School Ambassadors. After this step the school receives the Certification Kit, the symbol of its status as a Blue School. This is the stage where schools begin interacting with the Blue School partner network, and the two-year period of their challenge begins to count.

During the two-year period, schools are free to modify and adapt their project and planning, as new partners enter the Blue School network making new activities available, or simply because students show a more marked interest in a specific topic shifting the project dynamics. Twice a year schools are asked to fill a survey where the Coordination can access how the implementation is developing.

Over the last three years, synergies with other national/international programmes have been created, either by the Coordination or by the Blue Schools themselves, while developing their activities. The Eco-schools Programme, the UNESCO Associated Schools' Network, Educational Passages and the Portuguese Ministry of Education's Nautical Sports Training Centres are programmes/projects that can overlay

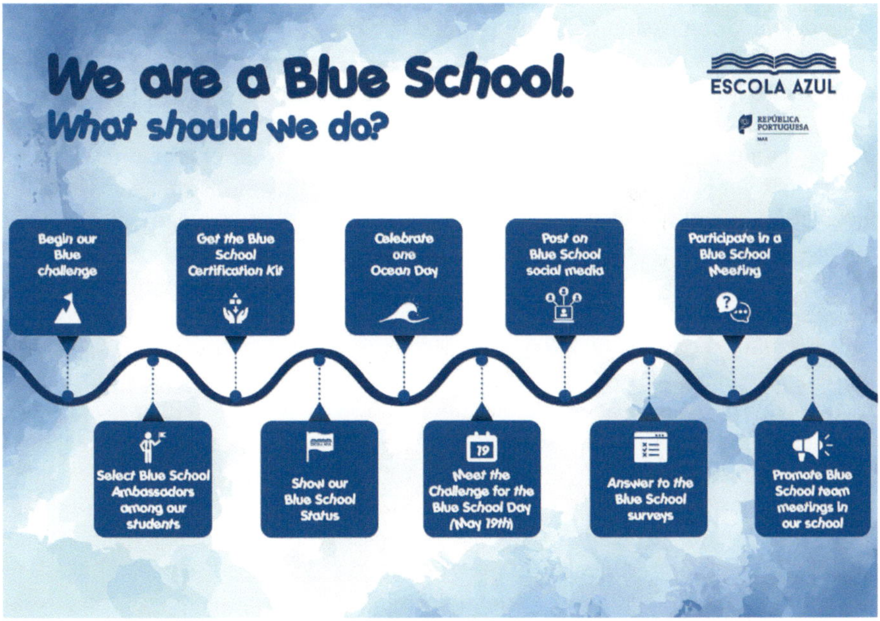

Fig. 10 Once a school becomes a Blue School these are the requirements every school must accomplish during the two project years

with the Blue School programme in specific areas such as themes worked, intended objectives or educational strategies used. This is the reason for which many schools end up combining these programmes with Blue School, thus being able to take advantage of the best points in each one.

Monitoring and Follow-Up

Schools are encouraged to share their activities on Blue School social media, either on the Blue School Facebook Group, where teachers, students, partners and families can express their opinion, discuss issues and share related content, or on Blue School Instagram, where students are more active.

But the monitoring of Blue Schools is not made exclusively through online interaction. Every year the national coordination tries to visit some schools to assess their projects. This personal interaction is key to the success of our programme since teachers need the opportunity to share their work, as well as their problems and difficulties, and pride themselves on showing the work done by their students. These visits to schools are also an opportunity for the Coordination to directly monitor the success of the application of the Blue School model in schools.

From 2019 on, due to the increase in the number of Blue Schools, these face-to-face visits to all schools became impossible and the Coordination chose to organize Regional Meetings. Regional Meetings are held once a year in each region and include Blue School teachers, regional partners, municipalities, the Scientific and Pedagogical Commission, as well as potential Blue School partners and schools. In these meetings the problems of implementing the programme in each school are addressed, but they also serve as an important moment for sharing experiences and strengthening relations.

This is one of the main objectives of the Blue School programme: to evolve as a community. More than an educational programme, the aim is for Blue School to become a national ocean literacy community, where schools, sea-related entities, partners and municipalities support each other in the development of projects and initiatives that promote an effective and affective bond between young people and the ocean.

The organization of annual Regional Meetings, and more recently the scheduling of triannual phone calls to Coordinating Teachers for follow-up, does not cancel visits to schools. Whenever it is considered important to visit a school, either to encourage the teaching and school staff to take part in the programme or because the school is unable to implement its plan, the Coordination schedules a visit.

Another important moment in the acknowledgement of the schools' work is the Blue School National Meeting. This two-day meeting is held annually and includes partners, teachers and students. It represents an excellent opportunity to link all the players involved and get them to share experiences and discuss their perspective on the programme.

Not only are these meetings an extraordinary learning platform for the Coordination (getting direct feedback from the participants) but they are also an essential moment to build the sense of community and belonging in teachers, students and partners alike. These meetings work as a very effective communication action.

Finally, on May 19, 2020, the 1st Blue School Day was celebrated. On the Blue School Day every Blue School must respond to a challenge released by the National Coordination. When 2020 suddenly became the most atypical of school years and schools were closed, the challenge was adapted so that it could be done from home and shared on social media. It was a huge success with more than 1000 people posting their feelings about being unable to see or simply be by the ocean. In such a particular time, this challenge acted as a community reinforcer bringing students, teachers, school staff, families, partners, sea-related entities and government together.

The 3rd Year, Fidelity Application

Schools ending the first biannual project must fill a permanence application to maintain their Blue School status. This step was a big challenge for the Coordination. It was decided to simplify the application since these schools had already given proof of being able to achieve all the Blue School criteria. Schools are encouraged to define a new challenge for the next two-year period and must comply with the same annual requirements of every other Blue School.

All the founder schools, which ended their first biannual project in 2019, applied for permanence in a clear sign of the effectiveness of the programme.

Blue School Data and Results

Currently, the Blue School programme includes 216 schools working on ocean literacy, distributed throughout Portugal's mainland, the Azores and Madeira Archipelagos (Fig. 11), and one Blue School outside Portugal, in Norway. Currently there are 33 schools with applications under evaluation. Over the last 3 years 35 schools failed to finish the analysis process, thus not becoming Blue Schools, meaning 86% of the applicant schools were approved as Blue Schools.

All school levels are fairly engaged in the programme. Only high schools present a lower number of applications (Fig. 12) due to the focus on final exams at the end of compulsory education.

Most schools choose to explore ocean themes related mainly to sustainability, biodiversity, conservation and marine pollution (Fig. 13), but it should be noted that most schools encompass more than one theme when defining their Blue School two-year challenge.

As previously mentioned, all school projects are structured according to the criteria defined in the Blue School concept. When submitting their application, teachers are asked to correlate their challenge with the IOC-UNESCO's OL multi-perspective approach (Santoro et al. 2017) and most schools choose sustainability issues closely followed by a scientific perspective (Fig. 14). Within the scientific

Fig. 11 Geographic distribution of Blue Schools in Portugal in 2020

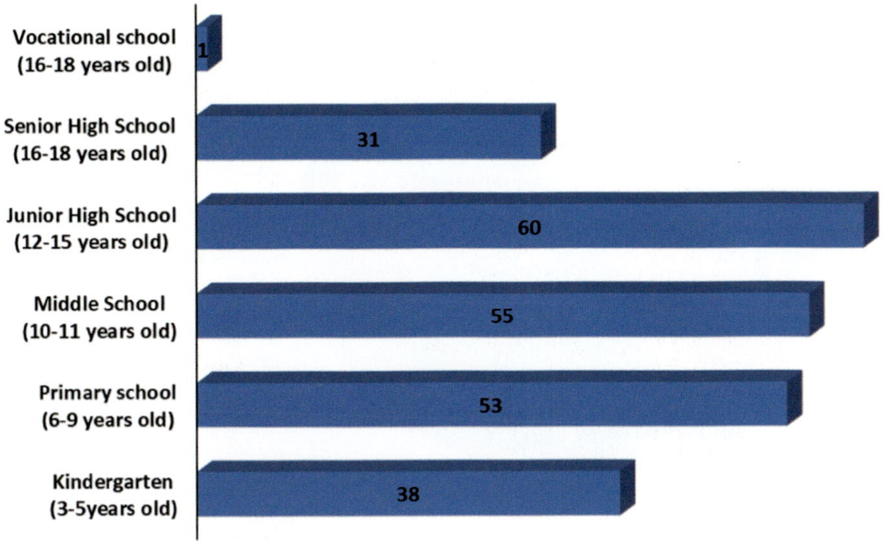

Fig. 12 Distribution of Blue Schools by school levels (absolute frequency)

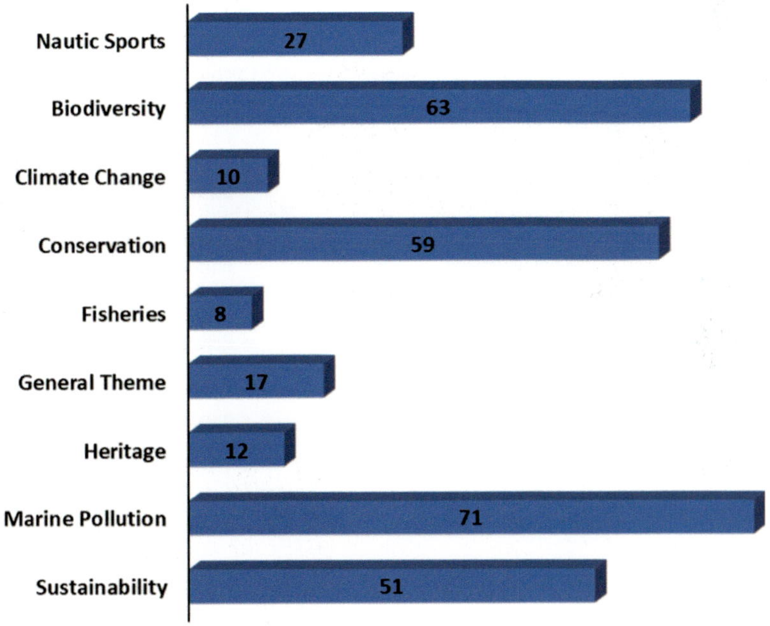

Fig. 13 Blue School challenges' themes (absolute frequency)

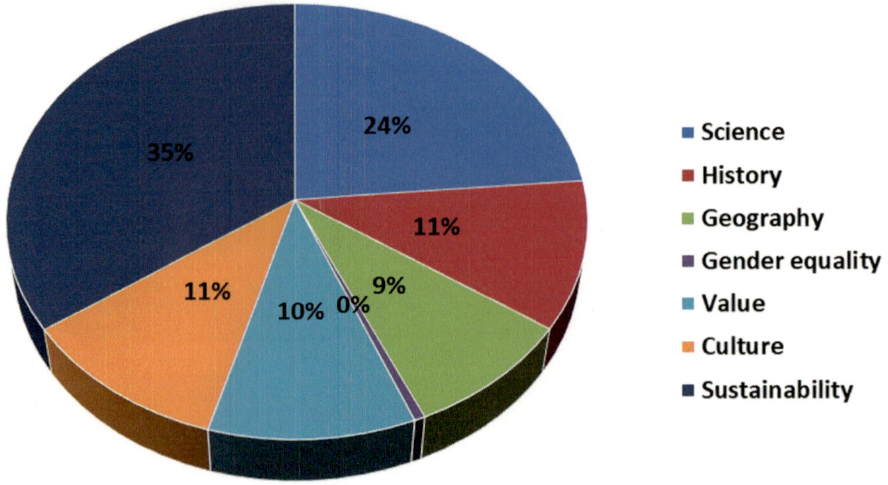

Fig. 14 Blue School challenges' correlation with IOC-UNESCO ocean literacy multi-perspectives

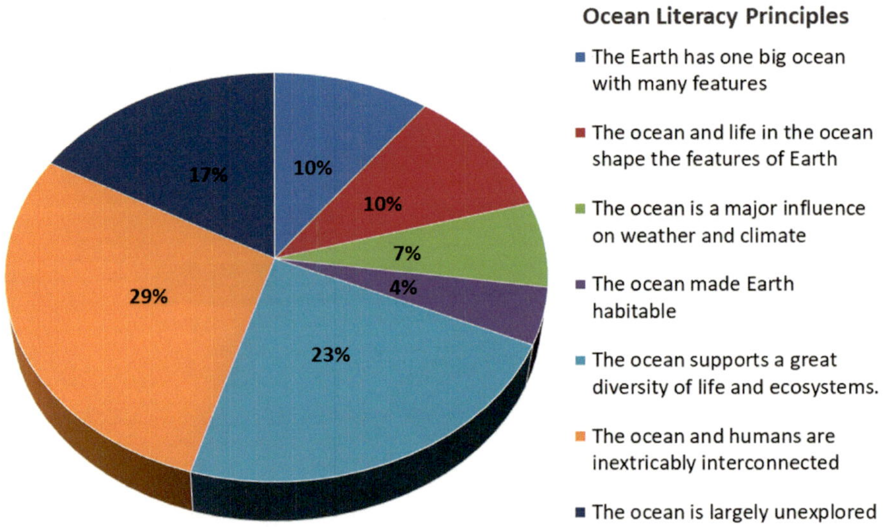

Fig. 15 Blue School challenges' correlation with Ocean Literacy Principles

perspective, teachers need to correlate their project with the Ocean Literacy Principles. Principles 5 ("The ocean supports a great diversity of life and ecosystems") and 6 ("The ocean and humans are inextricably interconnected") are the most explored by far (Fig. 15).

The variability of the activities available for Blue Schools from the partners' network, which comprises 89 institutions providing educational activities or mentoring in ocean-related issues, is displayed in Fig. 16. Ranging from different

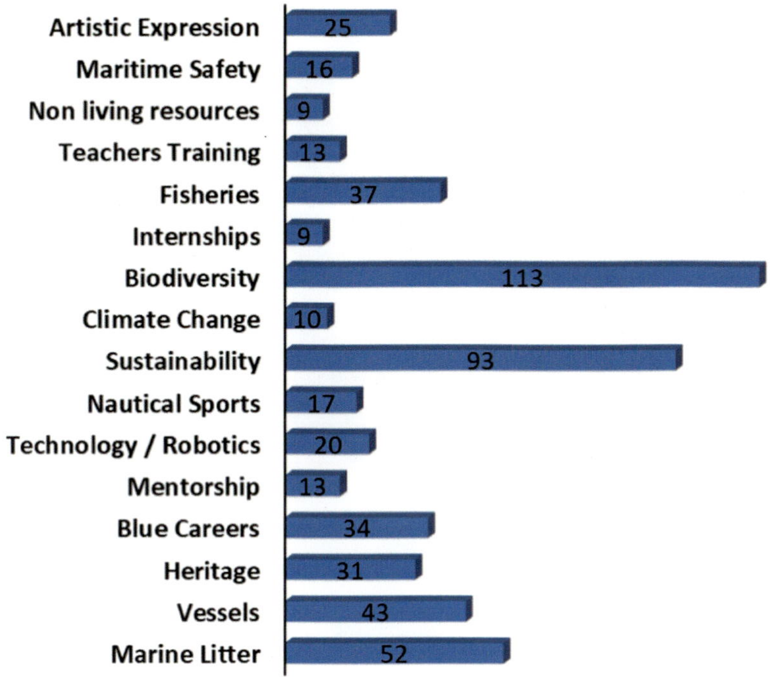

Fig. 16 Categorization of the educational activities available from the Blue School Partners' Network (absolute frequency)

backgrounds (maritime sector, education, sports, universities, companies, non-governmental organizations, museums and aquariums) these institutions contribute with 275 educational offers, available all over the country, covering different ocean-related topics (Fig. 16) and aimed at different school levels. This is very important data to consider if Blue School is to grow in the right direction, in order to respond to the major needs identified. Kindergarten activities are needed, and teacher training is lacking. Teacher training is paramount to achieve a successful ocean literacy implementation in schools. Most teachers need tools to feel confident when addressing ocean-related topics and adopting a problem-based learning approach in their classes.

While most of these activities are meant to be developed face-to-face, 2020 brought a new challenge in all education activities and teaching methods. In a collective effort by the Blue School Coordination and its partners' network, Blue School assembled more than 80 online offers in what was named "Blue School at Home". These offers are distributed in four categories and aim for different levels (Figs. 17 and 18). "Blue School at Home" represents not only a specific response to the new needs expressed by teachers in the scope of the COVID-19 pandemic, but also a long-term solution for inclusivity, allowing all schools to have direct access to educational opportunities about the ocean, even when far away from the coast or from the major urban centres. Nonetheless, the lack of specific educational offers for teachers and kindergarten persists.

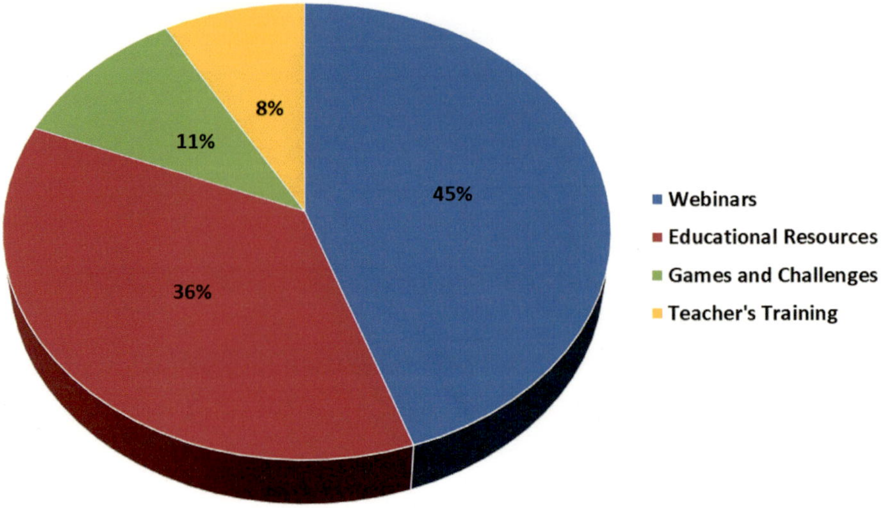

Fig. 17 "Blue School at home" online resources

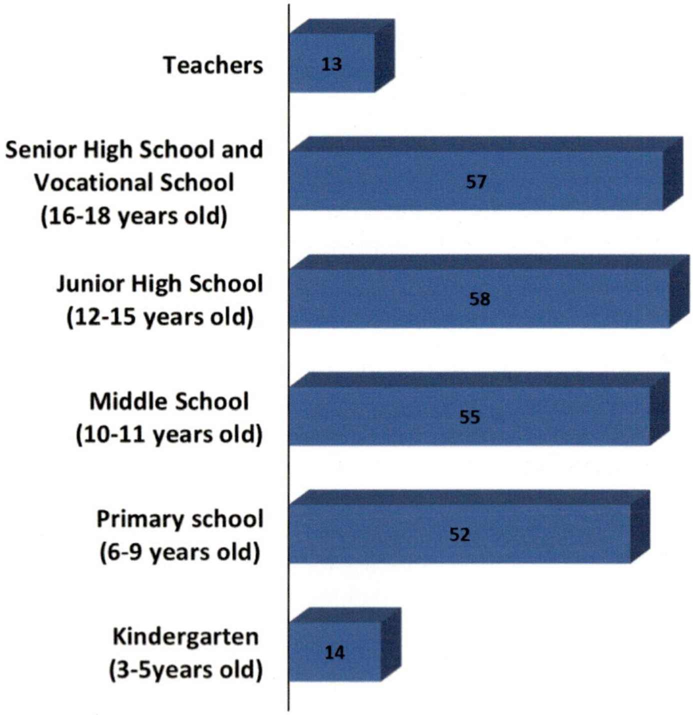

Fig. 18 "Blue School at home" online resources by school level (absolute frequency)

Discussion

In 2020, the Blue School programme comprised 215 Portuguese schools and 1 Norwegian school working on ocean-related themes (marine pollution and biodiversity being the most common topics addressed). Blue Schools mostly take on ocean literacy at a Junior and Elementary level. Their challenges or actions relate mainly to the Ocean Literacy Principles 5 ("The Ocean supports a great diversity of life and ecosystems") and 6 ("The Ocean and Humans are inextricably interconnected") and to the Sustainability and Science perspectives of the IOC-UNESCO multi-perspective approach to OL.

Providing more than 350 educational activities for Blue Schools, the programme's network of partners includes 89 institutions.

After three years of implementation, continuous improvement and constant feedback from teachers, students and partners, Blue School has proven a successful ocean literacy programme, raising teachers' awareness for the importance of teaching about the ocean and its adaptability to multiple subjects. Blue School is perceived as an inclusive, versatile and adaptable programme, respecting the most recent national educational standards as well as international guidelines for quality education. Furthermore, it opens opportunities and supports teachers to implement an interactive problem-based approach contrasting with traditional teaching methods that persist in many national schools. Blue School has been developing as an ocean literacy community where different partners, different schools, teachers, students and even municipalities cooperate towards the development of ocean-literate generations.

But the Blue School programme still faces shortcomings and barriers that need to be addressed. Most of them relate to the difficulty in committing schools and teachers to the increased effort that the Blue School programme requires. Teachers lack direct incentives (like reduced teaching hours or training actions with impact on their career progression), a problem which is commonly reflected on the Coordinating Teacher's difficulty in engaging other colleagues to participate in their Blue School challenge. Furthermore, most schools do not have logistic or financial support for their actions, which is particularly relevant in socially disadvantaged backgrounds. The need to fulfil the school curriculum established for each academic year, and its perceived incompatibility with innovative teaching approaches, also present difficult challenges for many schools.

Even if Blue School has accomplished positive results in garnering bigger awareness towards the ocean, one of the major upcoming challenges relies on assessing its contribution to the ability of teachers to work ocean topics and the impact of the programme on youngsters' attitudes and behaviours. That is the ultimate step for true ocean-literate citizens and therefore one of the ultimate goals of Blue School.

The Blue School concept and implementation model has been changing and evolving since its inception. Working collaboratively with its community of thoroughly engaged teachers, students, partners and municipalities, the Coordination constantly aims to improve the programme. Considering the success achieved in Portugal, it is increasingly important to discuss if this Blue School concept can be adapted to other countries and if Blue School does represent an ocean literacy strategy that adequately responds to the global obstacles in teaching about the ocean.

After three years of implementation and already marked as a success story by UNESCO (Santoro et al. 2017), the Blue School programme has been generating international interest. The first seeds for the internationalization of the concept have already been planted. Jessheim Upper Secondary School, a Norwegian school, became the first international Blue School back in January 2020. To feed the EU4Ocean Blue Schools EU network several proposals were submitted to Erasmus + for the implementation of sea basin Blue Schools networks. In the European Atlantic countries (Spain, France, Ireland and Portugal) the Portuguese Blue Schools will play a decisive role as mentors of the new international candidate schools, testing and adapting the Blue School PT concept in other EU countries.

The adaptation of the Portuguese Blue School concept to other countries is rapidly becoming a reality. And if in Europe the process is seemingly more advanced, the first steps are also in place in more distant latitudes like South America or Africa.

Conclusions

The Blue School model, as described in this chapter, aims to give an answer to the implementation of ocean literacy in formal education. By the end of the first three years of implementation, a blue community representing a true ocean literacy movement was created, and with much greater impact than originally planned.

The solidity of the Blue School programme, built through the last three years, is supported by the following foundations:

The Blue School concept and its eight associated criteria are in accordance with the international education guidelines;
The Blue School programme directly contributes to SDG 4, 14 and 17;
The Blue School programme involves all the stakeholders who have a relevant role in ocean literacy and blue growth, fostering relations and improving networking between them: government, UNESCO, public and private sea sector, municipalities, universities, NGOs, museums and schools;
The Blue School initiative contributes to increase the global awareness for ocean issues among younger generations and local communities, stimulating active citizenship towards the ocean;
The Blue School programme acts as an umbrella supporting other ocean literacy actions and projects;
The Blue School programme provides the ideal setting to create a dynamic blue community, where students, teachers and partners have an active voice;
The Blue School programme relies on a dedicated team to guide and support schools since the beginning.

But there are also challenges to be faced if we want to assure the effectiveness of the Blue School programme. Being this a new field of knowledge, the published research on ocean literacy is insufficient or dedicated to the development of educational resources only. There are no tools developed to analyse if initiatives like these can effectively reach their objectives and produce behavioural changes in new generations. Such tools may take several years to consolidate and validate and, at a time of great dynamics and fluidity around the ocean literacy concept, tools may fall short of what is necessary. To overcome this barrier, it is essential to boost research in ocean literacy.

Another challenge is the difficulty in committing schools and teachers to the increased efforts required by the Blue School programme. In order to support both schools and partners in developing effective ocean literacy actions, the Portuguese Ministry of Sea allocated resources and opened two calls granting funding for ocean literacy projects.

Adaptation of the Blue School model by other countries or communities should follow some common features for success: (i) adopting common Blue School criteria; (ii) selecting an entity to coordinate the programme nationally; (iii) gathering political support; (iv) creating a supportive network of partners from different economical and societal sectors; (v) creating a global identity for the programme and a formal recognition for the schools engaged in the programme. National implementation models appropriate to the educational realities of each country should be discussed and defined locally.

Becoming a global initiative, an international Blue School network would allow the exchange of ways of learning and teaching ocean literacy. An international community of Blue Schools, with its social, cultural and educational differences, brings diversity and allows for a global exchange of experiences that will favour the implementation of ocean literacy and contribute to quality education, inclusiveness and gender equality.

Acknowledgements This chapter was truly a collaborative work, with contributions from several sources. We would like to thank all the teachers and students, the scientific commission, partners and municipalities from the Portuguese Blue School network, particularly to teacher Cíntia Costa from Santa Maria Maior School.

The authors also acknowledge the support of Miguel Fonseca (DGPM) who was essential for the creation of the infographics presented in the chapter.

Finally, this work would not be possible without the continuous political and financial support granted by the Ministry of Sea for the Blue School programme.

LG was supported by national funds through FCT (Portuguese Foundation for the Science and Technology) within the scope of UIDB/04423/2020 and UIDP/04423/2020.

References

Ciência Viva (2020, July 6) Ciência Viva. Conhecer o Oceano. Ciência Viva – Agência Nacional para a Cultura Científica e Tecnológica. https://www.cienciaviva.pt/oceano/home/

COLC (2020, July 6) Canadian Ocean Literacy Coalition (COLC). https://colcoalition.ca/about/our-story/

Correia M, Azevedo IC, Peres H, Magalhães R, Oliva-Teles A, Almeida CMR, Guimarães L (2020) Integrated multi-trophic aquaculture: a laboratory and hands-on experimental activity to promote environmental sustainability awareness and value of aquaculture products. Front Mar Sci 7:156. https://doi.org/10.3389/fmars.2020.00156

Direção-Geral de Política do Mar (2009) Observatório Economia Azul. https://96594c97-1436-40ba-b257-d6d0d780b25f.filesusr.com/ugd/eb00d2_75476e45b61644fc9c981b5db94e2967.pdf

Direção-Geral de Política do Mar (2014) National Ocean Strategy 2013–2020. Uzina Books. ISBN 978-989-8456-64-9

Escola Azul (2020, July 6) Blue School. Ministério do Mar. Lisboa, Portugal. https://escolaazul.pt/en

European Commission (2018) Council recommendation on key competences for lifelong learning. Official J Eur Union 4(6). https://ec.europa.eu/education/sites/education/files/recommendation-key-competences-lifelong-learning.pdf

European Commission (2019, July 6) Tender: ocean literacy for all. Directorate-General for Maritime Affairs and Fisheries (DG MARE). https://etendering.ted.europa.eu/cft/cft-document.html?docId=50198

European Commission (2020, July 6) Network of European Blue Schools. European Union. https://webgate.ec.europa.eu/maritimeforum/en/frontpage/1485

EMSEA (2020, July 6) European Marine Science Educators Association (EMSEA). http://www.emsea.eu/

Fauville G, McHugh P, Domegan C, Mäkitalo Å, Friis Møller L, Papathanassiou M, Chicote C, Lincoln S, Batista V, Copejans E, Crouch F, Gotensparre S (2018) Using collective intelligence to identify barriers to teaching 12–19 year olds about the ocean in Europe. Mar Policy 91:85–96. https://doi.org/10.1016/j.marpol.2018.01.034

Gotensparre SM, Fauville G, McHugh P, Domegan C, Mäkitalo Å, Crouch F (2017) Meta-analysis of the consultation reports. EU Sea Change Project. Retrieved July 6, 2020, from http://seachangeproject.eu/images/SEACHANGE/SC_Results//Deliverable-3.3.pdf

Governo de Portugal (2017) Educação, Despacho n.º 5907/2017. Autonomia e Flexibilidade Curricular. Diário da República, 2ª Série (128), 13881–13890. https://www.dge.mec.pt/autonomia-e-flexibilidade-curricular

Governo de Portugal (2020) Lei n.º 3/2020 de 31 de Março. Diário da República, 1ª série (64), 337–460. https://dre.pt/application/conteudo/130893437

Innis L, Simcock A (2016) The first global integrated marine assessment: world ocean assessment I. United Nations regular process for global reporting and assessment of the state of the marine environment, including socioeconomic aspects. United Nations. https://wedocs.unep.org/bitstream/handle/20.500.11822/9795/-The_First_Global_Integrated_Marine_Assessment_World_Ocean_Assessment_I-2016WOA_2016.pdf.pdf?sequence=3&amp%3BisAllowed=

Martins GDO, Gomes CAS, Brocardo J, Pedroso JV, Camilo JLA, Silva LMU, Encarnação MMGAD, Horta MJDVC, Calçada MTCS, Nery RFV, Rodrigues SMCV (2017) Perfil dos alunos à saída da escolaridade obrigatória. Ministério da Educação, Direção-Geral da Educação. ISBN 978-972-742-416-0. http://www.dge.mec.pt/sites/default/files/Curriculo/Projeto_Autonomia_e_Flexibilidade/perfil_dos_alunos.pdf

Ministério da Educação (2017) Estratégia Nacional de Educação para a Cidadania. Ministério da Educação, Direção-Geral da Educação. http://www.dge.mec.pt/sites/default/files/Projetos_Curriculares/Aprendizagens_Essenciais/estrategia_cidadania_original.pdf

Ministério do Mar (2019) Soberania. In 1. Oceano, p 13

European Marine Board (2013) Linking oceans and human health: a strategic research priority for Europe. Position Paper 19. https://marineboard.eu/sites/marineboard.eu/files/public/publication/Oceans%20and%20Human%20Health-214.pdf

National Geographic Society, NOAA, COSEE, National Marine Sanctuary Foundation, National Marine Education Association, Co Exploration (2007) Ocean literacy—the essential principles of Ocean Sciences K-12. An ocean-oriented approach to teaching science standards. http://www.coexploration.org/oceanliteracy/documents/OceanLitConcepts_10.11.05.pdf

OECD (2018) The future of education and skills: education 2030, the future we want. Organisation for Economic Co-operation and Development (OECD). https://www.oecd.org/education/2030/E2030%20Position%20Paper%20(05.04.2018).pdf

Our Ocean (2018, October 29–30) Conference—our ocean commitments. Bali, Indonesia. https://ourocean2018.org/?l=our-ocean-commitments

Santoro F, Santin S, Scowcroft G, Fauville G, Tuddenham P (2017) Ocean literacy for all: a toolkit. Paris: UNESCO, p 136. ISBN:978-92-3-100249-6

Sea Change (2020, July 6) Sea change: our ocean, our health. EU-funded Project (Horizon 2020). https://www.seachangeproject.eu/

UNESCO (2014) Global citizenship education: preparing learners for the challenges of the 21st century. United Nations Educational, Scientific and Cultural Organization (UNESCO), Paris, France. ISBN 978-92-3-100019-5. http://unesdoc.unesco.org/images/0022/002277/227729E.pdf

UNESCO (2015) Education 2030 framework for action. Incheon declaration and framework for action: for the implementation of sustainable development goal 4: ensure inclusive and equitable quality education and promote lifelong learning opportunities for all. World Educators Forum. United Nations Educational, Scientific and Cultural Organization (UNESCO). http://uis.unesco.org/sites/default/files/documents/education-2030-incheon-framework-for-action-implementation-of-sdg4-2016-en_2.pdf

UNESCO (2020, July 6) The United Nations Decade of Ocean Science for Sustainable Development (2021–2030). https://www.oceandecade.org/

Two Ocean Aquarium Academy: An Introduction to Ocean Literacy Programmes and a Marine Sciences Curriculum

Russell A. Stevens

Abstract The Two Oceans Aquarium, a privately run institution, has quality education as a critical part of its vision. South Africa is an educationally challenged country with a highly diverse education landscape with extremes of both highly resourced and deeply impoverished schools. Over the past 20 years and against this backdrop, the Aquarium's Environmental Education Foundation has developed an extensive Aquarium school excursion and outreach programme and a subsequent suite of holiday/weekend enrichment courses. Out of this enrichment programme a customised national Grade 10–12 Marine Sciences Curriculum has been written, has been approved by the South African Parliament and is being implemented at pilot schools in 2020. This chapter outlines the background to the development of this curriculum. Highlighted topics include geology, chemistry, meteorology, marine biology, marine ecosystems and marine sciences applications such as aquaculture human interactions with the ocean and the education sustainable development goals (ESD). Its focus is to prepare scholars by introducing a marine sciences knowledge base in preparation for further tertiary studies in the field. In addition, it is designed as a stand-alone subject which adequately challenges and nurtures students in the way that similar Grade 10–12 three-year courses do. This strategy has achieved a high reputation among many international ocean literacy peers. This chapter describes the Aquarium Marine Sciences Academy programmes and the curriculum. It illustrates the process used as the curriculum was developed and how it morphed into its current form. Reference is made to widely researched international marine science texts and curricula in the development of the Marine Sciences Curriculum. A wide range of consultation was undertaken at tertiary levels, including the USA's Essential Ocean Literacy Principles and Fundamental Concepts and the emerging global Blue Economy. The onset of COVID-19 and subsequent lockdown has provided the necessity to offer several online courses.

Keywords Two Oceans Education Foundation · Marine education · Aquarium education · Marine sciences curriculum · Marine education facilities

---

R. A. Stevens (✉)
Two Oceans Aquarium Education Foundation, Cape Town, South Africa
e-mail: Russell@aquariumfoundation.org.za

K. C. Koutsopoulos and J. H. Stel (eds), *Ocean Literacy: Understanding the Ocean*, Key Challenges in Geography, https://doi.org/10.1007/978-3-030-70155-0_11

Introduction

The conservation strategy of the World Association of Zoos and Aquariums reports that a significant number of the 3 Billion annual Zoo and Aquarium visitors end their visit with an increased level of understanding of the meaning of biodiversity (WAZA 2005). In addition, visitors acquire a greater level of knowledge about biodiversity (Barongi et al. 2015). Aquariums therefore have a significant role to play in increasing awareness of their visitors' future actions and are well placed to educate about solutions to impacts of human activities to the ocean such as overfishing, impacts of plastic pollution such as ingestion or entanglement of marine animals, using exhibits, print and social media (Penning et al. 2009).

Many public aquariums purposefully highlight ocean awareness, conservation, research, education and ocean literacy programmes (Fig. 1). These are offered in the form of a wide range of free-choice learning options through signage or public gallery presentations (Falk et al. 2007). In addition—and on the increase—are formal public education programmes offered in discovery centres such as the Marine Sciences Academy at the Two Oceans Aquarium. These programmes are considered by leading aquariums as key solutions to increasing awareness and promoting student actions to secure a better future for the ocean. It is sobering to consider that of the population in sub-Saharan Africa 70% are functionally illiterate, the lowest level of literacy on Earth (PRB 2020). In such a context the Two Oceans Aquarium Education Foundation has the opportunity to make a significant contribution to the welfare of the ocean shores and the levels of literacy in the Southern African education landscape. This is important since the Two Oceans Aquarium displays marine organisms representative of those living off the two distinct east and west coasts of southern Africa. These coasts experience very different sea temperatures, climates, precipitation levels and nutrients. Consequently, there is a high level of endemism along the South African coastline constituting one of the most diverse on Earth. South Africa's formal government education departments are in great need of support from the non-governmental sector such as aquariums and science centres. Many South African school students

Fig. 1 Connecting students with Marine Life with experiential learning

go through their school career without exposure to ocean content. There is a critical need for aquariums to use their learning resources to reduce the gaps found in the current public education space—be they outreach programmes, science labs or online courses to increase the quality of formal education and help improve basic literacy. Aquariums, too, have the need to increase opportunities for marine sciences awareness and education in the formal school system. It is the moral imperative of aquariums to do so. The opportunity is there for every major aquarium worldwide to do the same.

In this publication, reference is made quite freely to Ocean Literacy and the excellent 7 Essential Principles developed in the USA (NOAA 2020). However, the term "Ocean Literacy" is used interchangeably with the terms "Ocean Awareness" and "Ocean Education". The term "Literacy" has a completely different meaning for the southern African educator audience. To refer to a brand of literacy, such as the "Scientific Literacy" brand or "Ocean Literacy" brand does not work. The experience of being African and working in Africa and among African people leads to a very strong resolve and motivation by Africans to find African solutions to African problems. In the process, we have learned that using the term Ocean Literacy could be seen as an elitist imposition from the "haves" in the Global North to the "have-nots" living in Africa. In this chapter, more neutral terminology is applied to facilitate meaningful communication. Where the term "Ocean Literacy" has truly honourable and values in the Global-North, it does appear patronising and consequently, inaccessible for the sub-Saharan African audience. Particularly, considering the "Rhodes must fall" and the "Decolonisation of Education" debates that currently rage. Notwithstanding, the term "Ocean Literacy" has traction and "validity" among the Marine Education community in many other parts of the world. Invaluable work has been done in the name of Ocean Literacy by international colleagues for which they should be commended (NOAA 2020).

The Two Oceans Aquarium Education Foundation and its current contribution to free-choice, informal and formal learning contributes significantly to the South African and the international education landscape (Fig. 2). Since opening in 1995, the Foundation has become a bastion of quality education as is presented in many case

Fig. 2 Aquarium teacher Chanelle Naidoo explains East and West Coast of South Africa

studies (Santoro et al. 2017). Through robust thinking and serendipitous evolution of practice, innovative education activities have developed mainly by responding to feedback and conducting formal interviews with visiting teachers to reflect and refine the best possible practice. Such endeavours have informed the development of a Grade 10–12 Marine Sciences Curriculum which was approved by the South African Parliament in August 2019 and has been implemented in January 2020.

Background to the Two Oceans Aquarium Education Efforts

The Two Oceans Aquarium, a private for-profit aquarium, is located in Africa's most visited tourist destination, Cape Town's Waterfront. Visited annually, in pre-COVID-19 times, by 500,000 visitors, no government, provincial nor municipality subsidises operational costs. In South Africa, a not-for-profit or NGO status is exceptionally difficult to obtain. The Aquarium therefore operates as a for-profit model, notwithstanding, 15% of its turnover or one-third of its profit financially supports education, conservation, awareness and research projects. After complex applications, the Two Oceans Aquarium Education Foundation was recently granted a not-for-profit status and the Aquarium Education Foundation became operational in March 2020. The Foundation fulfils the conservation programmes for turtles and Cape Fur Seals, plastic pollution awareness programmes, research and the extensive education operation (see links below).

A critical question asked at any aquarium is how to engage one's facilities, resources and opportunities to optimally serve educational purposes. The Aquarium's signage and gallery teams have developed valuable free-choice learning opportunities for guests. In addition, formal education programmes have been developed by teaching staff.

The Education team's vision informs programmes and activities offered both informally and formally to visiting school excursion groups.

> *Connect a wide audience with nature by nurturing ambassadors with integrity and empathy by promoting environmental activism through exciting, experiential learning.*

The Aquarium is committed to using these school group excursions to serve the needs of the South African schools and its formal curriculum (Fig. 3). Over 1.3 million school children have visited since 1995, making the Two Oceans Aquarium a significantly meaningful role player in the South African education landscape and literacy upliftment.

The Two Oceans Aquarium's vision is *Abundant and Healthy Oceans for life* and its mission statement to inspire action for the future well-being of our ocean; it well positions to serve an international audience within the marine sciences education arena. Its staff consider themselves as partners to any organisation with like-minded visions on both African and global scales.

Since March 2020, with the onset of COVID-19 and the closure of the Aquarium, online courses offered by aquarium teachers teaching from their homes have been sold

Fig. 3 Aquarium teacher Monwabisi Ncence presents to children at the Two Oceans Aquarium Kelp forest exhibit Where gallery presentation has a role in Aquariums focused lessons in Aquarium wet laboratories add greater value

as to generate income, retain staff and to sustain the Foundation financially. Online teaching by Aquarium teaching staff has been a reimagining of the Aquarium's purpose; given that the Two Oceans Aquarium has invested in a Learning Management System (LMS), online teaching is here to stay to serve the needs of an international audience.

Aquarium Teacher Experience

The Two Oceans Aquarium's education strategy has evolved in various ways, firstly by appointing fully qualified and experienced school teachers from both high and primary schools onto the Aquarium teaching team. The teachers in the employ of the Aquarium have earned a range of qualifications from a number of South African universities. In addition to these qualifications, they hold a range of high school, primary school, special needs and Early Childhood Development education qualifications. In addition, teaching staff have extensive teaching experience at primary, secondary and tertiary levels. Added to this teaching experience is a range of school management and school governance portfolios. The Aquarium teachers hail from a diversity of social, racial and faith communities. This richness of educational qualifications, social backgrounds with social insights, as well as a wide range of education experience enables the teaching staff to connect and communicate with highly diverse sectors of visitors from socioeconomic, religious and racial backgrounds. The team of sixteen aquarium teachers learns daily from each others' cultural backgrounds and develops an increased sensitivity to a very wide range of cultural diversity of visitors.

Increased Quality of School Group Visitor Experience

Overcrowding of Aquarium galleries during peak times is commonplace for school outing destinations such as zoos, museums, science centres and aquariums. During crowded times, the sheer volume of very excited and often educationally unfocussed scholars becomes overwhelming for Aquarium guest relations staff. In the final analysis such school group visits become superficial, and the level of learning and the quality of education experienced is seriously compromised. In such crowded spaces aquariums cannot live up to the expectations of their mission and vision. Regrettably many turn a blind eye to crowded galleries and such compromised educational experiences. The boosted high numbers of children in groups are useful and impressive when applying for grants or sponsorships or for including in annual board reports bragging of "apparent" education. The Two Oceans Aquarium has invested much effort in preventing such overcrowding and compromised education experiences (Fig. 4).

Key strategies to increase the quality of on-site experiences include a pre-booking for specific time slots and a variable entrance pricing structure similar to the bookings practice applied by the airline industry. An analysis of child group booking trends reveals high and low demand seasons for school group bookings. High demand times are charged at a premium price. The increased revenue at such times subsidises the lower season rates. These measures pace the flow of visiting school groups for the benefit of both the children and aquarium visitors. The lower pricing structure provides a financial incentive for schools, to visit during a less busy time. It spreads the school group visitor load and reduces crowding at peak times. To better manage the school group bookings, the number of school group children is capped at 400 school children per school day and no discounted groups can visit during school holidays or weekends.

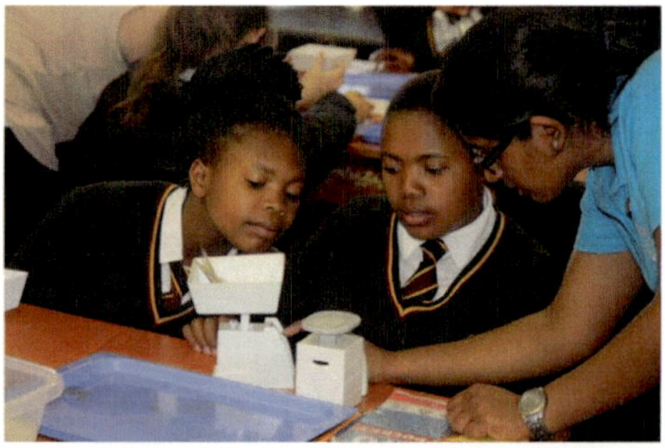

Fig. 4 Aquarium teacher Kirshia Govender Khombatis teaches Gansbaai students to weigh fish

Fig. 5 Aquarium teacher Xavier Zylstra teaching in the Wet Laboratory

Wet Laboratories

In the Aquarium galleries, the touch pool and microscope displays are the most hands-on, engaging and most effective way to authentically connect visitors to underwater marine life. For school children, touching sea animals and observing live animals under a microscope at school are not possible. Tapping into the Aquarium's air and seawater life support systems, the classrooms were converted to wet laboratories (Fig. 5). Equipped with storage and touch tanks live marine life is displayed. The Aquarium teachers supervise and facilitate the touching of the rocky shore invertebrates and small sharks. Water conservation and data collection lessons support the formal school curriculum. Smaller animals are placed under the microscope producing images displayed on large interactive screens, showing body parts of echinoderms, polychaetes and the feeding behaviour of barnacles. The wet laboratories enable children to engage with nature, make sense of their ideas of nature, be in awe of the beauty of nature and be challenged about their attitudes and values associated with nature. These activities facilitate the school group children to be diverted from the Aquarium galleries to increase the educational focus of their visit to the Aquarium.

By increasing the quality of meaningful activities, there has been an increased demand by school group facilitators wishing to pre-book a school group visit for their children. The Aquarium's school excursion programme has therefore grown from 20,000 children in 1999 to 76,000 in 2019. This includes students from other neighbouring provinces, countries as well as Europe and the Americas.

School Group Activities

The Aquarium's programmes and activities have progressively evolved. The Grade R to Grade 3 learners are offered puppet shows on environmental themes including marine food chains; environmentally smart living; turtle conservation, litter as well as the water cycle. Each puppet story has a reading book which reinforces the environmental and conservation messages and provides opportunities to increase levels of literacy among the learners.

Teachers of Grade 3 to Grade 12 select curriculum-based wet laboratory activities. These include lessons on rocky shore animals, water conservation, fish or shark activities as well as a mathematics data collection activity. Each activity has been developed by the Aquarium teaching staff in collaboration with school teachers to suit their curricular needs. Programmes are offered to first-and second-year zoology students from local universities. These students attend a programme with the live animals as opposed to observing animals in formalin in university laboratories. Group facilitators can meet with Aquarium staff prior to their visit and jointly align the activity with the school or university's course requirements. Special collections of animals are sometimes used in the wet laboratory during lectures and safely returned to the rocky shore thereafter.

The Aquarium has developed outreach programmes that visit schools in communities that cannot afford the transport or aquarium entrance costs. The "Oceans in Motions" programme takes live rocky shore animals to township schools. The second SMART Living outreach vehicle visits Grade 7 classes at schools offering a programme within the "environmental footprint" themes including biodiversity, energy saving, water saving and waste minimisation. A third outreach activity performs puppet shows to Kindergarten childhood.

In addition to the above-mentioned immediate learning goals for the school children, the Aquarium has a deliberate objective to demonstrate and display alternative teaching and experiential learning methodologies for visiting teachers. Regular feedback received from each accompanying teacher confirms that a significant number leave the Aquarium inspired by the demonstration of exciting, hands-on, experiential opportunities allowing students to make enquiry-based connections with nature.

Courses

After observing the value of the wet laboratory experience school teachers and education district subject advisors asked the aquarium teachers to develop enrichment courses for school students who showed an interest in marine sciences (Fig. 6). A week-long course increases the quantity and quality of engagement and the

Fig. 6 Aquarium teacher Bianca Engel shows students how to use a Secchi disk to measure turbidity

Aquarium's long-term vision and mission would be better achieved. The first Young Biologist (YB) course was offered to Grade 10 students in 2001 and over time, with reflexive practice, it has evolved into a course currently in great demand. Up to 400 applications are typically received for the 50 course positions. For some courses, applicants have applied from as far afield as Durban and Johannesburg (1,000 km from Cape Town) as well as from Vancouver, Canada and the Philippines from where the students travel to Cape Town and arrange their own accommodation for the week.

Since its opening, the Aquarium volunteer courses have been on offer to adults wishing to assist "Front of House" or "Behind the Scenes". Content and topics taught is similar to the volunteer content taught at peer aquariums across the globe. This material is tried and tested. It is followed that this be offered, as a Young Biologist course, to Grade 10 students wishing to job shadow and do community service at the Aquarium. It is affectionately called "The YB Course". The focus of this training is the invertebrates and sea plants displayed, in particular the touch pool and microscope displays.

Students learn content including interesting facts about the basic animal phyla and basic information about red, green and brown marine algae. A highlight of the hands-on activities is the cutting open of a kelp holdfast out of which tens if not hundreds of little animals such as tangle-worms and polychaetes emerge. These animals are then studied under the wet lab microscope. The last two days of the week-long course takes the form of a camp at an Environmental Education Centre. Once completed the students are given bridging training about the basics of volunteering.

Courses are advertised to school through the Education Department and the Aquarium's social media. The student applies by submitting a written page indicating

(a) why they believe themselves suitable to participate in the course,
(b) their prior experience relating to the course content,
(c) what they will do with the experience thereafter and
(d) their leadership aspirations.

The student also needs a teacher's letter of recommendation with a signature of endorsement from the school's principal.

Course Assessment

The Marine Sciences Academy courses are academic in nature, making best use of the animals on display in the galleries and applying experiential, hands-on learning methodologies aligned with the Aquarium's enquiry-based education philosophy. Marine sciences content is spread across the suite of courses and the introductory content is scaffolded from course to course culminating in the more challenging Grade 11 Marine Biology and Oceanography courses. For each course, pre-test assessments with predominantly knowledge-based questions are written before the course starts. This serves as an instrument of formative assessment guiding the Aquarium teachers' insight into the student's prior knowledge and their entry level to the content. Reflection on the assessment responses informs a presenter about her or his teaching pedagogy and the appropriateness of their teaching methods. The same test is completed as a post-course assessment to evaluate the level of learning. On each course-morning, an assessment of the previous days' activities is conducted. Scripts are marked and returned that day to guide students of the standard expected. Other post-course requirements include a report from the student providing evidence of follow-up work such as their presentations to eco-clubs at their school as well as images of a litter clean-up which they organised. Their submission forms part of the overall assessment requirement for the course completion. In the Environmentally SMART Living course, students return to the Aquarium and present their environmental actions in a PowerPoint presentation as part of a mini-symposium to their peers and parents. The average course test scores are used to determine the certificate level awarded to the student. Two Oceans-accredited certificates are awarded to the students on a range of levels: Bronze for attendance, Silver 70%, Gold 80% and Diamond 85%.

The Marine Sciences Academy Courses

Further requests resulted in the development of a suite of multi-tiered Grade 6 to Grade 11 weekend and during school holiday courses. Sponsorship allows the offer of these free of charge. Each course builds upon the content of previous courses. The table below shows a summary of courses (Fig. 7).

Course	Grade	Content supplementing CAPS and much more
Junior Biologist Marine Sciences 101 (4 days)	Gr 6	Fun introduction to marine science, basic marine sciences content
Smart Living Marine Sciences 102 (5 days)	Gr 7	Environmentally sensitive living biodiversity; water saving; electricity saving and waste minimisation

(continued)

Fig. 7 Marine Sciences students learning from Dr Hendricks, lecturer of Marine Sciences at two local universities

(continued)

Course	Grade	Content supplementing CAPS and much more
Marine Science Discoverer Course Marine Sciences 103 (5 days)	Gr 8	**Environmental, zoological and biological** content provide enrichment for students interested in marine life. Many experiments and hands-on activities provide career guidance throughout the programme
Marine Science Explorer Course Marine Sciences 104 (5 days)	Gr 9	Follows on from Marine Sciences 103, encouraging students to study Maths and Science from Gr 10 and to venture into marine sciences careers
Young Biologist Marine Sciences 202 (7 days)	Gr 10	The flagship course **preparing volunteers** to work at our touch pool, microscope and behind the scenes
Ocean Science Marine Sciences 301 (5 days)	Gr 11	Link to basic **oceanography** at university
Marine Biology Marine Sciences 302 (5 days)	Gr 11	**Shape and form of life**, linked to university zoology

The content for each course provides enrichment and complements the school curriculum. Each course aims to energise students to return to school and share newly discovered content. The students should be able to use the content to complete curricular requirements such as essays and school projects. For example, in a number of courses, biodiversity threats are taught using the HIPPO acronym which links to the curriculum [it represents the destruction of natural indigenous **H**abitats; the **I**nvasion by alien organisms; **P**ollutants; the increase in demand on natural habitats and biodiversity due to the increase in human **P**opulation and the destruction of biodiversity by **O**verexploitation]. Aquarium staff have been advised by colleagues teaching at school that the thorough teaching of these concepts and principles in the courses has played a positive role in students' ability to express themselves in Grade 8 through to 12 essay exam questions on environmentally related topics.

Marine Sciences Curriculum

Teachers, principals and education district subject advisors requested that the Two Oceans Marine Sciences Academy content be used to develop a Grades 10–12 Marine Sciences curriculum. Consultations with school teachers and university professors informed the curriculum development to bridge the gaps between the current school curriculum and the gaps identified while teaching marine biology and oceanography at university level.

An international Internet search revealed that versions of marine sciences curricula exist in New South Wales and Queensland in Australia, in some American states, the UK and a private high school on Vancouver Island BC, Canada. These documents

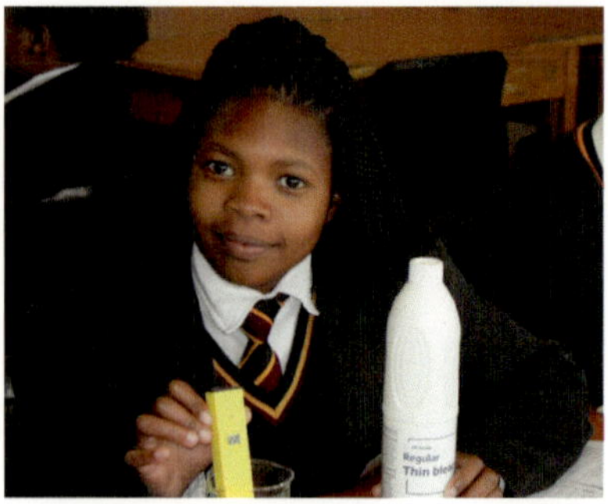

Fig. 8 A student from Gansbaai Academy using a hand held pH meter at a Marine Sciences Node School

showed a different approach to high school teaching in their respective contexts, including a less than three-year programme and others not being a formal academic course.

The Aquarium content considered in developing the curriculum framework was the Grade 11 Oceanography and the Marine Zoology course material. The course content is needed to complement the current South African curriculum (Fig. 8). Since the late '90s specialist South African schools, such as Simon's Town High School's Lawhill Maritime Centre, have offered Nautical Sciences and Maritime Economics as Grades 10–12 subjects. The introduction of the Marine Sciences curriculum coincides with the South Africa's initiative similar to Blue Economy initiatives in other parts of the Developing World, *Operation Phakisa.*

The proposal to write the curriculum was considered favourably by the Aquarium Board, as it would support the Aquarium's mandate to offer best quality education to the South African and international education landscape. There has been an increased demand from schools who have expressed a willingness to offer the subject at schools. On a National level the *Operation Phakisa* added traction to legislate marine sciences as it was seen as a subject that would support this national strategy at governmental level and the subject would complement the other Grade 12 maritime subjects. Maritime Economics and Nautical Sciences are offered at the specialised Simon's Town High School, Lawhill Maritime Centre.

The implementing of the curriculum involved much research into the processes required to introduce such a subject on a national level. We mostly had to invent a process in consultation with national role players. Unlike the practice in other countries, South African's Grade 12 certificate is based upon a national curriculum across all the provinces.

Authors of the Marine Sciences curriculum kept constant contact with university advisors, school teachers, committed school principals and district subject advisors. These colleagues needed to teach and support the curriculum when it was implemented. The "skeletal framework" of the course followed a similar path of already existing texts including:

- *An Introduction to the World's Oceans*; 10th edition; *by* Keith Sverdrup, Virginia Armbrust, Raphael Kudela; McGraw Hill; 2014.

- *Fundamentals of Oceanography* by Keith Sverdrup, Alyn Duxbury and Alison Duxbury 7th edition McGraw Hill 2000.
- *Oceanography: An Invitation to Marine Science*; by Tom S. Garrison, 2012.
- *Essentials of Oceanography* (12th Edition), by Alan P. Trujillo and Harold V. Thurman; Pearson, 2016.
- *Marine Science for Australian Students* by Bob Moffatt, Tim Ryan and Alan Zann; Marine Education Society of Austrasia, Wet Publications, 2003.
- *Marine Science: The Dynamic Ocean* by Meghan Marrero and Glen Schuster; Pearson, 2012.
- *Marine Science Marine Biology and Oceanography*; by Thomas Greene; Amsco School Publications; 2004.

Such books are characterised by very accessible language usage, the text readable and interesting, diagrams clear and concise together with comprehensive glossaries.

Once the basic content was scripted, a marine sciences advisory group of senior university professors and lecturers of oceanography and marine biology reviewed the content and helped the authors to keep the content relevant to the South African coastline and marine experience. More importantly direction was given about recent and relevant research. A key theme of such discussions was which gaps should be filled so that students entering university courses would receive a solid foundation of the fundamentals in the various topics at the school level.

A framework was developed and key concepts were written. Draft documents were distributed to colleagues at universities, district subject advisors and school teachers enquiring about the feasibility of teaching the course. Teaching of content was tested in terms of scope, cognitive demand and volume of content at school extra-mural marine sciences clubs. Feedback was received and included from partners of the process including scientists of the South African Network for Oceanographic and Coastal Research (SANCOR); specialists in education and school assessment. All comments were considered as the curriculum document was refined.

Objectives for Teaching Marine Sciences as a Subject

The authors designed this course with the intent that the students take direct control of their own learning journey, with the teacher being the facilitator. Our experience shows that excellent quality assessment scores are earned by students who learn by teaching themselves rather than being taught from a "podium". Self-learning has become a dominant approach in education, particularly in a post-COVID-19 context, the online learning of core content will dominate future pedagogy. So, no more blaming of a teacher for poor test scores! Internet searches produce many excellent quality video clips and BBC or National Geographic style documentaries, PowerPoints and documents that support the teaching of marine sciences content. Teachers facilitating the learning process teach their students to discern what material should be selected above the content which is inaccurate or simply fabricated.

In the learning process students learn core content by building upon previously developed mental constructs of marine sciences, assimilating their understanding of observations with the new cognitions (Fig. 9).

The Two Oceans Aquarium's teaching methodology is undergirded by the inductive and an enquiry-based learning approach to learning. This Socratic methodology encourages learning that develops an improved cognitive engagement with the natural world. Accompanying these learning methodologies is the enabling of students to complete the course with an increased capacity to ask the best possible questions in the field. Such thinking should develop the capacity to facilitate solutions through a questioning or inductive methodology. But, to do so, students need a thorough understanding of the core marine sciences content.

The subject aims to build upon the academic and scientific understanding in the minds and thinking of the students enabling them to hold their own in a conversation and to debate with confidence about any marine sciences topic. By the end of the programme, the students should be equipped with well-informed thinking skills and the ability to apply those skills within their future careers whether they become marine scientists, maritime lawyers or policy developers on a parliamentary portfolio committee. Students should gain an increased confidence in their ability to use their newly acquired thinking and knowledge and to critically evaluate and debate scientific processes and content. This new confidence enables students to provide informed perspectives.

Marine sciences students would develop an appreciation of the scientific method and scientific practices. Science teachers from around the world are increasingly concerned about students' lack of ability to apply the scientific method in designing an experiment, understanding the variables and the controls and meaningfully reflecting and developing conclusions about the data presented. Students will be encouraged to focus upon developing competencies in using the scientific method in research conducted and to substantiate their arguments in a debate.

Most current professionals, be they in medicine or in marine sciences, recognise the weakness of specialising in sub-fields. As such, professionals gather with like-minded individuals into silos and restricted ways of thinking. The different pillars of marine sciences, be they oceanography, zoology, botany or ecology all impact upon each other and no element of the field of marine sciences should be studied without

Fig. 9 A Kelp hold fasts reveals a little ecosystem of many different invertebrates to discover

due consideration for other disciplines within the field. Topics in marine sciences should thus not be taught in isolation from each other. The subject with its many sub-disciplines lends itself well to teaching and learning the multidisciplinary of the field. Each topic should be considered holistically whether it is about the physical abiotic elements of the field, the biotic elements or the functioning of ecosystems. All these elements and their interdependencies should make students aware of the delicate balances to be found in the ocean basins. The subject should increase student awareness and appreciation of the ocean's spectacular beauty, the sensitive nature of the ocean and give insight into the sensitive relationships between humans and the ocean, its organisms and ecosystems. The lungs of the earth!

Students should learn of the urgency for humans to institute migratory measures to reduce negative human impacts on the ocean. Students should be encouraged to develop a solutions-based form of thinking where their insight into the different marine sciences sub-disciplines leads to better understanding for developing future sustainable practices. They are encouraged to address the root causes of ocean problems, for example, it is preferable responding to the symptoms, be they over-exploitation, or short-term political and managerial decisions that lead compromised actions. Consideration is given to the sustainability of resource utilisation as opposed to the over-exploitation of resources.

Throughout the course, students are made aware of current and future marine sciences career opportunities and associated skills.

Conclusion

This chapter has outlined the Two Oceans Aquarium Education Foundation's strategic thinking and planning. It can be used as a forward thinking case study of how an Aquarium's education offering can make a significant contribution to marine education and the ocean literacy landscape both locally and abroad. Strategic thought has directed activities and programmes making the once-off visit meaningful by improving the quality of an Aquarium Educational experience. Outreach vehicles visit their schools and offer hands-on interactive programmes to unresourced schools (Fig. 10). Enrichment courses have been put in place for Grade 6 to Grade 11 students in the form of the Marine Sciences Academy. The Aquarium team engages proactively with the formal education sector including school principals, school district directors and senior education officials on a national and international level, to refine their strategic thinking.

The ultimate contribution on a national level has been the development of a Marine Sciences curriculum that is being taught in South African Schools and in an online course by the Two Oceans Aquarium staff. The first South African Marine Sciences National Examination examinations will be written at the end of the 2021 school calendar year. An enormous effort has been made by the Two Oceans Aquarium Education Foundation staff to best use of the education opportunities. This effort is the key justification for housing animals in captivity. It is envisioned that the Two

Fig. 10 A Marine Sciences Student learns how to use a refractometer salinity reading

Oceans Aquarium could partner other like-minded organisations in other parts of the world to participate in a similar way to contribute to marine sciences, ocean awareness and ocean education.

References

Barongi R, Fisken FA, Parker M, Gusset M (eds) (2015) Committing to conservation: the world zoo and aquarium conservation strategy. Gland, WAZA Executive Office, p 69

Falk JH, Reinhard EM, Vernon CL, Bronnenkant K, Deans NL, Heimlich JE (2007) Why zoos & aquariums matter: assessing the impact of a visit. Association of Zoos & Aquariums, Silver Spring, MD

National Oceanic and Atmospheric Administration (NOAA) (2020) Ocean literacy: The essential principles and fundamental concepts of ocean sciences for learners of all ages. Washington, DC

Penning M, Reid GMcG, Koldewey H, Dick G, Andrews B, Arai K, Garratt P, Gendron S, Lange J, Tanner K, Tonge S, Van den Sande P, Warmolts D, Gibson C (eds) 2009 Turning the tide: a global aquarium strategy for conservation and sustainability. World Association of Zoos and Aquariums, Bern, Switzerland. https://www.waza.org/wp-content/uploads/2019/03/Aquarium-strategy-EN.pdf

PRB 2020 Population Reference Bureau. https://www.prb.org/worldpopdata/

Santoro F et al (eds) 2017 Ocean literacy for all—a toolkit, IOC/UNESCO & UNESCO Venice Office, Paris (IOC Manuals & Guides, 80 revised in 2018)

WAZA 2005 Building a future for wildlife—The world zoo and aquarium conservation strategy. © WAZA 2005

Links

Best responsible Tourist Attraction. https://www.aquarium.co.za/blog/entry/two-oceans-aquarium-named-best-responsible-attraction-at-the-2019-african-r

https://www.aquarium.co.za/blog/entry/the-wild-ones-schools-aquarium-classroom-education-with-xavier-mr-zee

Two Oceans Aquarium Education Foundation Courses

https://www.aquarium.co.za/blog/entry/i-and-j-exploring-zoology-grade-11s-get-a-taste-of-advanced-marine-science

https://www.aquarium.co.za/blog/entry/young-biologists-graduate-thanks-to-the-support-of-de-beers-marine-2019
https://www.aquarium.co.za/blog/entry/america-international-school-cape-town-sponsor-marine-science-course-2018
https://www.aquarium.co.za/blog/entry/averda-sponsors-marine-science-explorers-grade-9-2018-aquarium
https://www.aquarium.co.za/blog/entry/the-wild-ones-environmental-education-teaching-two-oceans-aquarium
https://www.aquarium.co.za/blog/entry/apply-now-for-three-free-marine-sciences-academy-courses-happening-during-t
https://www.aquarium.co.za/blog/entry/new-marine-sciences-school-subject-for-grade-10-learners-in-2019

Index

K. C. Koutsopoulos and J. H. Stel (eds), *Ocean Literacy: Understanding the Ocean*, Key Challenges in Geography, https://doi.org/10.1007/978-3-030-70155-0

Printed by Printforce, the Netherlands